Diterpenes of Flowering Plants

F. Seaman F. Bohlmann
C. Zdero T.J. Mabry

Diterpenes of Flowering Plants

Compositae (Asteraceae)

Springer-Verlag
New York Berlin Heidelberg
London Paris Tokyo Hong Kong

Fred Seaman
Drug Dynamics Institute
College of Pharmacy
The University of Texas
Austin, Texas 78712, USA

Ferdinand Bohlmann
Institute of Organic Chemistry
Technical University
D-1000 Berlin 12
Federal Republic of Germany

Christa Zdero
Institute of Organic Chemistry
Technical University
D-1000 Berlin 12
Federal Republic of Germany

Tom J. Mabry
Department of Botany
The University of Texas
Austin, Texas 78712, USA

Cover: The plant illustrated is *Ageratina gypsophylla* B. Turner, a localized endemic species of Nuevo Leon, Mexico. *Ageratina* species produce both the acyclic geranylnerol-type and bicyclic *ent*-labdane diterpenes shown on the cover (in the background). The illustration is provided by B. Turner.

Library of Congress Cataloging-in-Publication Data
Diterpenes of flowering plants/F. Seaman ... [et al.].
    p. cm.
  Contents: Compositae (Asteraceae)
  Includes bibliographical references.
  ISBN 0-387-97058-4 (alk. paper)
  1. Diterpenes. 2. Angiosperms—Composition. 3. Angiosperms—Classification. 4. Botanical chemistry. 5. Plant chemotaxonomy.
I. Seaman, F. (Fred)
QK898.T4D57 1989
582.13′04192—dc20   89-21877

Printed on acid-free paper

Typeset by Technical Typesetting Incorporated, Baltimore, Maryland.
Printed and bound by R. R. Donnelley & Sons, Harrisonburg, Virginia.
Printed in the United States of America.

9 8 7 6 5 4 3 2 1

ISBN 0-387-97058-4 Springer-Verlag New York Berlin Heidelberg
ISBN 3-540-97058-4 Springer-Verlag Berlin Heidelberg New York

Contents

CHAPTER 1

Introduction

More than 1200 diterpenes have been identified from approximately 550 Compositae taxa (Figure 1 [pp. 219–384], Tables 1–3 [pp. 7–218]). The annual output of such reports rose sharply during the past several years, a trend that was also reflected by other major lipophilic constituents of the Compositae: acetylenes (*63, 465*), sesquiterpene lactones (*266, 458*), and benzofurans and chromenes (*429*). Unlike these other natural products, no review of the diterpene chemistry of the Compositae has been published. Given the bulk of data currently on hand, a review of this topic is clearly overdue.

Several major goals influenced the preparation of this review. The first was to provide a useful reference to guide terpenoid chemists to the literature of known compounds. The published reports for 1200 structures provide a sizable body of spectroscopic data applicable to the investigation of new compounds.

The study of diterpenes has been hampered by the absence of an all-inclusive nomenclature and classification system for the various skeletal types. Consequently, a second goal was to organize structural data into a classification scheme designed to group compounds by biogenetic homology, and to standardize such features as diterpene nomenclature and numbering. Pursuant to this goal, the published names (Table 1) and numbering systems (Table 2) for all compounds are included.

The third goal was to compare the distribution of diterpenes to the classification scheme used to subdivide the Compositae into its subfamilies, tribes, subtribes, genera, etc., and to seek chemical distribution patterns that paralleled the plant taxonomic boundaries. As diterpenes constitute a substantial body of raw taxonomic data, they may hold some promise as novel characters for use in plant taxonomy.

The fourth goal was to critically review the quality of published structure-elucidation arguments. As will be discussed in detail (Chapter six), eagerness to describe novel compounds in absolute stereochemical terms has led to much confusion, particularly in the description of new clerodane structures. The lengthy review of this topic hopefully will aid in the proper assessment of stereochemical assignments.

The fifth goal was to develop a computerized diterpene database from which this review's tables could be extracted. Major topics dealt with in each chapter represent some facet of programmed access to the diterpene database. Accomplishing this final goal simplifies the future updating of this review by automating the process of compiling up-to-date tabular information (e.g., Tables 1–3).

Diterpene data for the Compositae are included in Tables 1–3 (Chapter 2). In Table 1, the compound names and numbers are grouped according to membership in one of 103 different skeletal types and listed alphabetically within the skeletal-type group. Table 2 lists for each skeleton the compound numbers and associated molecular descriptions, while Figure 1 displays the corresponding structures. Table 3 lists all compound numbers and the Compositae taxa from which the compounds were reported together with the reference numbers corresponding to these reports. Distributional data were extracted from the literature and from the authors' unpublished results. Alphabetized lists of compound names and species appear in the Index together with citation page numbers.

In the discussion of diterpenoid biogenesis (Chapter 3), each skeletal type is described in terms of its likely mode of origin along one of two routes, the *ent*-labdane-based route or the normal-labdane-based route. Hypothetical biogenetic schemes for most skeletons are illustrated. Next, compound distribution is discussed in terms of patterns of structural variability found at different plant taxonomic levels (Chapter 4). Interspecific variation is detailed using six genera (*Brickellia*, *Stevia*, *Baccharis*, *Helichrysum*, *Helianthus* and *Montanoa*) representing the major diterpene-producing tribes of the Compositae. The next chapter (five) describes the biological activities associated with different diterpene skeletal types. The last chapter reviews the analytical methods used to identify diterpenes and provides an interpretation of spectroscopic data for several types, principally the clerodanes.

CHAPTER 2

Compound Names, Structures, and Sources

II.1. Names of Reported Compounds

The names (Table 1 [pp. 7–59]) and structures (Figure 1 [pp. 219–384]) of diterpenes reported from the Compositae are organized according to skeletal type following the sequence indicated in Figure 2 (p. 392). If a name was assigned to a structure at the time the compound was reported, that name was included in Table 1. Otherwise, the structure was given a name compatible with the names previously assigned to related compounds. If more than one name was applied in the literature to the same structure, usually the first-published name was included in Table 1, followed by additional names listed within brackets. The compound number listed with each name was used throughout this treatment when reference to the compound was made. Comments about a reported compound were also included in brackets on a line below the compound name. Usually these comments describe situations in which the name applied to the compound was incompatible with either the illustrated structure or the published spectroscopic data.

II.2. Structures of Reported Compounds

An alternative database-generated method of diterpene structure representation introduced in Table 2 (pp. 60–149) departs from the traditional approach using the "R-group" system for differentiating related compounds. The rationale for developing an alternative method is that by using it, a computer-generated set of structures (e.g. an expanded Table 2 contained within a future update of this review) can easily be produced from the diterpene database, even when the database has grown to many times its present size. The alternative insertion of many "R-group"-style structure illustrations into an updated version of Figure 1 would be prohibitively time consuming. Thus, illustrating structures in updates of this review will be handled by this alternative method.

II.2.1. Structure Description Codes

In Table 2, diterpene structures are identified by compound number and grouped with related structures belonging to the same skeletal type. The structural information given for each molecule (Table 2), together with the accompanying carbocyclic (or in some cases heterocyclic) skeleton, are sufficient to derive the compound's constitution and stereochemistry. By superimposing on the skeleton the substituents listed in the molecular description code, a molecule can be constructed with as much structural information as was included in the original report. In the molecular description, each substituent code is separated by semicolons within a string that begins with double bond substituents, proceeds through various oxygen substituents, and ends with hydrogens:

I.   Double Bonds
Double bonds are always indicated by a two-center code (e.g., 3(4), 9(11)) that identifies the two carbons linked by the double bond. If double bonds occur within the ring system, the steric relations around the double bond are defined by the skeleton illustrated with the description. Thus, for each cyclic double bond the steric relationship (Z or E) of the ring carbons attached to the doubly bonded carbons is defined by the relationship indicated in the illustrated skeleton.

For linear diterpenes or the linear segments of cyclic compounds with ambiguous double bond stereochemistries, the steric relationship about the double bond is indicated by a Z or E suffix (e.g., 13(14)E, 2(3)Z). The Z or E assignment results from the comparison of the sequence-rule-preference of the four atoms attached to the doubly bonded carbons (as specified in Rule E-2.2. of the IUPAC Rules for Nomenclature of Organic Chemistry). Several examples of Z and E assignment are discussed below in order to familiarize the reader with this method.

II.   Oxygen Functionalities
A.   Carboxyls: If present, the first-listed oxygen substituent is the carboxyl function (e.g., 19(=O,OH)). This code assigns a carbonyl group (=O) and a hydroxyl group (OH) to a specific center, carbon number 19. In compounds containing a lactone function, this carboxyl code is replaced by a single carbonyl group code (e.g. 19(=O)) and a separate string-entry indicating the position and orientivity of the oxido linkage (e.g., 19(—O—)6β).

B.   Ketones and Aldehydes: Ketones and aldehydes are denoted by the carbonyl group code, (=O), associated with a secondary (ketone) or primary (aldehyde) carbon of the skeleton (e.g. 3(=O)).

C.   Hydroxyls, Esters, etc.: Hydroxyls are indicated by a code, (OH), that is preceded by the number of the carbon to which it is attached and the orientation (α or β) of the attachment (e.g., 6β(OH)). Esters and other attached groups are specified in the same fashion, using approximately 80 different codes corresponding to the attached organic acid, etc. (e.g. 7α(OAng), 2β(OAc)). The Appendix lists the codes and structures for these attached groups. If attachment is through an oxygen, this is indicated in the code. For example, the codes for an acetate group and an acetyloxy substituent are respectively (Ac) and (OAc).

D.   Epoxides, Ether- or Oxido-Linkages, etc.: Epoxides and other oxido-bridges are indicated by a code (e.g. 3β(—O—)4β) specifying the two centers of attachment and, if known, the orientivity of attachment.

III.   Hydrogens

Hydrogens are included in order to define the orientivity (R or S) at a chiral center. For example, the entry, 8β(H), in reference to a labdane diterpene defines the orientation of the attached ring-methyl at C-8 as being alpha (α):

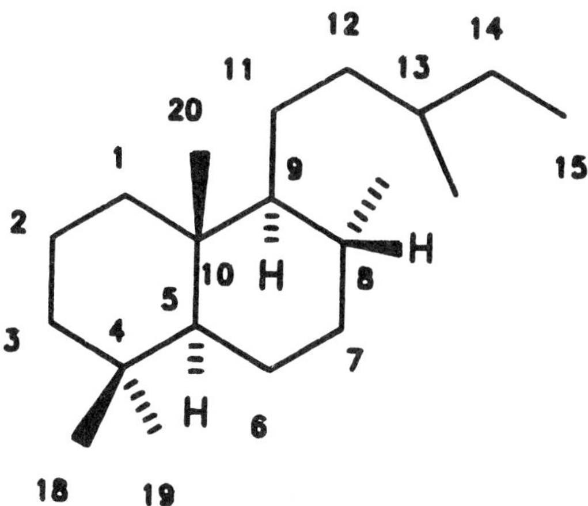

If no substituent is indicated for an available point of attachment at a skeletal carbon, then the presence of hydrogen(s) at that position is implicit.

2.2.2. Examples of Molecular Descriptions

I.   Linear diterpene example: 15,19-dihydroxy-14-oxo-geranylnerol
     Skeleton (No. 1. Geranylnerol):

Molecular description:

1004   2(3)Z;6(7)Z;10(11)E;14(=O);1(OH);15(OH);19(OH)

Because the skeleton immediately preceding the compound number 1004 description lacks double bonds, it is necessary to include in the molecular description the steric relationships about each of the three double bonds. The first double bond occurs at C-2-C-3 and according to sequence-rule-preference, the "Z" suffix stipulates that C-1 and C-4 occur on the same side of the double bond. The "Z" suffix of the C-6-C-7 double bond requires that the two sequence-rule-preferred groups on this double bond, the hydroxymethylene function (C-19) and C-5 (and its associated ligands), occupy the same side of the double bond. The "E" suffix of the 10,11-double bond indicates that, because C-9 and C-12 and their associated ligands are the sequence-rule-preferred groups, they must be on opposite sides of the double bond.

The remainder of the description indicates that a carbonyl group is located at C-14 and hydroxyl groups are located at C-1, C-15, and C-19.

II.   Bicyclic example: 14β-Carterothamnotriol-15-O-acetate
         Skeleton (No. 16. "Normal"-labdane):

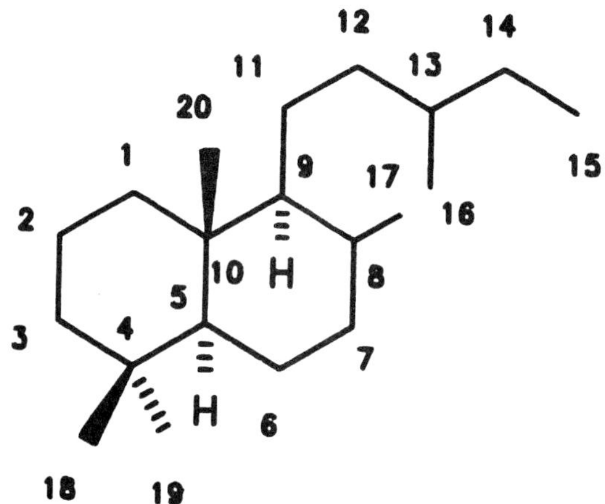

Molecular description:

95   12(13)E;9α(OH);14β(OH);15(OAc);8α(H)

The skeleton immediately above this description of compound number 95 possesses a six-carbon sidechain into which the single double bond must be inserted. Because this sidechain is not part of the cyclic portion of the molecule, the molecular description code, 12(13)E, includes an "E" defining the steric relationship around the double bond. Thus, sequence rule-preferred C-11 and C-14 are situated on opposite sides of the double bond.

The next three entries, 9α(OH), 14β(OH), and 15(OAc), indicate the placement of an α-oriented hydroxyl group at C-9, a β-oriented hydroxyl group at C-14 and an acetate at C-15. The presence of an α-oriented H-14 is implicit in the placement of a β-hydroxyl at C-14.

The code, 8α(H), defines the orientivity about C-8. In this case, it dictates β-orientation for the C-8 methyl group.

II.3. Plant Sources and References for Diterpenes

Table 3 (pp. 150–218) lists all diterpene compound numbers, the species of the Compositae from which they were reported and the reference numbers for the reports. Species names complete with authorities are listed alphabetically in the Species Index.

TABLE 1. Compound numbers and names of diterpenes from the Compositae (Asteraceae).

Compound:
No.   Name

I. Linear or Unicarbocyclic

1. Geranylnerol

1   Geranylnerol

2   Geranylnerol, 18-hydroxy-

3   Geranylnerol, 20-hydroxy-

4   Geranylnerol, 17,20-dihydroxy-

5   Geranylneral

6   Wyethic acid

7   Geranylnerol, 19-hydroxy-13-oxo-

8   Geranylnerol, 12,20-dihydroxy-16-oxo-

14   Viguieric acid [(2Z,6Z,10E)-3,15-dimethyl-7-carboxy-11-
      formyl-2,6,10,14-hexadecatetraen-1-ol]

15   Geranylnerol, 12,20-dihydroxy-19-acetoxy-

984   Geranylnerol, 19-acetoxy-20-hydroxy-

948   Geranylnerol, 8,12,19-trihydroxy-

957   Geranylnerol, 16,18,19-trihydroxy-

949   Geranylnerol-6,7-epoxide, 8,12,19-trihydroxy-

978   Geranylnerol, 19-hydroxy-12-oxo-

979   Geranylnerol, 6,7-epoxy-19-hydroxy-12-oxo-6,7-dihydro-

1002 Geranylnerol-19-oic acid

1003 Geranylnerol-19-oic acid, 1-acetyl-

1004 Geranylnerol, 15,19-dihydroxy-14-oxo-

1005 Geranylnerol acetate, 19-acetoxy-15-hydroxy-14-oxo-

1020 Geranylnerol, 20-acetoxy-10,11-epoxy-10,
      11-dihydro-9-hydroxy-

1095 Zoapatanol, pre-

1096 Tomexanthol, pre-

1109 Geranylnerol, 17,18-dihydroxy-

1110 Geranylnerol, 17,18,20-trihydroxy-

1111 Geranylnerol, 12-oxo-

TABLE 1. (*contd.*)
___

Compound:

No.   Name

1112 Geranylnerol, 18,19-diacetoxy-17,20-dihydroxy-

1113 Geranylnerol, 18,19,20-triacetoxy-17-hydroxy-

1114 Geranylnerol, 18-acetoxy-17,20-dihydroxy-

1115 Geranylnerol, 19-acetoxy-17,20-dihydroxy-

1116 Geranylnerol, 9,17-dihydroxy-

1117 Geranylnerol, 9,17,19-trihydroxy-

2. Geranylgeraniol

11   Geranylgeraniol, 12,19-dihydroxy-

12   Geranylgeraniol acetate, 12,19-dihydroxy-

13   Geranylgeraniol, 16-hydroxy-

16   Geranylgeraniol, 14,
      15-dihydro-4,14,15-trihydroxy-

17   Geranylgeranial

18   Geranylgeranial, 4-hydroxy-

19   Koanoadmantic acid

20   Geranylgeraniol,13-oxo-2,3-dihydro-

21   Geranylgeraniol, 19-hydroxy-12-oxo-10,11-dihydro-

22   Geranylgeraniol, 12,18-dihydroxy-6,7Z-

954   Hexa-deca-2E,6E,10E,14-tetraene, 1-hydroxy-11-carboxy-3,
      7,15-trimethyl-

1118 Geranylgeraniol, 17-hydroxy-18-acetoxy-

1119 Geranylgeranal, 12-oxo-

1120 Aspiliaparviol

3. Geranyllinalool

23   Geranyllinalool

24   Gerranyllinalool, 13-hydroxy-

25   Geranyllinalool, 13-acetoxy-

26   Geranyllinalool, 9-hydroxy-

27   Geranyllinalool, 9-acetoxy-

28   Geranyllinalool, 5-hydroxy-

TABLE 1. *(contd.)*

Compound:
No.   Name

29   Geranyllinalool, 9-acetoxy-5-hydroxy-

30   Geranyllinalool, 5,9-diacetoxy-

31   Geranyllinalool, 13-acetoxy-5-hydroxy-

32   Geranyllinalool, 12-oxo-

33   Geranyllinalool, 13-oxo-

34   Geranyllinalool, 13-oxo-14,15-epoxy-

35   Geranyllinalool, 15-hydroxy-13,14-dehydro-14,15-dihydro-

36   Geranylinalool, 14,15-dihydro-13,14,15,16-teradehydro-

37   Geranyllinalool, 14,15-dihydro-14,15-dihydroxy-

38   Geranyllinalool, 5,14,15-trihydroxy-14,15-dihydro-

39   Geranyllinalool, 14,15-dihydroxy-9-acetoxy-14,15-dihydro-

1121 Geranyl linlol, 13-acetoxy-

1122 Geranyl linalol, 12-oxo-10,11-dihydro-

1123 Geranyl linalol, 12-oxo-15-hydroxy-13,14E-dehydro-
      10,11,14,15-tetrahydro-

4.   Phytol-Derived

59   Neophytadiene

80. 14-Methylgeranylnerol

9   Geranylnerol, 12,19,20-trihydroxy-14-methylene-

10   Geranylnerol, 12,20-dihydroxy-19-acetoxy-14-methylene-

980   Smallantha-2Z,6Z,10E,13E-tetraene, 1,19-dihydroxy-

981   Smallantha-2Z,10E,13E-triene,
      1,19-dihydroxy-6,7-epoxy-12-oxo-

982   Smallantha-2Z,6Z,10E,14(21)-tetraene, 1,
      19-dihydroxy-12-oxo-

983   Smallantha-2Z,10E,14(21)-triene, 1,
      19-dihydroxy-6,7-epoxy-12-oxo-

1094 Tomentol, pre-

5.   Acanthoaustralane

TABLE 1. (*contd.*)

Compound:

No.   Name

40   Acanthoaustralide

41   Acanthoaustralide-1-O-acetate

42   Acanthoaustralide, 17-hydroxy-

43   Acanthoaustralide, 17-acetoxy-

1124 Acanthoaustralide, 1,6-diacetyl-

6. Acanthoaustralane-6,11-Epoxide

44   Melfusanolide, 1,10-dihydroxy-17-acetoxy-

45   Melfusanolide, 1,10,17-trihydroxy-

7. Isoacanthoaustralane

46   Isoacanthoaustralide, 17-hydroxy-

47   Isoacanthoaustralide, 17-acetoxy-

48   Isoacanthoaustralide-1-O-acetate

     [According to illustrated structures, **46** and **47** were assigned an orientation at 2(3) opposite to the one assigned earlier to **48**. No revision of **48** was indicated in the later publication.]

8. Geranylgeraniol-18,8-Lactone

49   Ichthyouleolide

50   Ichthyouleolide, 19-acetoxy-

51   Ichthyouleolide, 15-peroxy-13,14t-dehydro-14-15-dihydro-

52   Ichthyouleolide, 15-hydroxy-13,14t-dehydro-14-15-dihydro-

53   Ichthyouleolide, 14-peroxy-15,17-dehydro-14,15-dihydro-

54   Ichthyouleolide, 14-hydroxy-15,17-dehydro-14,15-dihydro-

9. Oxepane

55   Zoapatanol

56   Tomexanthin

57   Montanol

1097 Tomentol

TABLE 1. (*contd.*)

Compound:
No.   Name

1098 Tomexanthol

1099 Tomentanol

86. Melcanthane

1006 Melcantholide, 1,6,7-trihydroxy-17-acetoxy-

10. Mikanofurane

58   Mikanifuran

11. Centipedane

60   Centipedic acid

61   Conypododiol

1125 Furosolidagonone

1126 Furosolidagonol

1127 Furosolidagonone, 13-hydroxy-

1128 Furosolidagonone, 15-peroxy-13,14E-dehydro-14,15-dihydro-

12. Geranylgeraniol-1,20-Lactone

62   Dimeroperatic acid

63   Dimeroaperatate, methyl-

64   Dimerobrasiolide

1129 Furosolidagonone, 1-oxo-2,3-dehydro-1,2,3,20-tetrahydro-

1130 Gutiesolbriolide, 17-hydroxy-

1131 Gutiesolbriolide, 10E-17-hydroxy-

1132 Gutiesolbriolide, 17-acetoxy-

1133 Geranylgeraniol-20-acid lactone, 17,19-dihydroxy-18-acetoxy-
     6,7,10,11-tetrahydro-

1134 Geranylgeraniol-20-acid lactone, 17-hydroxy-18,19-diacetoxy-
     6,7,10,11-tetrahydro-

1135 Gutiesolbriolide, 17-hydroxy-iso-

1136 Gutiesolbriolide, 17-acetoxy-iso-

TABLE 1. *(contd.)*

Compound:
No.   Name

1137 Furosolidagonone, 20-oxo-1,2-epoxy-1,2,3,20-tetrahydro-

13. Geranylterpinene

65   Helicallen-16-ol

66   Helicallen-16-al

67   Helicallen-16-oic acid, 14,15-dihydro-

68   Geranyl-α-terpinene, 9-

69   Geranyl-α-terpinene, 10,11H-10,11-dihydroxy-9-

70   Geranylcurcumen,9-

71   Geranylcurcumen, 10,11H-10,11-dihydroxy-9-

      [Compounds **69** and **70** were named "11,12H-11,12-dihydroxy-..."
      derivatives, but this numbering was incompatible with
      illustrated structures and previously published homologous
      compounds.]

72   Geranylascaridol, β-10,11H-10,11-dihydroxy-9-

1101 Geranylascaridol, α-10,11H-10,11-dihydroxy-9-

1138 Geranyl-α-terpineol, 9-

14. Isocembrene

73   Isocembrene, 15-hydroxy-13,18H-

15. Geranylnerol-11,14-α-Epoxide

74   Ligantrol

75   Ligantrol-monoacetate

II. Normal-Bicarbocyclics and Derivatives

16. Normal-Labdane

76   Labda-8(17)-13-dien-15-ol

1103 Labd-13-ene, 8α,15-dihydroxy-

77   Acanthospermol-β-galactosidopyranoside

78   Labda-8(17),13-diene

TABLE 1. (*contd.*)

Compound:
No.   Name

80   Gymnospermin

79   Labdan-8α,15-diol

81   Labdan-8α,15-diol, 13-epi-

      [In published structure, C-5-H was shown as β-oriented.]

82   Carterothaminotriol

83   Carterothaminotriol-15-O-acetate

84   Carterothaminotriol-14,15-O-diacetate

85   Labd-8(17)-ene, 15-hydroxy-13,14-epoxy-

      [In the published name, this compound was called "...14,15-
      epoxy-...".]

86   Labd-7-ene, 13R,14R,15-trihydroxy-

87   Labd-(8,9)-ene, 13R,14R,15-trihydroxy-

88   Labd-8(17)-ene, 14,15-dihydroxy-

89   Labd-8(17)-ene, 13,14,15-trihydroxy-

90   Labd-8(17)-ene, 15-acetoxy-14-hydroxy-

993   Labd-7-ene, 3α,15-dihydroxy-

994   Labd-7-ene, 15-[2-methylbutyryloyloxy]-3α-hydroxy-

995   Labd-8(17)-ene, 3α,7α,15-trihydroxy-

91   Labd-8(17)-ene, 15-acetoxy-13,14-dihydroxy-

92   Labd-8(17)-ene, 14-acetoxy-15-hydroxy-

93   Labd-8(17)-ene, 15-acetoxy-13-hydroxy-14-propionyloxy-

94   Carterothamnotriol, 14β-

97   Brickellidiffusic acid angelate

98   Cativic acid, 2α,3α-[angeloyloxy-&
      2-hydroxy-2-methylbutyryloxy]-7α-trans-cinnamoyloxy-

99   Cativic acid, 2α,3α-[angeloyloxy- &
      2-hydroxy-2-methylbutyryloxy]-7α-cis cinnamoyloxy-

100   Cativic acid, 2-angeloyloxy-13,14Z-dehydro-

116   Labda-7,13Z-15-oic acid, 3α-hydroxy-

101   Cativic acid, 18-angeloyloxy-13,14-dehydro-

102   Cativic acid, 18-tigloyloxy-13,14-dehydro-

TABLE 1. *(contd.)*

Compound:

No.   Name

103   Cativic acid, 3α-angeloyloxy-2α-hydroxy-13,14Z-dehydro-

104   Cativic acid,3α-trans-cinnamoyloxy-2α-hydroxy-13,
14Z-dehydro-

105   Cativic acid, 3α-cis-cinnamoyloxy-2α-hydroxy-13,
14Z-dehydro-

106   Cativic acid, 3α-angeloyloxy-2α-hydroxy-(13Z)-13,
14-didehydro-

     [Compound 106 is identical to 103.]

107   Cativic acid, 2α-angeloyloxy-3α-hydroxy-(13Z)-13,
14-didehydro-

108   Cativic acid, 3α-angeloyloxy-2α-hydroxy-(13Z)-13,
14-didehydro-8(17)-

109   Cativic acid, 2α-angeloyloxy-3α-hydroxy-(13Z)-13,
14-didehydro-8(17)-

110   Batudioic acid

111   Labda-8(17),13E-dien-15-oic acid

112   Labda-7,13E-dien-15-oic acid, 2α-tigloyloxy-

113   Sempervirenic acid [3β-acetoxy-labda-7,13-dien-15-oic acid]

114   Labda-[7.13E-dien]-15-oic acid, 3α-hydroxy-

115   Labda-[7.13E-dien]-15-oic acid, 3β-hydroxy-

958   Labdan-15-oic acid, 8α-hydroxy-

117   Labda-7,13E-dien-15-oic acid, 3-oxo-

118   Cativic acid

991   Cativic acid, 3α-hydroxy-

992   Cativic acid methyl ester, 3α-hydroxy-

119   Labda-7-en-15-oic acid, 3α-hydroxy-

120   Labda-7-en-15-oic acid, 3β-hydroxy-

360   Labdane, (-)-7β,8α,15-trihydroxy-　　[Gymnospermin]

121   Labdanol acid

871   Labd-7-en-2α,15-diol, 13 -

872   Labd-7-ene, 2α-hydroxy-15-[3,4-dihydroxycinnamoyloxy]-13 -

873   Labd-7-ene,2α-hydroxy-15-
[4-hydroxy-3-methoxy-cinnamoyloxy]-13 -

TABLE 1. (*contd.*)

Compound:
No.   Name

874   Labd-7-ene, 2α-hydroxy-15-[4-hydroxy-cinnamoyloxy]-13 -

122   Cativic acid, 18-angeloyloxy-

123   Cativic acid, 18-tigloyloxy-

[Published names for 122 and 123 indicated that they were C-19-substituted, but the illustrated structures and reference to the substituents' axial orientation are compatible only with a C-18-substituted structure.]

124   Cativic acid, 2α,3α-dihydroxy-

125   Cativic acid, 2α-hydroxy-3α-[2-hydroxy-2-methylbutyryloxy]-

126   Cativic acid, 3α-angeloyloxy-2α-hydroxy- [Brickellia acid C]

127   Cativic acid, 3α-trans-cinnamoyloxy-2α-hydroxy-

128   Cativic acid, 3α-cis-cinnamoyloxy-2α-hydroxy-

129   Cativic acid, 3-oxo-

870   Labd-7-en-15-oic acid, 3β-hydroxy-2α-senecioyloxy-13 -

130   Agathenic acid, 8,17H-7,8-dehydro-

977   Labd-8(17)-en-15-al-19-oic acid

131   Agathenic acid, 8,17,13,14H-7,8-dehydro-

132   Biformene, 12,13E-

133   Biformene, 3α-hydroxy-12,13E-

134   Biformene, 3β-hydroxy-12,13E-

135   Biformene, 3β-acetoxy-12,13E-

136   Biformene, 3-oxo-12,13E-

959   Labd-8(17)-en-15-oic acid, 2β-acetoxy-

137   Communic acid

1104 Labd-8(17),12E,14-trien-19-oic acid

138   Communic acid, 7α-acetoxy-trans-

139   Abienol

140   Abienol, 7β-hydroxy-

141   Nidorellol

142   Nidorellol, 6α-hydroxy-

143   Nidorellol, 6α-angeloyloxy-

TABLE 1. *(contd.)*

Compound:

No.   Name

144   Abienol, 6β-hydroxy-12E-

145   Abienol, 7β-acetoxy-12E-

146   Abienol, 6β,7β-dihydroxy-12E-

147   Abienol, 6β,18-dihydroxy-12E-

148   Abienol, 7β-acetoxy-6β-hydroxy-12E-

149   Aristin-19-O-arachinate, 1β-acetoxy-

150   Aristin-19-O-coumarate, 1β-acetoxy-

151   Aristin-19-O-cis-coumarate, 1β-acetoxy-

152   Labda-12(E),14-diene, 9α,19-dihydroxy-

153   Sclareol, 6α-angeloyloxy-

      [Published structure contained a C-6β-oriented angelate,
      contradicting the published name.]

1102 Sclareol

996   (+)-Manool

1105 Manool, iso-

1106 Manool, 3β-hydroxy-iso-

1107 Manool, 3-oxo-iso-

154   Salicifoliol

      [Absolute stereochemistry not indicated.]

155   Abienol, iso-

156   Abienol, 7β-acetoxy-6β-hydroxy-12β-peroxiiso-

157   Abienol, 7β-acetoxy-6β-hydroxy-12α-peroxiiso-

158   Solidagenone

159   Lambertianic acid, 17-hydroxyiso-

160   Gutierrezianol acid-[2'-methylbutyrate]

161   Gutierrezianol acid-isobutyrate

162   Gutierrezianol acid-isovalerate

163   Lambertianic acid, 7α-hydroxy-

1021 Austroinulin-7-O-acetate, (12Z)-6-desoxy-

1022 Abienol, 6α-hydroxy-7β-acetoxy-12E-

TABLE 1. (*contd.*)

Compound:
No.   Name

1023 Austroinulin-7-O-acetate, (12Z)-

1024 Labdenol acid, (13Z)-

1139 Labdan-8α,15-diol, 13β-H-

1140 Cordobic acid [7α,18-Dihydroxylabd-8(17)-en-oic acid]

1141 Cordobic acid 18-acetate

1142 Cordobic acid, 7-epi-

1143 Salvic acid

1144 Salvic acid, 7α-acetyl-

1145 Discoidic acid [4β-hydroxymethyllabd-7-en-15-oic acid]

1147 Labd-8(17),12E,14-trien-19-oic acid

96   Gutierrezia sphaerocephala labdane-α-L-arabinopyranoside 1

97   Gutierrezia sphaerocephala labdane-α-L-arabinopyranoside 2

1148 Gutierrezia sphaerocephala labdane-β-D-arabinopyranoside 3

1149 Gutierrezia sphaerocephala labdane-β-D-arabinopyranoside 5

1150 Gutierrezia sphaerocephala labdane-β-D-arabinopyranoside 6

1151 Abienol, 3β-acetoxy-iso-

1152 Gnaphala-13(16),14-diene, 8α-hydroxy-

17. Grindelane

165   Grindelic acid, 7α-hydroxy-7,8-dihydro-8(17)-dehydro-

166   Grinde-6,8(17)-dienic acid

167   Grindelic acid

168   Grindelic acid, 1α-hydroxy-

169   Grindelic acid, 3β-hydroxy-

170   Grindelic acid, 6α-hydroxy-

172   Grindelic acid, 6β-hydroxy-

173   Grindelic acid, 17-hydroxy- [Oxygrindelic acid]

171   Grindelic acid, 6α-formyloxy-

174   Grindelic acid, 17-acetoxy-

175   Grindelic acid, 17-propionyloxy-

TABLE 1. *(contd.)*

Compound:

No.   Name

176 Grindelic acid, 17-isobutyryloxy-

177 Grindelic acid, 17-isovaleryloxy-

178 Grindelic acid, 17-[2-methylbutyryloxy]-

179 Grindelic acid, 17-methoxy-

180 Grindelic acid, 19-hydroxy-

181 Grindelic acid, 19-acetoxy-

182 Grindelic acid, 19-isovaleryloxy-

183 Grindelic acid, 19-isobutyryloxy-

184 Grindelic acid, 19-[2-methylbutyryloxy]-

185 Grindelic acid, 19-succinyloxy-

186 Grindelic acid, 19-oxo-

187 Grindelic acid, 19-oic-

188 Grindelic acid, 18-hydroxy-

189 Grindelic acid, 6-oxo-

190 Grindelic acid, 19-acetoxy-6-oxo-

191 Grindelic acid, 7β,8β-epoxy-7,8-dihydro-

192 Grindelic acid, 7α,8α-epoxy-7,8-dihydro-

955 Grindelic acid, iso-

956 Grindelic acid, 17-grindeloxy-

193 Solidagenone, 9-dehydroxy-9α,13α-epoxy-13,16-dihydro-

194 Solidagenone, 13-epi-dehydroxy-9α,13α-epoxy-13,16-dihydro-

1035 Grindelic acid, 3α-hydroxy-

1036 Grindelic acid, 7β-hydroxy-8(17)-dehydro-7,8-dihydro-

1060 Grindelic acid-15-O-arabinoside

1153 Grindelate, methyl-

1187 Grindelic acid 15-O-arabinoside

     [Incorrectly illustrated as an <u>ent</u>-labdane]

18. Normal-Labdane-8,13-Epoxide

272 Manoyl oxide

TABLE 1. (*contd.*)

Compound:
No. Name

195 Jhanol

196 Jhanol acetate

197 Jhanidiol

198 Jhanidiol-19-monoacetate

199 Jhanidiol-diacetate

200 Manoyl oxide, 3β-hydroxy- [18-deoxylazarcardenasol]

201 Manoyl oxide, 18-benzoyloxy-3β-hydroxy-
     [18-benzoyloxylarcardenasol]

964 Manoyl oxide, 15-oxo-14,15-dihydro-

273 Manoyl oxide, 13-epi-

202 Manoyl oxide, 18-hydroxy-13-epi-

     [Published structure was shown with a C-18 substituent but
     given a name indicating C-19 substitution. Reference to the
     substituent's axial orientation indicates a C-18
     substituent.]

19. Normal-Labdane-8,12-Epoxide

203 Carterochaetol, 16-hydroxy-

204 Carterochaetol, 16-acetoxy-

205 Carterochaetol acetate, 16-acetoxy-

206 Carterochaetol, 16-oxo-

207 Carterochaetal, 16-oxo-

208 Carterochaetol, 16-hydroxy-13,14H-13α,14α-epoxy-

209 Carterochaetol, 16-hydroxy-13,14H-13β,14β-epoxy-

210 Carterochaetol

211 Carterochaeta acid

212 Carterochaetal, 16-oxo-13,14H-12,13-dehydro-

213 Carterochaetal, 16-acetoxy-14-oxo-13,14H-12,13-dehydro-

20. Normal-Labdane-15,16-Lactone

164 Agathenic acid-16-lactone, 8,17H-7,8-dehydro-

965 Labda-7,13-dien-15-acid-16-lactone, 17-hydroxy-

TABLE 1. (*contd.*)
____

Compound:

No. Name

966 Labda-7,13-dien-15-acid-16-lactone, 17-oxo-

21. Normal-Labdane-20,12-Lactone

214 Nidorella lactone

22. Normal-Labdane Spiroketal

215 Brickellidiffusic acid spiro ketallactone,
     2α-angeloyloxy-17,O-dihydro-

216 Brickellidiffusic acid spiro ketal lactone,
     2α-angeloyloxy-17α-hydroxy-17,O-dihydro-

217 Brickellidiffusic acid spiro ketal lactone,
     2α-angeloyloxy-17β-hydroxy-17,O-dihydro-

218 Brickellidiffusic acid spiro ketal lactone, 2α-hydroxy-17,
     O-dihydro-

88. Normal-Labdane-15,12-Lactone

938 Evillosin

23. Normal-Labdane, 7,8-Seco-

219 Koanolabda-12E,14-diene, seco-

24. Normal-Labdane, 3,4-Seco-

220 Cativin-3,15-dioic acid, 3,4-seco-

25. Grindelane, 8,9-Seco-

223 Chrysothane   [Strictanonic acid]

89. Grindelane, 19-Nor-

221 Grindelic acid, 4α-hydroxy-19-nor-

222 Grindelic acid, 4α-formyloxy-19-nor-

1154 Grindelate, methyl-4β-hydroxy-6-oxo-19-nor-

TABLE 1. (*contd.*)

Compound:

No.   Name

26. Grindelane, 8,17-Bis-Nor-8,9-Seco-

224   Grindelistrictic acid

27. Normal-Labdane-20,12-Lactone, 5,10-Seco-

225   Nidorella lactone, seco-

226   Nidorella lactone-6-O-angelate, seco-

227   Nidorella lactone-6-O-isobutyrate, seco-

228   Nidorella lactone-6-O-[2-methylbutyrate], seco-

229   Nidorella lactone-6-O-isovalerate, seco-

230   Nidorella lactone-6-O-angelate, 5-deoxy-5-hydroxy-5H-seco-

100. Normal-Labdane, 14,15-Bis-Nor-

1155 Labden-13-one, 14,15-bisnor-8-hydroxy-11E-

1157 Sterebin A

1158 Sterebin B

1159 Sterebin C

1160 Sterebin D

101. Normal-Labdane, 13,14,15,16-Tetra-Nor-

1156 Norambreinolide

87. Hexanorlabdane

941   Jhanilactone

28. Relhaniane

231   Relhania acid

232   Koanophyllic acid B

233   Koanophyllic acid A

234   Koanophyllic acid C

TABLE 1. *(contd.)*

Compound:
No.   Name

235   Koanophyllic acid D

    [Absolute stereochemistry for the koanophyllic acid series was unproven.]

29. <u>trans</u>-Normal-Clerodane

(No proven structures of this type)

30. Clerodane, 5,10-Seco-

255   Nidoreseda acid, seco-   [Conyzic acid, strictic acid]
    [Revised to 9α-methyl, = 573]

256   Nidoreseda acid methyl ester, seco-

    [Revised to 9α-methyl, = 573 methyl ether]

31. <u>cis</u>-Normal-Clerodane

238   Solidago alcohol, 8-epi-

    [Reported as 5β,10α-<u>trans</u>-clerodane]

253   Nidorellalactone, iso-

    [Reported as 5β,10α-<u>trans</u>-clerodane]

254   Nidorellalactone

    [Reported as 5β,10α-<u>trans</u>-clerodane]

257   Clerodan-15-oic acid, 3,4-dehydro-cis-

258   Solidagolactone II   [Elongatolide C]

259   Solidagolactone III

260   Elongatolide E   [Solidagolactone VII]

261   Solidagolactone III, 3α,4α-epoxy-3,4-dihydro-

262   Solidagolactone V

263   Solidagolactone II, 2α-hydroxy-   [Solidago virgaurea compound 4a]

264   Solidagolactone III, 2α-hydroxy-   [Solidago virgaurea compound 4c]

265   Solidagolactone V, 2β-hydroxy-

266   Solidagolactone II, 3β,4α-dihydroxy-3,4-dihydro-

267   Solidagolactone III, 3β,4α-dihydroxy-3,4-dihydro-

TABLE 1. (*contd.*)

Compound:

No. Name

268 Solidagolactone II, 3-oxo-3,4-dihydro-

269 Solidagolactone III, 3-oxo-3,4-dihydro-

545 Elongatolide A [SolidagolactoneIV;
     6β-hydroxy-solidagolactone]

546 Elongatolide B

551 Kolavenic-15-acid lactone, 2α,16-dihydroxy-3α,4α-epoxy-
     [Revised from <u>trans</u>-ent-clerodane]

552 Kolavenic-15-acid lactone, 6β-angeloyloxy-16-hydroxy-3α,
     4α-epoxy-
     [Revised from <u>trans</u>-ent-clerodane]

553 Elongatolide D [Solidagolactone VI(structure no.555)]

988 Solidagolactone VIII

239 Solidagoic acid A

240 Solidago aldehyde

241 Solidago alcohol

242 Solidago glycol

243 Solidago dialdehyde

244 Solidagoic acid B

246 Solidago lactol

247 Solidago epoxylactol

270 Guteriolide

556 Kolavenic acid lactone, 16-hydroxy-3,4-epoxy-
     [Structure was revised to 5α,10α-cis]

557 Kolavenic acid lactone, 7β-angeloyloxy-16-hydroxy-3,
     4-epoxy-

     [Revised to 5α,10α-cis]

1025 Kolavenool, 6β-acetoxy-2-oxo-

1062 Solidago gigantea compound I

1063 Solidago gigantea compound III

1064 Solidago gigantea compound IV

1065 Solidago gigantea compound V

1066 Solidago gigantea compound II

1067 Solidago gigantea compound VI

TABLE 1. (*contd.*)

Compound:
No. Name

1068 Solidago gigantea compound VII

271 Chrysolic acid

1083 Cleroda-3,13(14)-diene-15,16-olide, 6α,18-dihydroxy-cis-

1084 Cleroda-3,13(14)-diene-15,16-olide, 18,19-dihydroxy-cis-

1085 Cleroda-3,13(14)-diene-15,16:18,19-diolide, cis-

1086 Cleroda-3,13(14)-diene-15,16-olide,
     18,19-epoxy-19α-hydroxy-cis-

1087 Cleroda-13(14)-ene-15,16-olide, 3α,4:18,19-diepoxy-18β,
     19α-dihydroxy-cis-

1088 Cleroda-13(14)-ene-15,16:18,19-diolide,
     3α,4-epoxy-19α-hydroxy-cis-

1089 Cleroda-13(14)-ene-15,16-olide,
     3α,4:18,19-diepoxy-19α-hydroxy-cis-

1090 Cleroda-13(14)-ene-15,16-olide, 3α,4β,19α-trihydroxy-18,
     19-epoxy-cis-

1091 Cleroda-3,13(14)-diene-15,16-olide-19-oic ester,
     19-O-α-L-arabinopyranosyl-cis-

1092 Cleroda-3,13(14)-diene-15,16:18,6α-diolide,
     2β,6α-dihydroxy-cis-

90. Tricycloclerodane

947 Tricyclosolidagolactone

III. Normal-Tricarbocyclics and Derivatives

32. Sandaracopimarane (Normal-Pimarane)

274 Sandaracopimaradiene

927 Sandaracopimar-8(14),15-diene, 7α-hydroxy-

928 Sandaracopimar-8(14),15-diene, 1β,7α-dihydroxy-

929 Sandaracopimar-8(14),15-diene, 7α-hydroxy-1β-acetoxy-

930 Sandaracopimar-8(14),15-diene, 1β-hydroxy-7α-acetoxy-

931 Sandaracopimar-8(14),15-diene, 1β,11α-dihydroxy-

932 Sandaracopimar-8(14),15-diene, 1β,11α-diacetoxy-

TABLE 1. (*contd.*)

Compound:
No.　Name

933　Sandaracopimar-8(14),15-diene, 1β,
　　11α-dihydroxy-7α-acetoxy-

934　Sandaracopimar-8(14),15-diene, 7α-hydroxy-1β,
　　11α-diacetoxy-

935　Sandaracopimar-8(14),15-diene, 1β,7α,11α-triacetoxy-

936　Sandaracopimar-8(14),15-diene, 1β,11α-diacetoxy-6,7-epoxy-

937　Sandaracopimar-8(14),15-diene, 1β,11α-diacetoxy-7-oxo-

275　Sandaracopimaradiene, iso-

276　Sandaracopimar-15-ene, 8β-hydroxy-

277　Sandaracopimar-15-en-6β,8β-diol

278　Sandaracopimar-15-en-8β-ol, 6β-acetoxy-

279　Sandaracopimar-15-en-8β,11α-diol

280　Sandaracopimar-15-ene, 8β-hydroxy-11α-senecioyloxy-

281　Sandaracopimar-15-ene, 8β-hydroxy-11α-tiglinoyloxy-

282　Sandaracopimar-15-ene, 8β-hydroxy-11α-acetoxy-

283　Sandaracopimar-15-en-11-one, 8β-hydroxy-

284　Sandaracopimar-15-ene, 8β-hydroxy-12β-acetoxy-

285　Sandaracopimar-15-en-3β,18-diol

286　Sandaracopimar-15-ene, 8β,20-dihydroxy-

287　Sandaracopimar-15-ene, 20-hydroxy-8β,20-oxido-

288　Sandaracopimar-15-ene, 8β,11α,12β-trihydroxy-

289　Sandaracopimar-15-en-8β,12β-diol, 11α-acetoxy-

290　Sandaracopimar-15-ene, 8β,12β-dihydroxy-11α-senecioyloxy-

291　Sandaracopimar-15-ene, 8β,12β-dihydroxy-11α-tiglinoyloxy-

292　Sandaracopimar-15-en-8β,11α-diol, 12β-acetoxy-

293　Sandaracopimar-15-en-8β,11α-diol,
　　12β-(p-hydroxycinnamoyloxy)-

294　Sandaracopimar-15-en-8β-ol, 11α,12β-diacetoxy-

295　Sandaracopimar-15-en-8β,
　　12α-diol, 11β-acetoxy-

296　Sandaracopimar-15-en-11-one, 8β,
　　12α-dihydroxy-

TABLE 1. (*contd.*)

Compound:

No. Name

297 Sandaracopimar-15-en-11-one, 12β-acetoxy-8β-hydroxy-

298 Sandaracopimar-15-en-6β,8β,11α-triol

299 Sandaracopimar-15-en-8β,11α-diol, 6β-acetoxy-

300 Sandaracopimar-8(14),15-dien-18-oic acid

301 Sandaracopimar-7,15-dien-18-oic acid

302 Sandaracopimar-8(14),15-dien-18-oic acid, 3β-acetoxy-

807 Trinervinol [Isopimara-8(14)-en-3β,15,16,17-tetrol]

303 Sandaracopimar-7,15-dien-18-oic acid, 3β-acetoxy-

921 Sandaracopimar-8(14)-en-19-oic acid, 6β,15,16-trihydroxy-

304 Palarosan, 3α-hydroxy-13-epi-

305 Palarosan, 9α-hydroxy-13-epi-

891 Pimar-15-en-8β-ol, 13-epi-

306 Sandaracopimar-7-ene, 3α,15,16-trihydroxy-13-epi-
     [Revised to rosane structure, = **313**; 168]

307 Sandaracopimar-7-ene, 9α,15,16-trihydroxy-
     [Revised to rosane structure, = **314**; 168]

308 Julslimtetrol

922 Sandaracopimar-8(14)-en-6β-D-glucoside, 15,16-dihydroxy-

923 Sandaracopimar-8(14)-ene, 6α,15,16,18-tetrahydroxy-

     [Compounds **922** and **923** were illustrated with β-oriented C-10 methyl groups.]

34. Cassane

310 Osteomuricone, 2α-hydroxy-

311 Osteomuricone, 1β,2α-dihydroxy-

312 Osteomuricone, 1β,2α-dihydroxy-15,16-dihydro-

35. Rosane

306 Palarosane, 3α-hydroxy

313 Palarosane, 3β-hydroxy-

314 Palarosane, 3β,18-dihydroxy-   [Jesromotetrol]

TABLE 1. (*contd.*)

Compound:
No.   Name

315   Palarosane, 3-oxo-

881   Rimuen-5β-ol, ent-

882   Rimuen-5β,18-diol, ent-

883   Rimuen-18-yl acetate, 5β-hydroxy-ent-

884   Rimuen-3α,5β-diol, ent-

885   Rimuen-18-yl tiglate, 5β-hydroxy-ent-

886   Rimuen-3α,5β,18-triol, ent-

887   Rosa-5,15-dien-3α,18-diol

888   Rosa-5,15-dien-3α-yl acetate, 18-hydroxy-

889   Rosa-5,15-dien-18yl acetate, 3α-hydroxy-

890   Rosa-1(10),15-dien-18-oic acid

      [Published structure indicated a C-19 acid, while the
      published name and other data indicated a C-18 substituent.]

1161 Jesromotrol, 3β,19-diacetoxy-

1162 Jesromotrol, 3β-acetoxy-

36. Sandaracopimarane, 19-Nor-13-Epi-

309. Julslimdiolone, nor-

37. Normal-Abietane

316   Abietadien-2α-ol acetate

317   Abietadien-3β-ol, 7,13-

318   Abietadien-3-one, 7,13-

      [Compounds **316-317** are probably artefacts of 8(9),13(14)-
      abietadiene-type compounds.]

319   Ferruginol

320   Inuroyleanol

321   Royleanone

322   Royleanone, dehydro-

323   Royleanone, 7-acetoxy-

324   Royleanone, 7-keto-

TABLE 1. (*contd.*)
____

Compound:
No.   Name

1072 Junceanol W

1073 Junceanol X

1074 Junceanol Y

1163 Turbinatone

94. Stevisalane

1026 Stevisalicinone

IV. Normal-Tetracarbocyclics

73. Normal-Stachane

325   Cupressen-18-oic acid
      [= 774?]

V. ent-Bicarbocyclics and Derivatives

38. ent-Labdane

326   ent-Labda-8(17),13-dien-15-ol

848   ent-Labda-8(17),13-dien-15-ol, 2α-hydroxy-

327   ent-Labda-8(17),13-dien-15-O-bernsteinoate

849   ent-Labda-8(17),13-dien-15-ol, 2-oxo-

328   ent-Labda-8(17),13,14E-dien-15-ol, 7β-acetoxy-

844   ent-Labda-8(17),13E-diene, 3α-angeloyloxy-15,18-dihydroxy-

329   ent-Labda-8(17),13E-dien-15-ol-18-oic acid
      [Viscidic acid A]

330   ent-Labda-8(17),13E-dien-15-acetoxy-18-oic acid
      [Viscidic acid B]

331   ent-Labda-8(17),13E-dien-15-al, 2α-hydroxy-

332   ent-Labda-8(17),13Z-dien-15-al, 2α-hydroxy-

333   ent-Labda-8(17),13E-dien-15-al, 2-oxo-

334   ent-Labda-8(17),13Z-dien-15-al, 2-oxo-

335   ent-Labda-8(17),13E-dien-15-oic acid [E-Copalic acid]

336   ent-Labda-8(17),13Z-dien-15-oic acid   [Z-Copalic acid]

337   ent-Labda-8(17),13Z-dien-15-oic acid, 3α-hydroxy-

TABLE 1. (*contd.*)

Compound:

No. Name

338 ent-Labda-8(17),13Z-dien-15-oic acid, 3α,7α-dihydroxy-

339 ent-Labda-13Z-en-15-oic acid, 3α-hydroxy-17-oxo-8β-H-

340 ent-Labda-13Z-en-15-oic acid, 3α-hydroxy-17-oxo-8α-H-

341 ent-Labda-13-en-15-oic acid, 3,17-dioxo-8β-H-

342 ent-Labda-13Z-en-15-oic acid, 3,17-dioxo-8α-H-

343 ent-Labda-13Z-en-15-oic acid, 3α,8α-dihydroxy-17-oxo-8β-H-

 [Published name, "3α,9β-dihydroxy-17-oxo-8β-H-ent-labda-13Z-
 en-oic acid", was inconsistent with the structure and
 published data.]

344 ent-Labda-13Z-en-15,17-dioic acid, 8β-H-

345 ent-Labda-13E-en-15,17-dioic acid, 3β-H-

346 ent-Labda-13Z-en-15-oic acid, 3-oxo-8,17-epoxy-

347 ent-Labda-8(17),13-dien-15-O-bernsteinoate, 8,17-dihydro-8β,
 17-epoxy-

348 ent-Labda-13(14)E-en-15-ol, 8-hydroxy-

349 ent-Labda-13(14)E-en-15-al, 8-hydroxy-

350 ent-Labda-13(14)Z-en-15-al, 8-hydroxy-

351 ent-Labda-7,8,13,14(Z)-diene, 6β,15-dihydroxy-

352 ent-Labda-7,8,13,14(Z)-dien-6-one, 15-hydroxy-

353 ent-Labda-7,13-diene, 3α,15-diacetoxy-11-hydroxy-

354 ent-Labda-7,13E-dien-15-oic acid

355 ent-Labda-7,13-dien-15-oic acid, 2β-angeloyloxy-
 [Dendroidinic acid]

356 ent-Labda-6,8(17)-dien-15-oic acid

357 ent-Labda-8(17)-ene, 2α,15-dihydroxy-

358 Eperuic acid, 2α-iso-valeroyloxy-

 [C-13 orientivity conflicts with earlier assigned
 stereochemistry for this center in eperuic acid.]

857 ent-Labda-8(17)-en-15-oic acid, 2α-hydroxy-

858 ent-Labda-8(17)-en-15-oic acid, 2α-angeloyloxy-

859 ent-Labda-8(17)-en-15-oic acid, 2α-tigloyloxy-

860 ent-Labda-8(17)-en-15-oic acid, 2α-[2-methylbutyryloxy]-

TABLE 1. *(contd.)*

Compound:

No.   Name

861   ent-Labda-8(17)-en-15-oic acid, 2-oxo-

862   ent-Labdan-15-oic acid, 2α-angeloyloxy-8β,17-epoxy-

863   ent-Labdan-15-oic acid, 2α-[2-methylbutyryloxy]-8β,17-epoxy-

359   ent-Labda-8(17)-en-15-ol, 2-oxo-

360   Haplopappus pectinatus ent-labdane 16

361   ent-Labda-7-en-15-oic acid, 2β-angeloyloxy-
         [13,14-Dihydrodendroidinic acid]

362   ent-Labda-7(8),13E-diene, 15,16,17-trihydroxy-

363   ent-Labda-6,13-diene, 15,16-diacetoxy-

364   ent-Labda-6,13-diene, 15,16,17-triacetoxy-

365   ent-Labda-6,8(17),13-trien-15-oic acid, 16-hydroxy-

366   ent-Labda-7,13E-dien-15-oic acid, 16-hydroxy

367   ent-Labda-7,13E-dien-15-oic acid, 16,17-dihydroxy-

368   ent-Labda-7,13Z-dien-15-oic acid, 12-hydroxy-16-oxo-

369   ent-Labda-7,13E-dien-15-oic acid, 16-hydroxy-17-oxo-

370   ent-Labda-7,13E-dien-15-oic acid, 17-hydroxy-16-acetoxy-

371   ent-Labda-7,13-dien-15-oic acid, 16-acetoxy-17-oxo-

372   ent-Labda-7,13E-dien-15-oic acid,
         16-acetoxy-6α-hydroxy-17-oxo-

373   ent-Labda-7,13E-dien-15-oic acid,
         16-acetoxy-6β-hydroxy-17-oxo-

374   ent-Labda-7,13Z-dien-15-oic acid, 16-acetoxy-6-oxo-

375   ent-Labda-7,13Z-dien-15-oic acid, 6,16-dioxo-

376   ent-Labda-7,13E-dien-15-oic acid, 16-hydroxy-6-oxo-

377   ent-Labda-7,13E-dien-15-oic acid, 16-acetoxy-6-oxo-

378   ent-Labda-7,13E-dien-15-oic acid, 16-methoxy-6-oxo-

379   ent-Labda-6,13E-dien-15-oic acid, 8α,16-dihydroxy-

380   ent-Labda-6,13E-dien-15-oic acid, 16-acetoxy-17-oxo-

381   ent-Labda-6,13E-dien-15,17-dioic acid, 16-acetoxy-

382   ent-Labda-8(17),13-diene, 7β,15,16-trihydroxy-

383   ent-Labda-8(17)-en-15-oic acid, 16-oxo-

TABLE 1. (*contd.*)

Compound:

No. Name

384 ent-Labda-6,8(17),13-trien-15-oic acid lactone, 16-hydroxy-

385 ent-Labda-8(17),13-dien-15-oic acid lactone, 16,19-dihydroxy-

386 ent-Labda-7,13-dien-15-oic acid lactone, 16-hydroxy-

387 ent-Labda-7,13-dien-15-oic acid lactone, 12,16-dihydroxy-

388 ent-Labda-7,13-dien-15-oic acid lactone, 16,16α-dihydroxy-

389 ent-Labda-7,13-dien-15-oic acid lactone, 16,16β-dihydroxy-

390 ent-Labda-7,13-dien-15-oic acid lactone, 16,12,16-trihydroxy-

391 ent-Labda-7,13-dien-15-oic acid lactone, 16,12,16α-trihydroxy-

392 ent-Labda-7,13-dien-15-oic acid lactone, 16,12,16β-trihydroxy-

393 ent-Labda-7,13-dien-15-oic acid lactone, 16-hydroxy-6-oxo-

394 ent-Labda-7,13-dien-15-oic acid lactone, 16,16-dihydroxy-6-oxo-

395 ent-Labda-7,13-dien-15-oic acid lactone, 16,16α-dihydroxy-17-oxo-   [Acritolongifolide A]

396 ent-Labda-7,13-dien-15-oic acid lactone, 16,16β-dihydroxy-17-oxo-   [Acritolongifolide B]

397 ent-Labda-8(17),13(14)-dien-16-oic acid lactone, 7β,15-dihydroxy-

843 Nivenolide

398 ent-Labda-7,13-dien-16-oic acid lactone, 15-hydroxy-

399 ent-Labda-7,13-dien-16-oic acid lactone, 15,12,15β-trihydroxy-

400 ent-Labda-7,13-dien-16-oic acid lactone, 15,12,15α-trihydroxy-

401 Polyalthine, 7β-hydroxy-   [Austrochaparol]

402 Austrochaparol acetate

403 Daniellol

404 Psiadiol

839 Daniellic acid

840 Daniellic acid, 3α-hydroxy-

841 Daniellol, 18-hydroxy-

TABLE 1. (*contd.*)

Compound:

No. Name

842 Daniellol, 3α,18-dihydroxy-19-deoxy-

974 ent-Labda-8(17)-ene, 3α-angeloyloxy-18-hydroxy-13-furyl-

975 ent-Labda-8(17)-ene, 18-angeloyloxy-3α-hydroxy-13-furyl-

830 Polyalthic acid

831 Polyalthic acid, 3α-hydroxy-

832 Polyalthic acid, 3α-angeloyloxy-

833 Polyalthic acid, 3α-[2-methylbutyryloxy]-

834 Polyalthic acid, 3α-isobutyryloxy-

835 Polyalthic acid, 19-hydroxy-

836 Polyalthic acid, 8α,17-epoxy-8,17-dihydro-

837 Polyalthic acid, 8β,17-epoxy-8,17-dihydro-

838 Polyalthic acid, 17-oxo-8β,17-dihydro-

405 Conyzic acid

406 Austrofolin-12-ol, 12-desoxo-

407 Austrofolin

408 Austrochaparol acetate, 18-hydroxy-

409 Austrochaparol acetate, 19-acetoxy-

410 Austrochaparol, 19-oxo-

411 Austrochaparol acetate, 18-oxo-

412 Austrofolin, 3α-hydroxy-

413 Austrofolin, 3α,12-dihydroxy-12-desoxo-

414 Polyalthine, 17-hydroxy-iso-

415 ent-Labda-7,13(16),14-trien-15,16-epoxy-18-oic acid,
     5β,9βH,10α-

416 ent-Labda-13(16),14-diene, 3α-angeloyloxy-8-hydroxy-15,
     16-epoxy-

417 Austrofolin-15α-ol, 15,16-dihydro-

418 Austrofolin-15β-ol, 15,16-dihydro-

421 Pumiloxide, 6β-hydroxy-

     [Published structure indicated a C-7β-hydroxyl group, but
     name and data contradicted structure.]

TABLE 1. (*contd.*)

Compound:

No.   Name

422   ent-Biformene

423   ent-Biformene, 18-oxo-

424   Ozic acid, cis-

425   Ozic acid, 12,13E-
[ent-Labda-8(17),12E,14-trien-18-oic acid]

426   Ozic acid, 2α-hydroxy-12,13Z-

427   Ozic acid, 2-oxo-12,13Z-

428   Austroinulin

429   Austroinulin, 6-O-acetyl-   [Stevinsol]

808   Austroinulin, 6-angelyl-7-acetyl-

430   ent-Biformene, 12,13E-

431   ent-Labda-8(17),12E,14-triene, 19-hydroxy-

432   ent-Labda-8(17),12E,14-triene, 3α,19-dihydroxy-

433   ent-Labda-8(17),12E,14-triene, 18-hydroxy-

434   ent-Labda--8(17),12E,14-triene, 18,19-dihydroxy-

435   Ozic acid, 1β-acetoxy-trans-   [1β-acetoxy-ent-Labda-8(17),
12E,14-trien-19-oic acid]

[Published structure contained a C-1α-oriented acetate,
while the name indicated that the acetate was β-oriented.
Although the compound was named as an ozic acid derivative,
the published structure contained an α-oriented carboxyl
function rather than the β-oriented carboxyl group typical
of ozic acid.]

436   ent-Abienol, 3β-hydroxy-

809   ent-Abienol, 6β,7α-dihydroxy-

810   ent-Abienol, 6β-acetoxy-7α-hydroxy-

811   ent-Abienol, 6β-angeloyloxy-7α-acetoxy-

437   ent-Labda-8(17),14-dien-18-oic acid, 12α,13αZ-epoxy-

438   ent-Labda-8(17),14-dien-18-oic acid, 12β,13βZ-epoxy-

439   ent-Labda-8(17),14-dien-18-oic acid, 12α,13αE-epoxy-

440   ent-Labda-8(17),14-dien-18-oic acid, 12β,13βE-epoxy-[12,
13-Dehydro-12,13α-epoxy-ozic acid]

441   ent-Labda-13(16),14-diene, 8α-hydroxy-

442   ent-Manool

TABLE 1. *(contd.)*

Compound:

No.   Name

443   ent-Manool, 2β-hydroxy-

444   ent-Manool, 2α-hydroxy-

445   ent-Manool, 2β-succinyloxy-

446   ent-Manool, 2-oxo-

447   ent-Labda-7,14-diene, 13-hydroxy-

448   ent-Labda-7,14-diene, 3,13-dihydroxy-

449   ent-Labda-7,14-diene, 13-hydroxy-3-oxo-

450   ent-Sclareol, 8-epi-

460   ent-Labda-13-Z-en-15-oic acid, 6,16-dioxo-8,12-oxido-

461   Erigerol [1,6β-dihydroxy-7,8β-epoxydihydrogrendelyl ester
      of 3,3-dimethylacrylic acid]

1012 ent-Labda-7(8),13E-diene, 15,18-diacetoxy-2β-hydroxy-
      [or enantiomer]

1013 ent-Labda-7(8),13E-diene, 18-acetoxy-2β,15-dihydroxy-
      [or enantiomer]

1014 ent-Labda-8(17),13E-diene,3α-acetoxy-15-hydroxy-18-
      tigloyloxy- [or enantiomer]

1015 Abienol, 18-hydroxy-

1016 ent-Labd-8,14-diene, 13,18-dihydroxy-7-oxo-ent-

1017 ent-Labd-14-ene, 18-hydroxy-7-oxo-9β,13β-epoxy-

1018 ent-Labd-14-ene, 7,18-dioxo-9β,13β-epoxy-

1019 ent-Labd-14-en-18-oic acid, 7-oxo-9β,13β-epoxy-

1037 ent-Labda-8(17),13(16),14-triene,
      18-hydroxy-3α-angeloyloxy-15,16-epoxy-

1038 ent-Labda-8(17),13(16),14-triene,2β,18-dihydroxy-
      3α-angeloyloxy-15,16-epoxy-

1039 ent-Labda-8(17),13Z-diene, 2β,15,16,
      18-tetrahydroxy-3α-angeloyloxy-

1040 ent-Labda-8(17),13E-dien-15-oic acid, [E-Copalic acid]

      [Structure is the same as **335.**]

1041 ent-Labda-8(17),13E-dien-15-oic acid, 18-hydroxy-

1042 ent-Labda-7(8),13E-dien-15-oic acid, 6β-hydroxy-18-oxo-

1043 ent-Labda-7(8),13E-dien-15,
      18-dioic acid, 6β-isovaleryloxy-

TABLE 1. (*contd.*)

Compound:
No.   Name

1044 ent-Labda-13E-en-15-oic acid, 5β,8β-peroxy-

1045 ent-Labda-7,13E-dien-15-oic acid, 6β,18-dihdroxy-

1046 ent-Labda-7-en-15-oic acid, 17-hydroxy-

1047 ent-Labda-7-en-15-oic acid, 17-acetoxy-

1048 ent-Labda-7-en-15-oic acid, 17-oxo-

1049 ent-Labda-7-en-15-oic acid, 2α,17-dihydroxy-

1050 ent-Labda-7-en-15-oic acid, 2α-hydroxy-17-acetoxy-

1051 ent-Labda-7-en-15-oic acid, 2α-hydroxy-17-oxo-

1052 ent-Labda-7-en-15,17-dioic acid

1053 ent-Labda-7-en-15-oic acid, 19-hydroxy-17-acetoxy-

1164 ent-Labd-8(17),13E-dien-15,18-dioic acid   [Guamaic acid]

1165 ent-Labd-7,13E-dien-2-O-β-fucopyranoside, 3α-angeloyloxy-
     2β,15-dihydroxy-

1166 ent-Labd-7,13E-dien-2-O-β-[fucopyranoside-4'-O-acetate],
     3α-angeloyloxy-2β,15-dihydroxy-

1167 ent-Labd-7,13E-dien-2-O-β-[fucopyranoside-3'-O-acetate], 3α-
     angeloyloxy-2β,15-dihydroxy-

1168 ent-Labd-7,13E-dien-2-O-β-[fucopyranoside-4'-O-acetate],
     3α-angeloyloxy-2β,15,16-trihydroxy-

1169 ent-Labd-7,13(16)-dien-2-O-β-[fucopyranoside-4'-O-acetate],
     3α-angeloyloxy-2β,14,15-trihydroxy-

1170 ent-Labd-7,13E-dien-2-O-[rhamnopyranoside-4'-O-acetate],
     3α-angeloyloxy-2β,15-dihydroxy-

1171 ent-Labd-7,13E-dien-2-O-β-xylopyranoside,
     3α-angeloyloxy-2β,15-dihydroxy-

1172 ent-Labd-13E-ene-8-O-β-xylopyranoside, 8β,15-dihydroxy-

1173 ent-Labd-8(17)-en-19-oic acid, 15-hydroxy-

1174 ent-Labd-13E-ene, 2β,8β,15-trihydroxy-

1175 Haplopappus pectinatus ent-labdane 13

1176 Haplopappus pectinatus ent-labdane 14

1177 Haplopappus pectinatus ent-labdane 15

1178 Haplopappus pectinatus ent-labdane 16

1179 ent-Labd-7-en-15-oic acid, 2-hydroxy-17-methoxy-

TABLE 1. *(contd.)*

Compound:
No.   Name

1180 ent-Labd-7,13(14)-diene-15-oic acid, 16-hydroxy-17-methoxy-

1181 ent-Labd-8(17),13-dien-15,16-olide, 2β,18-dihydroxy-

1182 ent-Labd-8(17),13-dien-15,16-olide, 3α-angeloyloxy-18-
      hydroxy-

1183 ent-Labd-8(17),13-dien-15,16-olide, 3α-angeloyloxy-18β-
      glucopyranosyloxy-

1184 ent-Labd-8(17),13-dien-15,16-olide, 3α-angeloyloxy-2β,18β-
      dihydroxy-

1185 ent-Labd-8(17),12E,14-triene, 18,19-diacetoxy-

1186 ent-Biformene, 3α,14,15-trihydroxy-12E,14,15-dihydro-

1188 ent-Labda-7,13(16),14-trien-2β,3α-diol, 15,16-epoxy-

1189 ent-Labd-8(17),13E-dien-15-oic acid, 3α,18-dihydroxy-

1190 ent-Labd-8(17),13E-dien-15-oic acid, 3α-angeloyloxy-18-
      hydroxy-

1191 ent-Labd-8(17),13Z-diene, 2β,15,16,18-tetrahydroxy-3α-
      angeloyloxy-

1192 ent-Labd-7,13-dien-2β,15-diol, 3α-angeloyloxy-

1193 ent-Labd-7,13E-diene, 2β,3β,15-trihydroxy-

1194 Gutierrezia spathulata ent-labdane 4

1195 Gutierrezia spathulata ent-labdane 5

1196 ent-Cativic acid, 2β,3β-dihydroxy-

1197 ent-Labd-7-ene, 2β,3β,15-trihydroxy-

39. ent-Labdane-19,6-Lactone

462   Hartwrightia acid

40. ent-Labdane-8,13-Epoxide-15,16-Hemiacetal

463   Schkuhriadiol, cyclo-

464   Schkuhriadiol-16-O-acetate, cyclo-

97. ent-Labdane Dimer

419   Acritopappus lactone A

TABLE 1. *(contd.)*

Compound:

No. Name

420 Acritipappus lactone B

98. ent-Manoyl Oxide, 13-Epi-

451   ent-Manoyl oxide, 13-epi- [ent-8,13β-Epoxy-14-labdene]

452   ent-Manoyl oxide, 3α-hydroxy-

453   ent-Manoyl oxide, 3-oxo-13-epi-   [ent-8,
      13β-epoxy-14-labden-3-one]

454   ent-Manoyl oxide, 2-oxo-13-epi-

455   ent-Manoyl oxide, 19-hydroxy-13-epi-

456   ent-Manoyl oxide, 2-oxo-19-hydroxy-13-epi-

457   Schkuhrianol acetate   [16-acetoxy-14,15-epoxy-14,
      15-dihydro-13-epi-ent-Manoyl oxide]

1198 ent-Manoyl oxide, 3α-acetoxy-

1199 ent-Manoyl oxide, 19-cinnamoyloxy-3α-hydroxy-

99. ent-Manoyl Oxide

458   ent-Manoyl oxide

459   ent-Manoyl oxide, 3β-hydroxy-

      [This structure is apparently identical to a compound from
      Coespeletia lutescens named 3β-Hydroxy-13-epi-ent-manoyl
      oxide for which the published structure did not indicate a
      13-epi-orientation.]

41. ent-Labdane, 17-Nor-

465   ent-Labda-6,13E-dien-15-oic acid,
      16-acetoxy-8β-formyl-8-desmethyl-

466   ent-Labda-6,13E-dien-15-oic acid,
      16-acetoxy-8-oxo-8-desmethyl-

467   Ayapanone acid, 3α-hydroxy-

468   Ayapanone acid, 3-oxo-

1201 ent-Labda-13Z-en-15-oic acid, 8-oxo-8-desmethyl-

1202 ent-Labda-13Z-en-15-oic acid, 3,8-dioxo-

TABLE 1. *(contd.)*

Compound:
No.   Name

42. <u>ent</u>-Labdane, 15-Nor-

469   ent-Labda-8(17),12E-diene, 14-oxo-nor-15-

43. <u>ent</u>-Labdane, 14,15-Bis-Nor-

470   ent-Labda-8(17)-ene, 13-oxo-bis-nor-14,15-

1203 ent-Labd-7-en-2-O-[fucopyranoside-4'-O-acetate],
      14,15-nor-

44. <u>ent</u>-Labdane, 13,14,15,16-Tetra-Nor-

471   ent-Labda-7-en-aldehyde, nor-   [Tetra-nor-13,14,15,
      16-12-oxo-ent-labda-7-ene]

472   Acritoconfertic acid   [Tetra-nor-13,14,15,
      16-17-hydroxy-ent-labda-7-en-12-oic acid]

45. <u>ent</u>-Labdane, 7,8-Seco-

473   Athrixic acid, seco-

74. <u>ent</u>-Labdane, 18-Nor-

828   ent-Labdane triol 1a, nor-18-

829   ent-Labdane triol 2a, nor-18-

1205 Austroeupatorione, 12-desoxo-

1206 Austroeupatorione, 12-desoxo-2-desacetyl-

1207 Austroeupatorione

95. <u>ent</u>-Friedolabdane

1059 ent-Labda-5,13E-dien-15-oic acid,
      18-hydroxy-10-desmenthyl-9α-methyl-8β(H)-

1208 [= **1059**]

1209 ent-Friedolabd-1(10),13E-dien-15-ol, 2-oxo-

1210 ent-Friedolabd-1(10),13E-diene, 2α,15-dihydroxy-

1211 Friedolabdaturbinic acid

TABLE 1. (*contd.*)

Compound:

No.   Name

1212 Chiliolide, 3α,5α-dihydroxy-

1213 Chiliolide, 3α-hydroxy-5,6-dehydro-

1214 Chiliolide, 3α-hydroxy-5β,10β-epoxy-

102. <u>ent</u>-Friedolabdane, 3,4-Seco-

1215 Chiliolide aldehyde, seco-

1216 Chiliolide acid, seco-

1217 Chiliolide, 19-hydroxy-seco-

1218 Chiliolide acid methyl ester, 19-hydroxy-

1219 Chiliolide lactone, seco-

1220 Chiliotrin, seco-

103. <u>ent</u>-Friedolabdane, Rearranged 3,4-Seco-

1221 Chiliolide lactone, iso-

46. <u>trans</u>-<u>ent</u>-Clerodane

236 Bedfordia diterpene alcohol

248 Solidago-18-oic acid, 8-epi-1,2-dehydro- [Reported as
     5β,10α-<u>trans</u>-clerodane]

249 Solidago-18-oic acid, 8-epi- [Reported as 5β,10α-<u>trans</u>-
     clerodane; probably (-)-hardwickiic acid]

250 Solidago-18-oic acid, 1,2-dehydro- [Reported as 5β,10α-
     <u>trans</u>-clerodane with C-8β-methyl;probably = 248]

251 Solidago-18-oate, methyl 8-epi-1,2-dehydro- [Nidoreseda
     acid methyl ester; Reported as 5β,10α-<u>trans</u>- clerodane]

245 Solidago-18-oic acid [Reported as 5β,10α-<u>trans</u>-clerodane
     with C-8β-methyl; probably the same as 249]

475 Kolavenol

476 Kolavenol, 17-hydroxy-13E-

477 Kolavenol acetate, 2α-hydroxy-3β,4β-epoxy-3,4-dihydro-

478 Kolavenol acetate, 2α,16-dihydroxy-3β,4β-epoxy-3,4-dihydro-

TABLE 1. *(contd.)*

Compound:
No.   Name

1010 Kolavenol arachidate, 17-hydroxy-13E-

1011 Kolavenol behenate, 17-hydroxy-13E-

479   Kolavelool   [Kolavenool]

480   Kolavelool, 6β-angeloyloxy-   [6β-Angeloyloxy-kolavenool]

481   Kolavenol, 17-hydroxy-13,14-dihydro-

482   Kolavenol arachidate, 17-hydroxy-13,14-dihydro-

483   Kolavenol behenate, 17-hydroxy-13,14-dihydro-

484   Kolavenol, 18-succinyloxy-13,14-dihydro-

485   Kolavenol, 2α-hydroxy-3α,4α-epoxy-13,14-dihydro-

486   Kolavenol, 17-oxo-13,14-dihydro-

487   Gochnatol-17-O-acetate

488   Kolav-3-en-15-ol, 6α,18-dihydroxy-17-acetoxy-

489   Kolav-3-en-15-ol, 6α,18-dihydroxy-17-phenylacetoxy-

490   Kolav-3-en-15-ol, 6α-hydroxy-17-phenylacetoxy-18-oxo-

491   Kolav-3-en-15-ol, 6α-hydroxy-17-acetoxy-18-oxo-

492   Kolaven-15-al, 17-hydroxy-13,14-dihydro-

493   Kolavenoic acid, 13,14Z-

494   Kolavenoic acid, 2β-acetoxy-13,14Z-

495   Kolavenoic acid [13,14E-Kolavenoic acid]

496   Kolavenoic acid, 6-acetoxy-   [6β-Acetoxy-13,
      14E-kolavenoic acid]

497   Kolavenoic acid, 6-angeloyloxy-   [6β-Angeloyloxy-13,
      14E-kolavenoic acid]

498   Kolavenic acid, 6-angeloyloxy- [6α-Angeloyloxy-13,
      14E-kolavenoic acid]

499   Kolavenic acid, 6-tigloyloxy-   [6α-Tigloyloxy-13,
      14E-kolavenoic acid]

500   Solidagonic acid   [7α-Acetoxy-13,
      14E-kolavenoic acid]

962   Tucumanoic acid

501   Kolavenoic acid, 6α,18-dihydroxy-17-acetoxy-

502   Kolavenoic acid, 6α,18-dihydroxy-17-phenylacetoxy-

TABLE 1. (*contd.*)

Compound:	
No.	Name
503	Kolav-3-en-15-oic acid, 13,14-dihydro-
504	Kolav-3-en-15-oic acid, 18-acetoxy-
961	Kolavenic acid, 2-oxo-
505	Haplociliatic acid
506	Kolavan-2-on-15-oic acid, 3β,4β-epoxy-
507	Kolav-3,13(16),14-triene, 15,16-epoxy-
508	Bacchalineol malonate
509	Bacchotricuneatin D
846	Kingidiol [Barticulidiol]
510	Barticulidiol, 18-acetoxy-19-malonyloxy-
967	Cleroda-3,13(16),14-trien-15,16-oxide, 17-oxo-
511	Hardwickiic acid, (-)- [= 517]
953	Hardwickiic acid, 12α-[2-methylbutyryloxy]-(-)-
512	ent-cleroda-1,3,13(16),14-tetraen-18-oic acid, 15, 16-epoxy-
911	Hautriwaic acid, 19-O-acetyl-1,2-dehydro-
513	ent-Kolav-3,13(16),14-trien-18-oic acid, 2β-hydroxy-15, 16-epoxy-
514	Imbricatol-isovalerate, 3-hydroxy-
515	Imbricatol-α-methylbutyrate, 3-hydroxy-
516	Imbricatol-angelate, 3-hydroxy-
518	Kolav-3,13(16),14-trien-18-oic acid, 15, 16-epoxy-19-hydroxy-ent- [Hautriwaic acid]
519	Kolav-3,13(16),14-trien-18-oic acid, 15,16-epoxy-2α, 19-dihydroxy-ent-
900	Hautriwaic acid, 19-acetyl-
901	Hautriwaic acid, 10-O-methyl
902	Hautriwaic acid, 19-O-angelyl-
903	Hautriwaic acid, 19-O-isovaleryl-
904	Hautriwaic acid, 19-oxo-
524	ent-Kolav-3,13(16),14-trien-20-oic acid, 15, 16-epoxy- [Junceic acid]

TABLE 1. (*contd.*)

Compound:

No.   Name

520   Hautriwaic acid lactone

950   Hautriwaic acid-19-lactone, 12α-hydroxy-

951   Hautriwaic acid-19-lactone, 7α,12α-dihydroxy-

521   Bacchofertin

522   Gochnatoic acid-17-O-phenylacetate

523   Hardwickiic acid lactone, 7-oxo-6α-hydroxy

525   Conycephaloide

526   Conycephaloide, 7-hydroxy-17-oxo-7,8-dehydro-8,17-dihydro-

527   Bacchotricuneatin A

528   Bacchotricuneatin C

969   Conycephaloide, 12-epi-

970   Conycephaloide, 12-epi-8β,17-dihydro-

971   Salviarin, 6α-angeloyloxy-1-oxo-2,3-dihydro-

972   Bacchotricuneatin A, 6β-hydroxy-7,8-dehydro-

973   Salviarin

529   Bacchotricuneatin B

968   Bacchotricuneatin B, 2α-angeloyloxy-

530   Bacchotricuneatin B, 7β-angeloyloxy-

531   Bacchotricuneatin, 2α-angeloyloxy-3α-senecioyloxy-3,4βH-

532   Bacchotricuneatin, 2α-senecioyloxy-3α-angeloyloxy-3,4βH-

533   Bacchotricuneatin,2α-angeloyloxy-3α-[2-methylbutyryloxy]-
      3,4βH-

534   Kolaven-15-al, 17-hydroxy-13Z-

535   Kolaven-15-al, 17-hydroxy-13E-

536   Symphyoreticulic acid, 13Z-

537   Symphyoreticulic acid, 13E-

865   Kolava-3,13E-dien-15-oic acid, 16-hydroxy-

962   Solidagonic acid

538   Kolavenic acid, 16,18-dihydroxy-

539   Kolavenic acid, 16-acetoxy-18-hydroxy-

TABLE 1. (*contd.*)

Compound:
No.   Name

540   Kolavenic acid, 16-hydroxy-18-oxo-

541   Kolavenic acid, 16-acetoxy-18-oxo-

542   Kolav-3-en-15-oic acid, 16-oxo-

543   Solidagolactone   [Soldagolactone I; 16-Hydroxy-kolavenic
      acid lactone]

544   Kolavenic acid lactone, 2α,16-dihydroxy-

864   Kolava-3,13Z-dien-15-oic acid lactone, 16,16-dihydroxy-

550   Kolavenic acid lactone, 16,18-dihydroxy-

551   Kolavenic-15-acid lactone, 2α,16-dihydroxy-3α,4α-epoxy-

      [Revised to <u>cis</u>-normal-clerodane]

552   Kolavenic-15-acid lactone, 6β-angeloyloxy-16-hydroxy-3α,
      4α-epoxy-
      [Revised to <u>cis</u>-normal-clerodane]

558   Hardwickiic acid, 15,16H-15-oxo-

      [Compound was named as "16-oxo-...", but data and structure
      indicate otherwise.]

909   Hautriwaic acid, 16α-acetoxy-19-O-acetyl-15-oxo-15,
      16-dihydro-

910   Hautriwaiic acid, 16β-acetoxy-19-O-acetyl-15-oxo-15,
      16-dihydro-

559   Kolaven-15,
      18-dioic acid dilactone, 7α,16,19-trihydroxy-

560   Olearin

561   Kolaven-15,18-dioic acid dilactone,7-oxo-16,
      19-dihydroxy-13,14-dihydro-

562   Kolaven-15,18-dioic acid dilactone, 7α,16,19-trihydroxy-13,
      14-dihydro-

563   Kolaven-15,18-dioic acid dilactone, 16,19-dihydroxy-13,
      14-dihydro-

564   Kolaven-15-oic acid lactone, 16-hydroxy-13,
      14-dihydro-

565   Kolav-3-en-15-oic acid lactone, 16,16-dihydroxy-

963   Kolaven-15-oic acid lactone,
      16-hydroxy-16-methoxy-18-acetoxy-

566   Rugosolide

905   Hautriwaic acid, 15α-hydroxy-16-oxo-15,16-dihydro-

TABLE 1. *(contd.)*

Compound:

No.   Name

906   Hautriwaic acid, 15β-hydroxy-16-oxo-15,16-dihydro-

907   Hautriwaic acid, 14α,15α-epoxy-16-oxo-13,14,15,
16-tetrahydro-

908   Hautriwaic acid, 14β,15β-epoxy-16-oxo-13,14,15,
16-tetrahydro-

567   Kolaven-16-oic acid lactone, 2β,15-dihydroxy-

568   Kolaven-16-oic acid lactone, 15-hydroxy-2-oxo-

569   Kolav-3-ene, 18-malonyloxy-15-hydroxy-15,16-epoxy-

570   Stephalic acid

1027 Kolavenic acid, 6α-isobutyryloxy-7β-acetoxy-13-Z-

1028 Kolavenic acid, 6α-angeloyloxy-7β-acetoxy-13Z-

1029 Kolavenic acid, 6α,7β-diacetoxy-13Z-

1030 Kolavenic acid, 6α,7β-dihydroxy-2-oxo-13E-

1031 Kolavenic acid, 6α-isobutyryloxy-7β-angeloyloxy-13,
14-dihydro-

1033 Hautriwaic Acid

     [= 518]

1034 Hautriwaic acid, 2β-hydroxy-

1069 Bacchotricuneatin A, 1α-hydroxy-

1070 Hautriwaic acid, 1,2-dehydro-

1071 Junceic acid epoxide

1075 Kolav-13Z-en-15-al, 3α,4α-epoxy-

1076 Kolav-13(16),14-diene, 3α,4α,15,16-diepoxy-

1077 Kolav-13E-en-15-al, 3α,4α-epoxy-

1078 Kolav-13Z-en-15-al, 3-oxo-

1079 Kolav-13E-en-15-al, 3-oxo-

1080 Kolav-13-en-15-oic acid lactone, 16-hydroxy-3α,4α-epoxy-

1081 Kolav-13-en-15-oic acid lactone, 16-hydroxy-3-oxo-

1082 Kolav-13-en-15-oic acid lactone, 16,3α,4β-trihydroxy

1222 Bincatriol

1223 Salvicin [19-Carboxy-6α,15-dihydroxy-<u>trans</u>-cleroda-
3,13(14)-diene]

TABLE 1. (*contd.*)

Compound:

No.   Name

1224 Bacchasalicyclic acid

1225 Bacchasalicyclic acid-15-O-acetate, 16-hydroxy-

1226 Bacchasalicyclic acid-15-O-[1'-β-xylopyranoside]

1227 Baccharis incarum neoclerodane 1

1228 Kolavenol acetate, 16-hydroxy-2α-acetoxy-3β,4β-epoxy-3,4-
     dihydro-

1229 Baccharis rhomboidalis neoclerodane 5

1230 Baccharis rhomboidalis neoclerodane 6

1231 Kingidiol, 19-O-malonyl-

1232 Bacchalineol

1233 Barticulidiol

1234 Salvin [15,16-Epoxy-<u>trans</u>-cleroda-3,13(16),14-trien-19,6β-
     olide]

1235 Salvinin [7-Hydroxy-15,16-epoxy-<u>trans</u>-cleroda-3,13(16),14-
     trien-19,6β-olide]

1236 Baccharis incarum neoclerodane 2

1237 Articulin

1238 Articulin acetate

1239 Baccharis grandicapitulata neoclerodane 16

1240 Baccharis incarum neoclerodane 3

1241 Baccharis rhomboidalis neoclerodane 1

1242 Baccharis rhomboidalis neoclerodane 3

1243 Baccharis rhomboidalis neoclerodane 4

1244 Bacchomagellin A

1245 Bacchomagellin B

1257 Bacchotricuneatin A, 1α-acetyl-

1258 Bacchotricuneatin A, 7α-hydroxy-

47. Printziane

571 Printzianic acid

TABLE 1. *(contd.)*

Compound:

No.   Name

572   Printzianic acid, iso-

913   Printziaic acid, 5α-hydroxy-5,10-dihydro-

914   Printziaic acid, 5α-methoxy-5,10-dihydro-

915   Printziaic acid, 5α-hydroxy-1,2-dehydro-5,10-dihydro-

916   Printziaic acid lactone, 5α-hydroxy-5,10-dihydro-

76.Conyscabrane

917   Conyscabraic acid, 5,6-dehydroiso-

918   Conyscabraic acid, 5,6-dehydro-

919   Conyscabraic acid, 5α-hydroxyiso-

920   Conyscabraic acid, 5α-hydroxy-

48. ent-Clerodane, 5,10-Seco-

573   Strictic acid

912   Nidoresedaic acid, 19-acetoxyseco-

985   Pulicaria angustifolia compound 1a

986   Pulicaria angustifolia compound 2a

987   Pulicaria angustifolia compound 3a

952   Strictic acid, 12α-[2-methylbutyryloxy]-

96. trans-ent-Clerodane, 17-Nor-

1061 Bacchascoparone

49. trans-ent-Clerodane, 13,14,15,16-Tetra-Nor-

574   Acritopappusol

575   Acritopappus acid

1032 Kolavenic acid, 6α,7β-dihydroxy-nor-

1093 Hautriwaic acid, 13,14,15,16-tetranor-1,2-dehydro-

TABLE 1. (*contd.*)

Compound:
No. Name

Undetermined

237 Bedfordia diterpene alcohol, iso- [Clerodane of unknown
     absolute stereochemistry]

50. cis-ent-Clerodane

578 Cleroda-3,13(16),14-triene, 15,16-epoxy-cis-

579 Cleroda-3,13(16),14-triene, 18-hydroxy-15,16-epoxy-cis-

580 Cleroda-3,13(16),14-triene, 18-acetoxy-15,16-epoxy-cis-

576 Haplopappic acid

577 Haplopappate, monomethyl-

581 Cleroda-3,13(16),14-triene, 6α,18-dihydroxy-15,16-epoxy-cis-

582 Cleroda-3,13(16),14-triene, 6α-hydroxy-18-acetoxy-15,
     16-epoxy-cis-

583 Cleroda-3,13(16),14-trien-18,6α-olide, 15,16-epoxy-cis-

584 Cleroda-3-en-15,16,18,6α-diolide, cis-

585 Cleroda-3-en-15,16,18,6α-diolide, 2β-hydroxy-cis-

1055 Cleroda-3,13E-dien-15-oic acid, 18-acetoxy-cis-

1056 Cleroda-3,13E-dien-15-oic acid, 18-hydroxy-cis-

1057 Cleroda-3-en-15-oic acid, 18-hydroxy-cis-

1058 Cleroda-3-en-15-oic acid, 19-hydroxy-

1246 Chiliomarin

82. Solidagonane

960 Solidagonal acid

VI. ent-Tricarbocyclics and Derivatives

51. ent-Labdane, Cyclopropyl-

586 Gnaphalene, 8α,13α-dihydroxy-

77. Gutierreziane

845 Gutierrezial

TABLE 1. *(contd.)*

Compound:

No.   Name

52. ent-Pimarane

587   Sandaracopimar-15-en-8β-ol [Published structure incorrectly
      drawn = Cmpd No. 276]

588   Sandaracopimar-15-en-8β,11α-diol, 12β-acetoxy-
      [Structure incorrectly drawn = Cmpd No. 292]

589   Kirenol

590   Pimara-8(14),15-dien-19-oic acid, ent-

820   Pimara-7,15-dien-19-oic acid, ent-

821   Pimara-8,15-dien-19-oic acid, ent-7-oxo-

823   Pimara-9(11)-en-19-oic acid, ent-12β-acetoxy-

591   Darutigenol

592   Darutigenol-3α-glucoside

593   Pimara-8(14)-en-6β-O-glucoside, 15β,16-dihydroxy-

594   Pimara-8(14)-ene, 6β,15β,16-trihydroxy-

595   Pimara-8(14)-ene, 2β,15β,16,19-trihydroxy-

944   Pimara-8(14)-ene, ent-6β,15,16,18-tetrahydroxy-

599   Pimar-9,11(15)-dien-19-ol, (-)-

600   Pimar-9,11(15)-dien-19-oic acid, (-)-

1247 ent-8(14),15-Pimaradiene-3β,19-diol

1248 ent-8(14),15-Pimaradiene-3β,18-diol

1249 ent-Pimar-8(9),15-diene, 20-hydroxy-7-oxo-

1250 ent-Pimara-9(11),15-dien-19-ol

1251 ent-Pimara-9(11),15-diene, 18-hydroxy-

1252 ent-Pimar-9(11),15-diene malonate, 18-hydroxy-

1253 ent-Pimar-9(11),15-diene malonate, 19-hydroxy-

1254 ent-Pimar-9(11),15-diene, 3α,19-dihydroxy-

53. ent-Pimarane-8,15-Tetrahydrofuran

596   Pimara-3β,12α,16-triol C-15-epimer 9a, ent-8,15-epoxy-

597   Pimara-3β,12α,16-triol C-15 epimer 10a, ent-8,15-epoxy-

TABLE 1. (*contd.*)

Compound:

No.   Name

824   Pimar-9(11)-en-19-oic acid, ent-8(R),
      15(S)-epoxy-12α-acetoxy-

598   Pimara-12α,16-diol, ent-8,15R-epoxy-3-oxo-

54. Cleistanthane (Cleistanthol)

601   Cleistanth-12-en-11-one, 15,16-epoxy-

602   Cleistanth-12-en-17-al, 15,16-epoxy-11-oxo-

603   Cleistanth-12-en-11-one, 17-acetoxy-15,16-epoxy-

604   Cleistanth-12-en-11-one, 3α-angeloyloxy-15,16-epoxy-

605   Cleistanth-12-en-11-one, 17-acetoxy-14β-hydroxy-15,16-epoxy-

606   Cleistanth-12-en-11-one,
      17-acetoxy-3α-angeloyloxy-15,16-epoxy-

607   Cleistanth-12-en-17-al, 11-oxo-8,9,15,16-diepoxy-

608   Cleistanth-12-en-11-one, 14β,15β-dihydroxy-16,17-oxido-

85. Cleistanthane, Iso-

609   Cleistanth-12-en-11-one, 17-acetoxy-15,16-epoxy-iso-

610   Cleistanth-12-en-11-one, 17-acetoxy-3,4,15,16-diepoxy-iso-

55. <u>ent</u>-Abietane

611   Missourienol A [initially isolated as dehydration product:
      3-oxo-abieta-7,13-diene]

612   Missourienol B [initially isolated as dehydration
      product:3β-hydroxy-abieta-7,13-diene]

613   Missourienol C [initially isolated as dehydration
      product:2β-acetoxy-abieta-7,13-diene]

614   Abieta-7,13-diene

615   Abieta-7,13(14)-diene, 5-acetoxy-

616   Abieta-7,13(14)-dien-18-oic acid

1054 Abieta-7,13-diene, 19-succinyloxy-

56. Acritoconfertane

617   Acritoconfert-7-en-15-oic acid, 16-acetoxy-17-hydroxy-

TABLE 1. (*contd.*)

Compound:
No. Name

57. Acritoconfertane-Acetal

1100 Acritoconfertic acid, 8,16,16,17-bisoxido-

83. Erythroxane

976 Erythroxa-3,15-dien-18-oic acid

VII. ent-Tetracarbocyclics and Derivatives

58. ent-Kaurane

618 Kaur-16-ene, ent-

619 Kaur-16-ene, 19-hydroxy-ent-

997 Kauranol, ent-   [(-)-kauranol]

620 Kaur-16-ene, 19-acetoxy-ent-

621 Kaurene, 19-formyloxymethylenoxy-ent-

622 Kaurene, 19-[p-hydroxyhydrocinnamoyloxy]-ent-

623 Kaur-16-en-3α,19-diol, (-)-

626 Kaur-16-ene, 18-acetoxy-ent-

624 Kaur-16-ene, 19-oxo-ent-

625 Kauren-19-oic acid, ent-

926 Kauren-18-oic acid, ent-

     [Although referred to as kaurenic acid, the published
     structure of this compound from Melampodium perfoliatum
     contained a C-4β-oriented carboxyl group.]

939 Kaurenic acid, 18-angeloyloxy-
     [Although referred to as a kaurenic acid derivative, the
     published structure of this compound from Melampodium
     perfoliatum contained a C-4β-oriented carboxyl group and a
     C-4α-oriented ester function.]

627 Kaur-16-en-19-oic acid glycol ester, ent-

628 Kaur-16-en-19-oic acid, 3α-hydroxy-ent-

629 Kaur-16-en-19-oic acid, 3α-acetoxy-ent-

630 Kaur-16-en-19-oic acid, 3α-angeloyloxy-ent-

631 Kaur-16-en-19-oic acid, 3α-tigloyloxy-ent-

632 Kaur-16-en-19-oic acid, 3α-cinnamoyloxy-ent-

TABLE 1. (*contd.*)

Compound:

No.   Name

633   Kaur-16-en-19-oic acid, 3α-senecioyloxy-ent-

634   Kaur-16-en-19-oic acid, 3α-isovaleryloxy-ent-

635   Kaur-16-en-18-al, ent-

636   Kaur-16-en-19-oic acid, 9β,13β-dihydroxy-15α-
      angeloyloxy-ent-

637   Kaur-16-en-19-oic acid, 11β,15α-dihydroxy-ent-

638   Kaurane, 18-hydroxy-16α,17-epoxy-ent-

1000 Kaur-16-en-19-oic acid; 7β-hydroxy-(-)-

639   Stenolobin

640   Kaur-16-en-19-oic acid, 11β-acetoxy-ent-

641   Kaur-16-en-19-oic acid, 12α-hydroxy-ent-

642   Kaur-16-en-19-oic acid, 12α-acetoxy-ent-

1108 [= **642**]

643   Kaur-16-en-19-oic acid, 18-angeloyloxy-ent-

644   Kaur-16-en-19-oic acid, 18-senecioyloxy-ent-

645   Kaur-16-en-19-oic acid, 18-isovaleryloxy-ent-

646   Grandifloric acid   [15α-hydroxy-ent-kaur-16-en-19-oic acid]

647   Kaur-16-en-19-oic acid, 15β-hydroxy-ent-

648   Kaur-16-en-19-oic acid, 15α-acetoxy-ent-

649   Kaur-16-en-19-oic acid, 15α-angeloyloxy-ent-

1007 Stenolobin, 15α-angeloyloxy-
      [methyl-15α-angeloyloxy-ent-kaur-16-en-19-oate]

650   Kaur-16-en-19-oic acid, 15α-tigloyloxy-ent-

651   Kaur-16-en-19-oic acid, 15α-senecioyloxy-ent-

652   Kaur-16-en-19-oic acid, 15α-isovaleryloxy-ent-

653   Kaur-16-en-19-oic acid, 15α-[2-methylacryloyloxy]-ent-

654   Kaur-16-en-19-oic acid, 15α-isobutyroyloxy-ent-

655   Kaur-16-en-19-oic acid, 15α-cinnamoyloxy-ent-
      [cinnamoyloxygrandifloric acid]

656   Kauren-16-en-19-oic acid, 15α-[2,
      3-epoxy-2-methylbutyloxy]-ent-

657   Kaur-16-en-19-oic acid, 15α-benzoyloxy-ent-

TABLE 1. *(contd.)*

Compound:

No.   Name

818   Kauren-16-en-19-oic acid,15α-[2',
      3'-dihydroxy-2-methylbutyryloxy]-ent-

892   Kaur-16-en-19-oic acid, 15α[2'-methylbutyryloxy]-ent-

658   Kaur-16-en-19-oic acid, 15β-angeloyloxy-ent-

659   Kaur-16-en-19-oic acid, 15β-tigloyloxy-ent-

660   Kaur-16-en-19-oic acid, 15β-senecioyloxy-ent-

661   Kaur-16-en-19-oate, methyl-15α-hydroxy-ent-   [grandifloric
      acid methylester]

662   Kaur-16-en-19-oic acid, 3α-angeloyloxy-9β-hydroxy-ent-

663   Kaur-16-en-19-oic acid, 3α-cinnamoyloxy-9β-hydroxy-ent-

664   Kaur-16-en-19-oic acid, 9β,15α-dihydroxy-ent-
      [grandifloric acid, 9β-hydroxy]

665   Kaur-16-en-19-oic acid, 9β-hydroxy-15α-angeloyloxy-ent-

666   Kaur-16-en-19-oic acid, 9β-hydroxy-15α-tigloyloxy-ent-

667   Kaur-16-en-19-oic acid, 9β-hydroxy-15α-acetoxy-ent-

668   Kaur-16-en-19-oic acid, 9β-hydroxy-15α-senecioyloxy-ent-

669   Kaur-16-en-19-oic acid, 9β-hydroxy-15α-isovaleryloxy-ent-

670   Kaur-16-en-19-oic acid, 9β-hydroxy-15α-cinnamoyloxy-ent-

819   Kaur-16-en-19-oic acid, 9β-hydroxy-
      15α-[3'-hydroxy-2'-methylbutyryloxy]-ent-

671   Kaur-16-en-19-oic acid, 11β,15β-dihydroxy-ent-

812   Kaur-16-en-19-oic acid, 17-carbomethoxy-ent-

672   Kaur-16-en-19-oic acid, 11β-hydroxy-15β-acetoxy-ent-

673   Kaur-16-en-19-oic acid, 15β-hydroxy-11-oxo-ent-

674   Kaur-16-en-19-oic acid, 11β-hydroxy-15-oxo-ent-

1260 Kaur-16-en-19-oic acid, 11β,12β,15β-trihydroxy-ent-

675   Kaur-16-en-19-oic acid, 3β,9β-dihydroxy-15α-angeloyloxy-ent-

869   Kaur-16-en-19-oic acid,
      9β,11β-dihydroxy-15α-angeloyloxy-ent-

676   Kaur-16-en-19-oic acid, 3β,
      9β-dihydroxy-15α-senecioyloxy-ent-

677   Kaur-16-en-19-oic acid, 3β,9β-dihydroxy-15α-tigloyloxy-ent-

TABLE 1. (*contd.*)

Compound:

No.   Name

945   Doronicoside D

678   Carboxyatractyloside

679   Wedeloside

680   Wedeloside, L-rhamnopyranosyl-

681   Atractyloside

895   Paniculoside I

896   Paniculoside II

897   Paniculoside III

898   Paniculoside IV

899   Paniculoside V

682   Steviolbioside

683   Stevioside

684   Rebaudioside B

685   Rebaudioside A

686   Dulcoside A

687   Dulcoside B

688   Rebaudioside D

689   Rebaudioside E

690   Kaur-9(11),16-dien-19-oic acid, ent-

691   Kaur-9(11),16-dien-19-oic acid, 2β-hydroxy-ent-

692   Kaur-9(11),16-dien-19-oic acid, 3β-hydroxy-ent-

693   Kaur-9(11),16-dien-19-oic acid, 7β-hydroxy-ent-

694   Kaur-9(11),16-dien-19-oic acid, 12β-hydroxy-

998   Kaur-9(11),16-dien-19-oic acid, 12β-ethoxy-ent-

695   Kaur-9(11),16-dien-19-oic acid, 12-oxo-ent-

696   Kaur-9(11),16-dien-19-oic acid, 15α-hydroxy-ent-

697   Kaur-15-ene, 17-hydroxy-ent-   [17-hydroxy-isokaurene]

698   Kaur-15-ene, 17,19-dihydroxy-ent-

699   Kaur-15-ene, 17-hydroxy-19-[p-hydroxy-hydrocinnamoyloxy]-
      [17-hydroxy-19-[p-hydroxy-hydrocinnamoyloxy]-isokaurene]

TABLE 1. (*contd.*)

Compound:

No.   Name

946   Kaur-15-en-19-al, (-)-

700   Kaur-15-en-19-oic acid, ent-

701   Kaur-15-en-19-oic acid, 17-hydroxy-ent-
      [17-hydroxy-ent-isokaur-15(16)-en-19-oic acid]

702   Kaur-15-en-19-oic acid, 17-oxo-ent-

703   Kaur-15-en-17,19-dioic acid, ent-

704   Kaur-9(11),15(16)-dien-19-oic acid, 3α-isovaleryloxy-
      [3α-isovaleryloxy-polymnia acid]

705   Kaur-9(11),15(16)-dien-19-oic acid, 3α-isobutyryloxy-ent-
      [3α-isobutyryloxy-polymnia acid]

706   Kaur-9(11),15(16)-dien-19-oic acid, 3α-tiglyloxy-ent-
      [3α-tiglyloxy-polymnia acid]

707   Kaur-15(16)-en-19-oic acid, 3α-isovaleryloxy-9β-hydroxy-ent-

708   Kaur-15(16)-en-19-oic acid, 3α-isobutyryloxy-9β-hydroxy-ent-

733   Kaurane, 16α,17-epoxy-ent-

709   Kaurane, 3α-angeloyloxy-16α,17-epoxy-ent-

734   Kauran-19-ol, 16α,17-epoxy-ent-

710   Kaurane, 19-angeloyloxy-16α,17-epoxy-ent-

711   Kaurane, 3α-angeloyloxy-19-oxo-16α,17-epoxy-ent-

735   Kauran-19-al, 16α,17-epoxy-ent-

712   Kauran-19-al, 16β,17-epoxy-ent-

713   Kauran-19-oic acid, 16α,17-epoxy-ent-

714   Kauran-19-oic acid, 16β,17-epoxy-ent-

715   Kauran-19-oic acid, 15α-angeloyloxy-16,
      17-epoxy-ent-   [Perymenium acid]

716   Kauran-19-oic acid, 15α-acetoxy-16,17-epoxy-ent-

717   Kauran-19-oic acid, 15β-angeloyloxy-16,17-epoxy-ent-

718   Kauran-19-oic acid, 15β-tigloyloxy-16,17-epoxy-ent-

719   Kauren-19-oic acid, 12β-hydroxy-16α,17-epoxy-16,
      17-dihydro-9(11)-dehydro-ent-

720   Kaurane, 16α-hydroxy-ent-   [(-)-kauranol]

1001 Kauran-16β-ol, (-)-

721   Kaurane, 3β-acetoxy-16α-hydroxy-ent-

TABLE 1. *(contd.)*

Compound:	
No.	Name

722 Kaurane, 16α-hydroxy-19-acetoxy-ent-

723 Kauran-19-oic acid, 16α-hydroxy-ent-

724 Kaur-11-en-19-oic acid, 16α-hydroxy-ent-

725 Kaurane, 16α,17-dihydroxy-ent-   [17-hydroxy-ent-kauranol]

726 Kaurane, 16α,17,19-trihydroxy-ent-

727 Kaurane, 16α,17-dihydroxy-3-oxo-ent-   [Abbeokutone]

851 Kaurane, 16β-hydroxy-ent- [ent-kauran-16β-ol]

852 Kaurane, 16β,19-dihydroxy-ent- [ent-kauran-16β,19-diol]

728 Kauran-19-oic acid; 16α,17-dihydroxy-ent-

924 Kauran-19-oic acid, 17-hydroxy-16α-H-ent-

925 Kauran-17,19-dioic acid, 16α-H-ent-

822 Kauran-19-oic acid, 16β,17-dihydroxy-ent-

729 Kauran-17-al, 16β-H-ent-

730 Kauran-17-al, 19-hydroxy-16β-H-ent-

731 Kauran-17-oic acid, 16β-H-ent-

732 Kauran-17-oic acid, 19-hydroxy-16β-H-ent-

736 Kaurane, 16α-acetyl-19-hydroxy-16-desmethyl-ent-

737 Kauran-17,19-dial, 16β-H-ent-

738 Kauran-19-oic acid, 17-hydroxy-16β-H-ent-

739 Kaur-9(11)-en-19-oic acid, 16α,17-dihydroxy-ent-

989 Kauran-19-al-17-oic acid, 16α(-)-

990 Kauran-17,19-dioic acid, 16α(-)-

740 Kauran-17-al, 18-[p-hydroxyhydrocinnamoyloxy]-ent-

741 Kauran-19-oic acid, 17-isobutyrloyloxy-16β-H-ent-

742 Kauran-19-oic acid, 17-[n-dodecanoyloxy]-16β-H-ent-

743 Kauran-19-oic acid, 17-[n-tetradodecanoyloxy]-16β-H-ent-

744 Kauran-19-oic acid, 17-[n-hexadecanoyloxy]-16β-H-ent-

745 Kauran-19-oic acid, 17-[n-octadecanoyloxy]-16β-H-ent-

746 Kauran-19-oic acid, 15α,16α-epoxy-17-hydroxy-ent-

747 Kauran-19-oic acid, 11β-hydroxy-15-oxo-16α-H-ent-

TABLE 1. (*contd.*)

Compound:

No.   Name

748   [= **815** = **747**]

942   Kaurenal, 9,11-dehydro-ent-

943   Kaur-16-en-19-oic acid, 9(11)-dehydro-15α-cinnamoyloxy-ent-

850   Kaur-16-en-19-oic acid thujanol ester, ent-

1255 Kauran-19-al, 16α-hydroxy-

1256 Ruilopezia acid   [16α-Hydroxy-kaur-9(11)-en-19-oic acid]

59. ent-Kaurane, 17-Nor-

749   Kauran-19-oic acid, 17-nor-16-oxo-ent-

60. ent-Kaurane, 18-Nor-

752   Kaur-16-ene, 18-nor-ent-

753   Kaur-16-ene, 4β-hydroxy-18-nor-ent-   [4-epi-ruilopeziol]

754   Athrixianone, 4-hydroxymethyl-

755   Athrixianone, 4-formyl-

756   Athrixianone, 4-carbomethoxy-

757   Athrixianone, 16,17-dihydro-16,17-epoxy-

758   Atractyligenin

813   Kaurene, 4β,19-epoxy-18-nor-ent-

61. ent-Kaurane, 19-Nor-

750   Kaur-16-ene, 19-nor-ent-

751   Kaur-16-ene, 4α-hydroxy-19-nor-ent-   [Ruilopeziol]

62. ent-Kaurane, 9,10-Seco-

759   Terminaloic acid

866   Wederegiolide, 15α-hydroxy-

867   Wederegiolide, 15β-hydroxy-

868   Wederegiolide, 15β-acetoxy-

760   Wedelia-seco-kaurenolide, 9-oxo-
       [wedelia-seco-kaurenolide]

TABLE 1. *(contd.)*

Compound:

No. Name

761 Wedelia-seco-kaurenolide, 9α-hydroxy-9-desoxo-

762 Wedelia-seco-kaurenolide, 9β-hydroxy-9desoxo-

763 Wedelia-seco-kaurenolide, 15β-hydroxy-

64. ent-Stachane (ent-Beyerane)

764 Beyer-15-en-3α-ol, ent-

765 Beyer-15-en-19-ol, ent-   [Erythroxylol-A]

766 Beyer-15-en-18-ol, ent-

767 Beyer-15-en-3α,19-diol, ent-

768 Beyer-15-en-3α,12β-diol, ent-

769 Erythroxylol A-malonate

770 Beyer-15-en-3α,17-diol, ent-

817 Beyer-15-en-12β,19-diol, ent-

771 Beyer-15-en-17,19-diol, ent-   [17-hydroxymonogynol]

772 Beyer-15-en-18,19-diol, ent-

774 Beyer-15-en-19-oic acid, ent-   [stach-15-en-19-oic acid;
     beyerenic acid]

773 Beyer-15-ene, 19-oxo-ent-   [20-oxostachene]

775 Beyer-15-en-19-oic acid; 3α-hydroxy-ent-

776 Beyer-15-en-19-oic acid, 3α-tigloyloxy-ent-

777 Beyer-9(11),15-dien-19-oic acid, ent-

778 Beyer-9(11),15-dien-19-oic acid, 3α-tigloyloxy-ent-

779 Beyeran-18-ol, 15β,16β-epoxy-ent-

780 Beyeran-19-ol, 15β,16β-epoxy-ent-   [erythroxylol A-14,
     15-oxide]

781 Nidoanomalin

93. ent-Stachane, 19-Nor-

782 Beyer-15-ene, 4α-hydroxy-19-nor-ent-

65. Atisirane (Atisane)

TABLE 1. *(contd.)*

Compound:
No.   Name

783   Atis-16-en-19-oic acid

853   Atisan-16α-ol, ent-

854   Atisan-16β-ol, ent-

814   Atis-16-en-19-oic acid, 13-hydroxy-

816   Atis-13-en-3α,16β-diol, ent-

784   Atis-16-en-19-oic acid, 7α-hydroxy-ent-   [occidentalic acid]

785   Atisirene acid, 11α-acetoxy-
      [11α-acetoxy-atisirene-20-acid]

786   Atisirene acid, 13α-angeloyloxy-

787   Atisirene acid, 13α-isovaleryloxy-

788   Atisirene acid, 13α-isobutyryloxy-

789   Atisirene acid, 7α-acetoxy-9,11-didehydro-   [7α-acetoxy-9,
      11-didehydroatisirene-20-acid]

790   Atisirene acid, 14β-acetoxy-9,11-didehydro- [14β-acetoxy-9,
      11-didehydro-atisirene-20-acid]

66. Trachylobane

855   Trachyloban-19-al, ent-

791   Trachyloban-19-oic acid

825   Trachyloban-19-oic acid, 15α-acetoxy-

826   Trachyloban-19-oic acid, 15α-angeloyloxy-

827   Trachyloban-19-oic acid; 15α-isovaleryloxy-

893   Trachyloban-19-oic acid, 15α-isobutyryloxy-

792   Trachyloban-19-oic acid, 7α-hydroxy-   [Ciliaric acid]

793   Trachyloban-19-oic acid, 11-oxo-

794   Trachylobanoic acid, 9,11-dehydro-

795   Trachyloban-19-oic acid,
      11α-[16'α-hydroxy-ent-kaur-11'-en-19'-oyloxy]-

856   Trachyloban-19-oic acid thujanol ester, ent-

1204 Tachyloban-19-oic acid methyl ester

1259 Trachyloban-19-oic acid, 7β-hydroxy-

67. Helifulvane

796   Helifulvan-19-ol

TABLE 1. (*contd.*)

Compound:
No.   Name

797   Helifulvan-19-oic acid

798   Helifulvan-19-oic acid, 11α-hydroxy-

799   Helifulvan-19-oic acid, 11α-acetoxy-

68. Tetrachyrane

800   Tetrachyrin   [Zoapatlin]

847   Zoapatlin, 15-oxo-

805   Eupatalbin

806   Eupatoralbin

78. Villanovane

875   Villanovane, 13α,17-dihydroxy-19-isovaleryloxy-

876   Villanovane, 17-acetoxy-13α-hydroxy-19-isovaleryloxy-

877   Villanovane, 13α,17-dihydroxy-19-(3-methylvaleryloxy)-

878   Villanovane, 17-acetoxy-13α-hydroxy-19-(3-methylvaleryloxy)-

879   Villanovane, 17-acetoxy-13α-hydroxy-3α-isovaleryloxy-

880   Villanovane, 17-acetoxy-13α-hydroxy-3α-(3-methylvaleryloxy)-

894   Villanovane, 3α,15α-dihydroxy-

69. Lycoctonine

801   Lycaconitine, methyl-

802   Lycoctonine   [royline; roylene]

803   Lycoctonine, anthranoyl-

VIII. Undetermined Stereochemistry

79. Inulaefane

804   Inulaefol

84. Hebeclinane (9,10-Seco-Labdane)

940   Jhanic acid

999   Hebeclinolide

TABLE 2. The Substitutional Patterns of the Diterpenes Reported from the Compositae (Asteraceae).

Compound: No.	Substituents

I. Linear or Unicarbocyclic

1. Geranylnerol

16 14 12 10 8 6 4 1
15 13 11 9 7 5 3 2
17   18   19   20

No.	Substituents
1	1(OH)
2	1(OH);18(OH)
3	1(OH);20(OH)
4	1(OH);17(OH);20(OH)
5	1(=O)
6	18(=O,OH);1(OH);16(OH)
7	13(=O);1(OH);19(OH)
8	16(=O);1(OH);19(OH);20(OH)
15	1(OH);12(OH);20(OH);19(OAc)
984	1(OH);20(OH);19(OAc)
948	1(OH);8(OH);12(OH);19(OH)
957	1(OH);16(OH);18(OH);19(OH)
978	12(=O);1(OH);19(OH)
1002	19(=O,OH);1(OH)
1003	19(=O,OH);1(OAc)
1109	1(OH);17(OH);18(OH)
1110	1(OH);17(OH);18(OH);20(OH)
1111	1(=O);12(=O)
1112	1(OH);16(OH);20(OH);18(OAc);19(OAc)
1113	1(OH);16(OH);18(OAc);19(OAc);20(OAc)
1114	1(OH);16(OH);20(OH);18(OAc)

TABLE 2. *(contd.)*

Compound:	
No.	Substituents
1115	1(OH);16(OH);20(OH);19(OAc)
1116	1(OH);9(OH);17(OH)
1117	1(OH);9(OH);17(OH);19(OH)

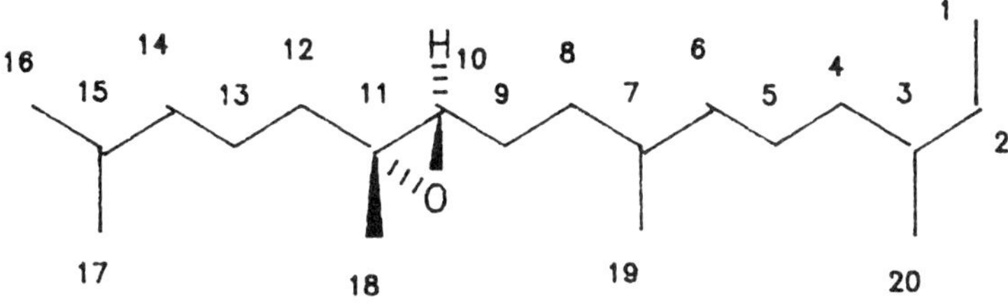

14	2(3)Z;6(7)Z;10(11)E;14(15);19(=O,OH);18(=O);1(OH)
949	2(3)Z;10(11)E;14(15);6(-O-)7;1(OH);8(OH); 12(OH);19(OH)
979	2(3)Z;10(11)E;14(15);12(=O);1(OH);19(OH);6(-O-)7
1004	2(3)Z;6(7)Z;10(11)E;14(=O);1(OH);15(OH);19(OH)
1005	2(3)Z;6(7)Z;10(11)E;14(=O);15(OH);1(OAc);19(OAc)
1095	2(3)E;14(15);12(=O);1(OH);20(OH);6(-O-)7
1096	2(3)E;13(14)E;12(=O);1(OH);15(OH);20(OH);6(-O-)7

1020	2(3)E;6(7)E;14(15);1(OH);9(OH);20(OAc)

2. Geranylgeraniol

TABLE 2. (*contd.*)

Compound: No.	Substituents
11	1(OH);12(OH);19(OH)
12	12(OH);19(OH);1(OAc)
13	1(OH);16(OH)
17	1(=O)
18	1(=O);4α(OH)
19	17(=O,OH);1(OAc)
954	18(=O,OH);1(OH)
1118	1(OH);17(OH);18(OAc)
1119	1(=O);12(=O)

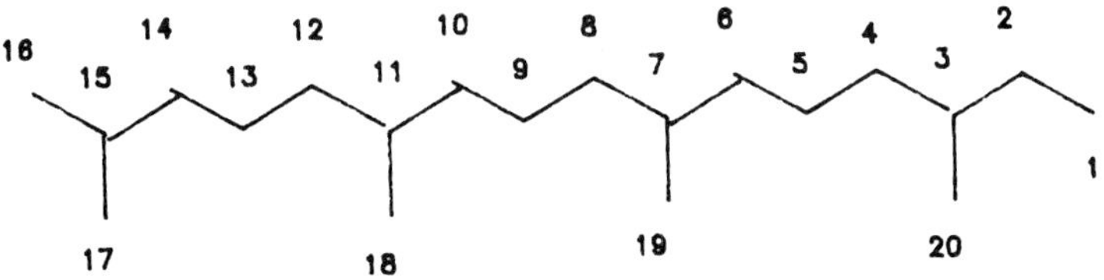

16	2(3)E;6(7)E;10(11)E;1(OH);4(OH);14(OH);15(OH)
20	6(7)E;10(11)E;13(=O);1(OH)
21	2(3)E;6(7)Z;14(15);12(=O);1(OH);19(OH)
22	2(3)E;6(7)Z;10(11)E;14(15);1(OH);12(OH);18(OH)
1120	2(3)E;6(7)E;10(11)E;14(15);18(=O);19(=O);1(OH)

3. Geranyllinalool

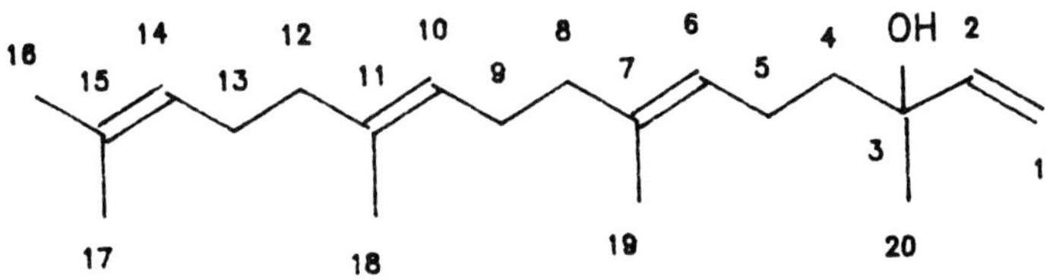

23	
24	13(OH)
25	13(OAc)

TABLE 2. (*contd.*)

Compound: No.	Substituents
26	9(OH)
27	9(OAc)
28	5(OH)
29	5(OH);9(OAc)
30	5(OAc);9(OAc)
31	5(OH);13(OAc)
32	12(=O)
33	13(=O)
1121	13(OAc)

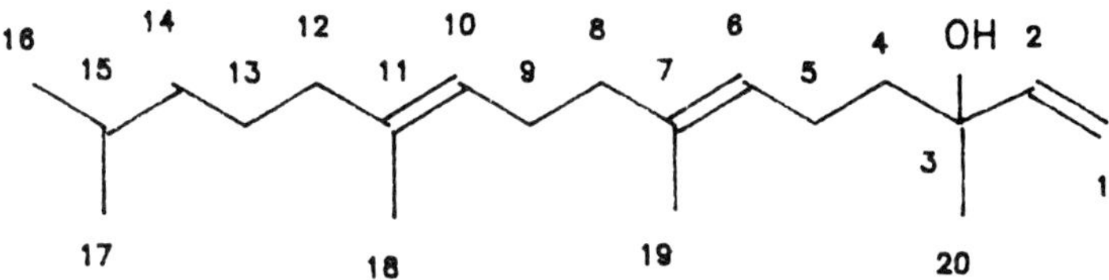

34	13(=O);14(-O-)15
35	13(14)E;15(OH)
36	13(14)E;15(17)
37	14(OH);15(OH)
38	5(OH);14(OH);15(OH)
39	14(OH);15(OH);9(OAc)

1122	6(7)Z;10(11)Z;14(15);12(=O)
1123	6(7)Z;13(14)E;12(=O);15(OH)

TABLE 2. (*contd.*)

Compound:	
No.	Substituents

4. Phytol-Derived

| 59 | 1(2);3(20) |

80. 14-Methylgeranylnerol

9	6(7)Z;10(11)E;14(21);1(OH);12(OH);19(OH);20(OH)
10	6(7)Z;10(11)E;14(21);1(OH);12(OH);20(OH);19(OAc)
980	6(7)Z;10(11)E;13(14)E;12(=O);1(OH);19(OH)
981	10(11)E;13(14)E;12(=O);6(-O-)7;1(OH);19(OH)
982	6(7)Z;10(11)E;14(21);12(=O);1(OH);19(OH)
983	10(11)E;14(21);12(=O);1(OH);19(OH);6(-O-)7
1094	14(21);12(=O);1(OH);20(OH);6(-O-)7

5. Acanthoaustralane

| 40 | 1(OH);6(OH) |

TABLE 2. *(contd.)*

Compound: No.	Substituents
41	6(OH);1(OAc)
42	1(OH);6(OH);17(OH)
43	1(OH);6(OH);17(OAc)
1124	1(OAc);6(OAc)

6. Acanthoaustralane-6,11-Epoxide

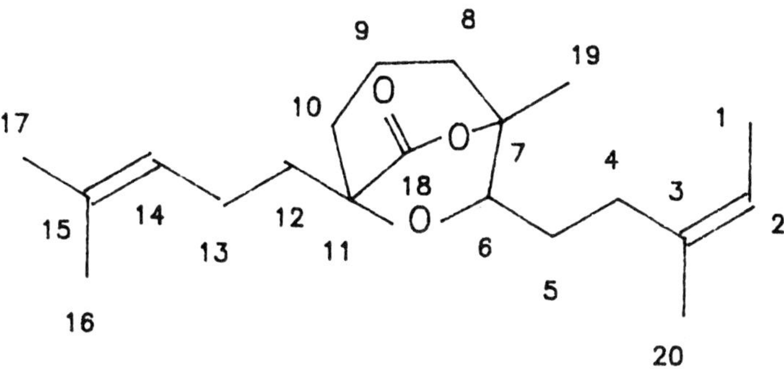

44	1(OH);10(OH);17(OAc)
45	1(OH);10(OH);17(OH)

7. Isoacanthoaustralane

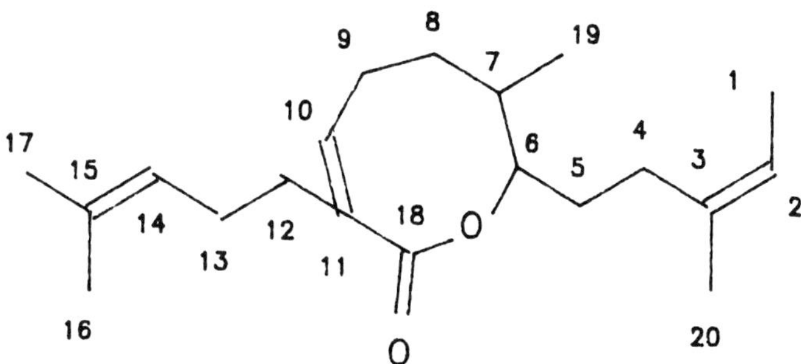

46	1(OH);7(OH);17(OH)
47	1(OH);7(OH);17(OAc)
48	7(OH);1(OAc)

TABLE 2. (*contd.*)

Compound:	
No.	Substituents

8. Geranylgeraniol-18,8-Lactone

49	14(15);1(OH)
50	14(15);1(OH);19(OAc)
51	13(14)E;1(OH);15(OOH)
52	13(14)E;1(OH);15(OH)
53	15(17);1(OH);14(OOH)
54	15(17);1(OH);14(OH)

9. Oxepane

55	14(15);1(OH);6α(OH)
56	13(14)E;6(OH);15(OH);1(OAc)
57	13(14)E;1(OH);6α(OH);14(Me)
1097	1(OH);6α(OH);14(CH2)
1098	13(14)E;1(OH);6α(OH);15(OH)
1099	15(16);1(OH);6α(OH);14(CH3)

TABLE 2. (*contd.*)

Compound:
No.   Substituents

86. Melcanthane

1006   2(3)Z;10(11)Z;14(15);1(OH);6(OH);7(OH);17(OAc)

10. Mikanofurane

58

11. Centipedane

60   18(=O,OH)

61   18(OH);19(OH)

1125   12(=O)

1126   12(OH)

1127   12(=O);13(OH)

TABLE 2. (*contd.*)

Compound:

No.   Substituents

1128   6(7)E;10(11)E;13(14)E;12(=O);15(OOH)

12. Geranylgeraniol-1,20-Lactone

No.	Substituents
62	14(15);17(=O,OH)
63	14(15);17(=O,OMe)
64	14(OH);15(OH)
1129	14(15);12(=O)
1130	14(15);18(=O,OH);17(OH)
1132	14(15);18(=O,OH);17(OAc)

No.	Substituents
1131	2(3);6(7)E;10(11)E;14(15);18(=O,OH);1(=O);17(OH)
1133	2(3);14(15);20(=O);17(OH);19(OH);18(OAc)
1134	2(3);14(15);20(=O);17(OH);18(OAc);19(OAc)
1135	2(3);6(7)E;10(11)Z;14(15);18(=O,OH);20(=O);17(OH)
1136	2(3);6(7)E;10(11)Z;14(15);18(=O,OH);20(=O);17(OAc)

TABLE 2. *(contd.)*

Compound:
No.   Substituents

1137   6(7)E;10(11)E;14(15);12(=O);1β(-O-)2β;20(=O)

13. Geranylterpinene

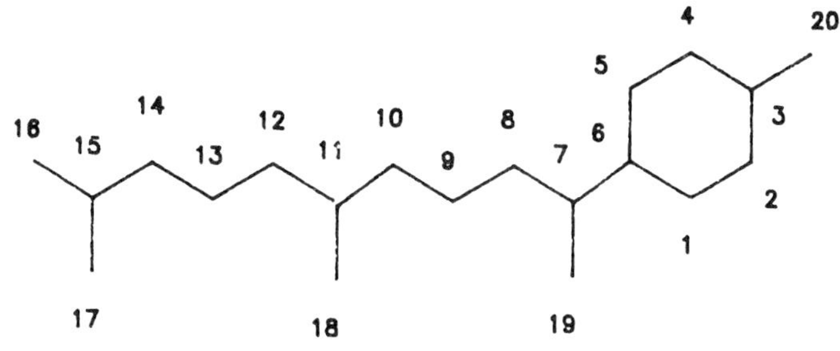

65   2(3);7(8)E;10(11)E;14(15)E;16(OH)

66   2(3);7(8)E;10(11)E;14(15)E;16(=O)

67   2(3);7(8)E;10(11)E;16(=O,OH)

68   1(6);2(3);10(11)E;14(15)

69   1(6);2(3);14(15);10α(OH);11(OH)

71   1(6);2(3);4(5);14(15);10α(OH);11(OH)

70   1(6);2(3);4(5);10(11)E;14(15)

72   14(15);3β(-O-O-)6β;10α(OH);11(OH)

1101   14(15);3α(-O-O-)6α;10α(OH);11(OH)

1138   2(3);10(11)E;14(15);7(OH)

14. Isocembrene

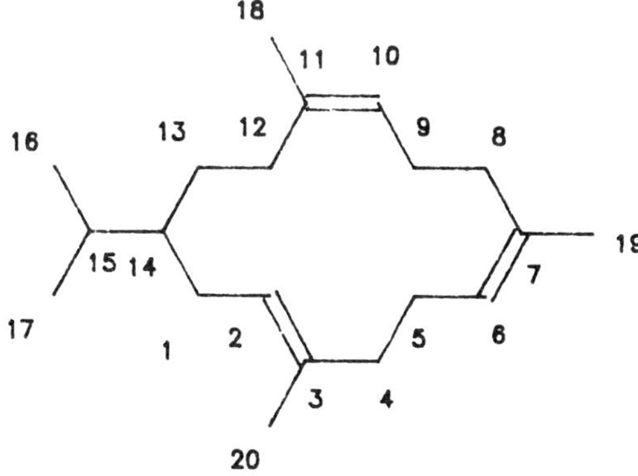

73   15(OH)

Table 2. (*contd.*)

Compound:

No.   Substituents

15. Geranylnerol-11,14-α-Epoxide

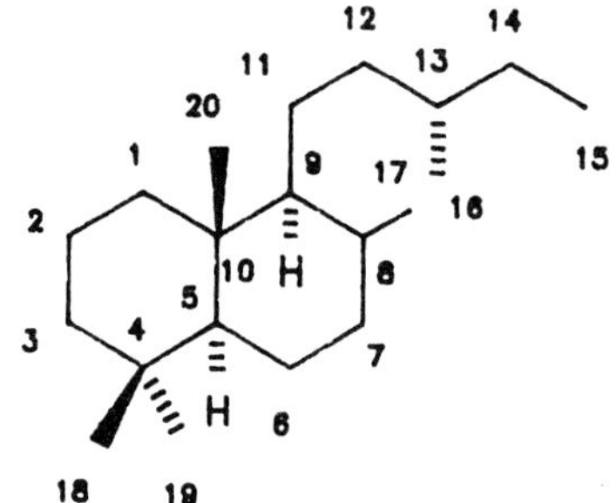

74	1(OH);10α(OH);15(OH);19(OH)
75	1(OH);10α(OH);15(OH);19(OAc)

II. Normal-Bicarbocyclics and Derivatives

16. Normal-Labdane

79	8α(OH);15(OH)
80	7β(OH);8α(OH);15(OH)
958	15(=O,OH);8α(OH)
1139	8α(OH);15(OH)

81	8α(OH);15(OH)

TABLE 2. (*contd.*)

Compound: No.	Substituents
86	7(8);13β(OH);14β(OH);15(OH);14α(H)
87	8(9);13β(OH);14β(OH);15(OH);14α(H)
124	7(8);15(=O,OH);2α(OH);3α(OH)
125	7(8);15(=O,OH);2α(OH);3α(OMebut-2-OH)
129	7(8);15(=O,OH);3(=O)
1140	8(17);15(=O,OH);7α(OH);19(OH)
1141	8(17);15(=O,OH);7α(OH);19(OAc)
1142	8(17);15(=O,OH);7β(OH);19(OH)

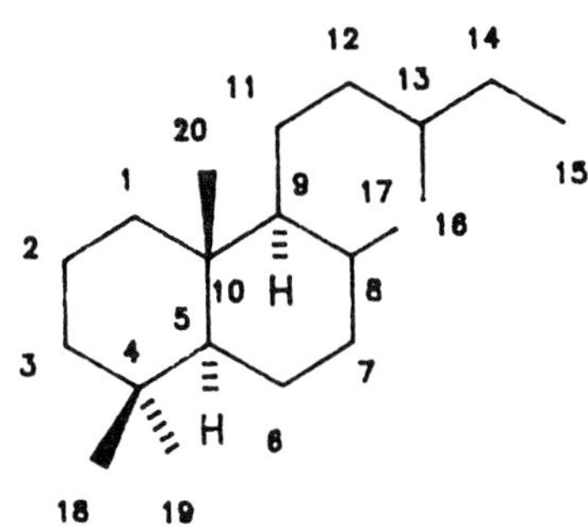

76	8(17);13(14)E;15(OH)
77	8(17);13(14)E;3α(OH);15(OH);7α(Oβ-Gal)
78	8(17);13(14)Z;2α(OH);15(OH)
1103	13(14)E;8α(OH);15(OH)
82	12(13)E;9α(OH);14α(OH);15(OH);8α(H)
83	12(13)E;9α(OH);14α(OH);15(OAc);8α(H)
84	12(13)E;9α(OH);14α(OAc);15(OAc);8α(H)
85	8(17);13(-O-)14;15(OH)
88	8(17);14(OH);15(OH)
89	8(17);13(OH);14(OH);15(OH)
90	8(17);14(OH);15(OAc)
993	7(8);3α(OH);15(OH);13β(H)
994	7(8);15(O2-Mebut);3α(OH);13β(H)
995	8(17);3α(OH);7α(OH);15(OH);13β(H)

TABLE 2. (*contd.*)

Compound: No.	Substituents
91	8(17);13(OH);14(OH);15(OAc)
92	8(17);15(OH);14(OAc)
93	8(17);13(OH);14(OPro);15(OAc)
94	12(13)E;9α(OH);14β(OH);15(OH);8α(H)
97	7(8);13(14)Z;15(=O,OH);12(=O);17(=O);2α(OAng);3α(OAng)
98	13(14)Z;15(=O;OH);8α(OH);7α(O-trans-Cinn);2(OAng or OMebut-2-OH);3(OAng or OMebut-2-OH)
99	13(14)Z;15(=O,OH);8α(OH);7α(O-cis-Cinn);2(OAng or OMebut-2-OH); 3(OAng or OMebut-2-OH)
100	7(8);13(14)Z;15(=O,OH);2(OAng)
116	7(8);13(14)Z;15(=O,OH);3α(OH)
101	7(8);13(14)Z;15(=O,OH);18(OAng)
102	7(8);13(14)Z;15(=O,OH);18(OTig)
103	7(8);13(14)Z;15(=O,OH);2α(OH);3α(OAng)
104	7(8);13(14)Z;15(=O,OH);2α(OH);3α(O-trans-Cinn)
105	7(8);13(14)Z;15(=O,OH);2α(OH);3α(O-cis-Cinn)
106	7(8);13(14)Z;15(=O,OH);2α(OH);3α(OAng)
107	7(8);13(14)Z;15(=O,OH);3α(OH);2α(OAng)
108	8(17);13(14)Z;15(=O,OH);2α(OH);3α(OAng)
109	8(17);13(14)Z;15(=O,OH);3α(OH);2α(OAng)
110	7(8);13(14)Z;15(=O,OH);19(=O,OH)
111	8(17);13(14)E;15(=O,OH)
112	7(8);13(14)E;15(=O,OH);2α(OTig)
113	7(8);13(14)E;15(=O,OH);3β(OAc)
114	7(8);13(14)E;15(=O,OH);3α(OH)
115	7(8);13(14)E;15(=O,OH);3β(OH)
117	7(8);13(14)E;15(=O,OH);3(=O)
118	7(8);15(=O,OH)
991	7(8);15(=O,OH);3α(OH);13β(H)
992	7(8);15(=O,OMe);3α(OH);13β(H)
119	7(8);15(=O,OH);3α(OH)

TABLE 2. (*contd.*)

Compound: No.	Substituents
120	7(8);15(=O,OH);3β(OH)
121	15(=O,OH);8α(OH)
871	7(8);2α(OH);15(OH)
872	7(8);2α(OH);15(OCoum)
873	7(8);2α(OH);15(O3-Me-Coum)
874	7(8);2α(OH);15(p-OH-Cinn)
122	7(8);15(=O,OH);18(OAng)
123	7(8);15(=O,OH);18(OTig)
126	7(8);15(=O,OH);2α(OH);3α(OAng)
127	7(8);15(=O,OH);2α(OH);3α(O-trans-Cinn)
128	7(8);15(=O,OH);2α(OH);3α(O-cis-Cinn)
870	7(8);15(=O,OMe);3β(OAc);2α(OSen)
130	7(8);13(14)E;15(=O,OH);18(=O,OH)
977	8(17);19(=O,OH);15(=O);13α(H)
131	7(8);15(=O,OH);18(=O,OH)
132	8(17);12(13)E;14(15)
133	8(17);12(13)E;14(15);3α(OH)
134	8(17);12(13)E;14(15);3β(OH)
135	8(17);12(13)E;14(15);3β(OAc)
136	8(17);12(13)E;14(15);3(=O)
959	8(17);15(=O,OH);2β(OAc)
137	8(17);12(13)E;14(15);18(=O,OH)
1104	8(17);12(13)E;14(15);19(=O,OH)
138	8(17);12(13)E;14(15);18(=O,OH);7α(OAc)
139	12(13)Z;14(15);8α(OH)
140	12(13)Z;14(15);7β(OH);8α(OH)
141	12(13)E;14(15);7β(OH);8α(OH)
142	12(13)E;14(15);6α(OH);7β(OH);8α(OH)
143	12(13)E;14(15);7β(OH);8α(OH);6α(OAng)

TABLE 2. *(contd.)*

Compound:

No.	Substituents
144	12(13)E;14(15);6β(OH);8α(OH)
145	12(13)E;14(15);8α(OH);7β(OAc)
146	12(13)E;14(15);6β(OH);7β(OH);8α(OH)
147	12(13)E;14(15);6β(OH);8α(OH);19(OH)
148	12(13)E;14(15);6β(OH);8α(OH);7β(OAc)
149	12(13)E;14(15);8α(OH);1β(OAc);19(OArac)
150	12(13)E;14(15);8α(OH);1β(OAc);19(OCoum)
151	12(13)E;14(15);8α(OH);1β(OAc);19(O-cis-Coum)
152	12(13)E;14(15);9α(OH);19(OH);8α(H)
153	14(15);8α(OH);13β(OH);6α(OAng)
1102	14(15);8α(OH);13β(OH)
996	8(17);14(15);13β(OH)
154	7(8);14(15);2(OH);13(OH)
1105	7(8);14(15);13β(OH)
1106	7(8);14(15);3β(OH);13β(OH)
1107	7(8);14(15);3(=O);13β(OH)
155	13(16);14(15);8α(OH)
156	13(16);14(15);6β(OH);8α(OH);7β(OAc);12β(O-OH)
157	13(16);14(15);6β(OH);8α(OH);7β(OAc);12α(O-OH)
158	7(8);13(16);14(15);6(=O);9α(OH);15(-O-)16
159	7(8);13(16);14(15);18(=O,OH);17(OH);15(-O-)16
160	7(8);13(16);14(15);17(=O,OH);6α(O2-Mebut);15(-O-)16
161	7(8);13(16);14(15);17(=O,OH);6α(Oi-But);15(-O-)16
162	7(8);13(16);14(15);17(=O,OH);6α(Oi-Val);15(-O-)16
163	8(17);13(16);14(15);19(=O,OH);7α(OH);15(-O-)16
1021	12(13)Z;14(15);8α(OH);7β(OAc)
1022	12(13)E;14(15);6α(OH);8α(OH);7β(OAc)
1023	12(13)Z;14(15);6α(OH);8α(OH);7β(OAc)
1024	13(14)Z;15(=O,OH);8α(OH)

TABLE 2. *(contd.)*

Compound:	
No.	Substituents
1143	8(17);15(=O,OH);7α(OH)
1144	8(17);15(=O,OH);7α(OAc)
1145	7(8);15(=O,OH);19(OH)
1146	7(8);15(=O,OH);3β(OH)
1147	8(17);12(13)E;14(15);19(=O,OH)
1151	13(16);14(15);8α(OH);3β(OAc)

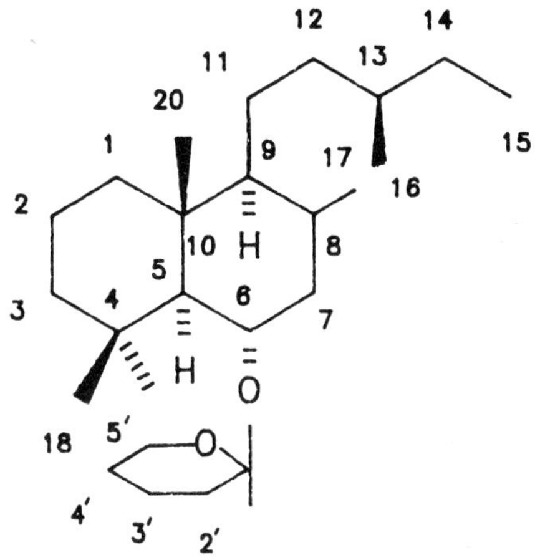

95	7(8);14(15);19(=O);13β(OH);2'α(OH);4'β(OH);3'β(OTig)
96	7(8);14(15);19(=O);13β(OH);2'α(OH);3'β(OH);4'β(OH)
1148	7(8);14(15);19(=O);13β(OH);3'β(OH);4'α(OH);2'α(OAc)
1149	7(8);14(15);19(=O,OMe);13β(OH);3'β(OH);4'α(OH);2'α(OAc)
1150	7(8);14(15);19(=O,OMe);13β(OH);2'α(OH);3'β(OH);4'α(OH)

1152	13(16);14(15);8α(OH)

TABLE 2. (*contd.*)

Compound:
No.   Substituents

17. Grindelane

No.	Substituents
165	8(17);7α(OH)
166	6(7);8(17)
167	7(8);13(S)
955	7(8);13(R)
168	7(8);1α(OH)
169	7(8);3β(OH)
170	7(8);6α(OH)
172	7(8);6β(OH)
173	7(8);17(OH)
171	7(8);6α(OCHO)
174	7(8);17(OAc)
175	7(8);17(OPro)
176	7(8);17(Oi-But)
177	7(8);17(Oi-Val)
178	7(8);17(O2-Mebut)
179	7(8);17(OMe)
180	7(8);19(OH)
181	7(8);19(OAc)
182	7(8);19(Oi-Val)
183	7(8);19(Oi-But)

TABLE 2. *(contd.)*

Compound:	
No.	Substituents
184	7(8);19(O2-Mebut)
185	7(8);19(OSucc)
186	7(8);19(=O)
187	7(8);19(=O,OH)
188	7(8);18(OH)
189	7(8);6(=O)
190	7(8);6(=O);19(OAc)
191	7β(-O-)8β
192	7α(-O-)8α
1035	7(8);3α(OH)
1036	8(17);7β(OH)
1153	7(8);19(=O,OH);15(=O,OMe)

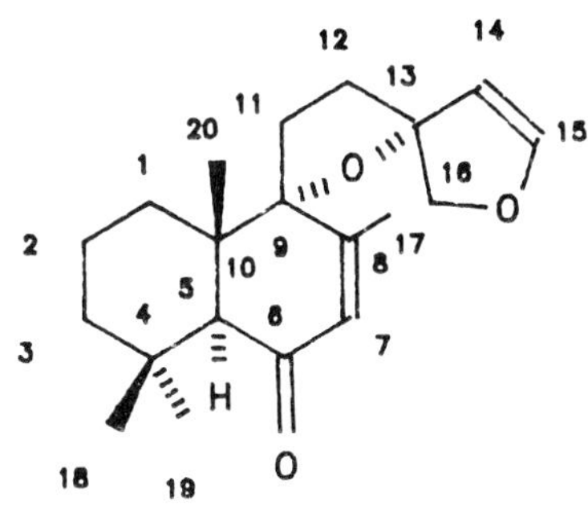

193

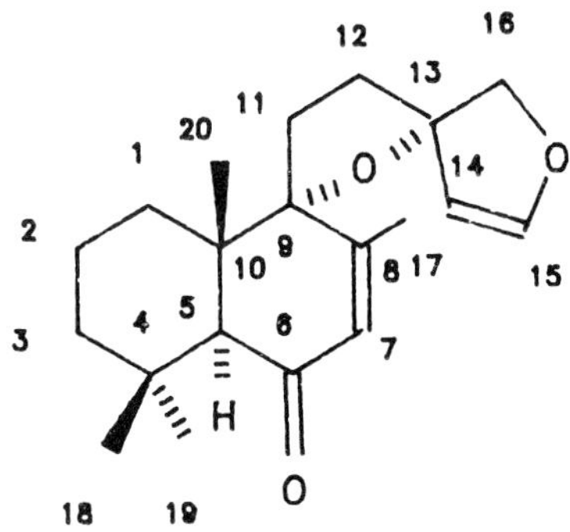

194

TABLE 2. *(contd.)*

Compound:

No.   Substituents

1060

956

1187   7(8);2'α(OH);3'β(OH);4'β(OH)

TABLE 2. (*contd.*)

Compound:

No.   Substituents

18. Normal-Labdane-8,13-Epoxide

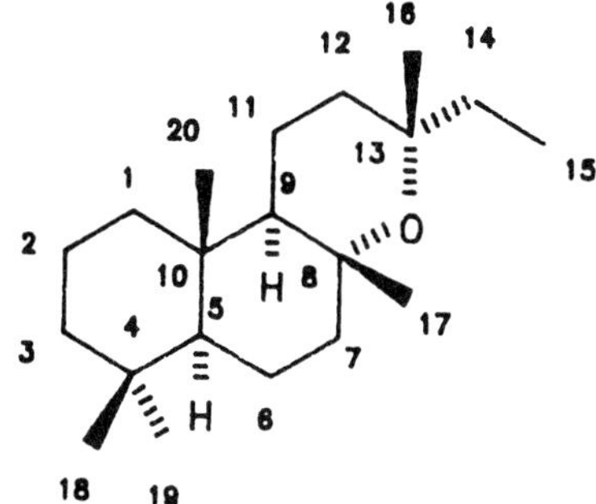

272	14(15)
195	14(15);19(OH)
196	14(15);19(OAc)
197	14(15);1β(OH);19(OH)
198	14(15);1β(OH);19(OAc)
199	14(15);1β(OAc);19(OAc)
200	14(15);3β(OAc)
201	14(15);3β(OH);18(OBenz)
964	15(=O)

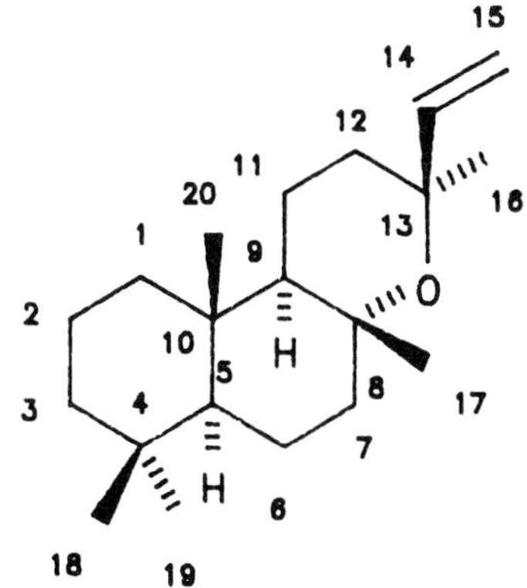

273	
202	19(OH)

TABLE 2. *(contd.)*

Compound:

No.	Substituents

19. Normal-Labdane-8,12-Epoxide

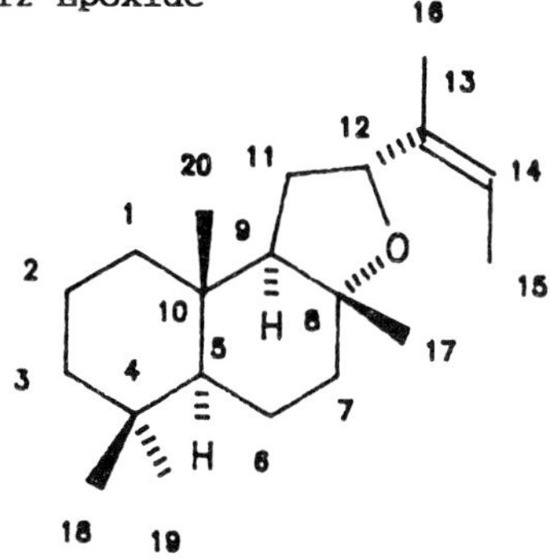

203	15(OH);16(OH)
204	15(OH);16(OAc)
205	15(OAc);16(OAc)
206	16(=O);15(OH)
207	15(=O);16(=O)
210	15(OH)
211	15(=O,OH)

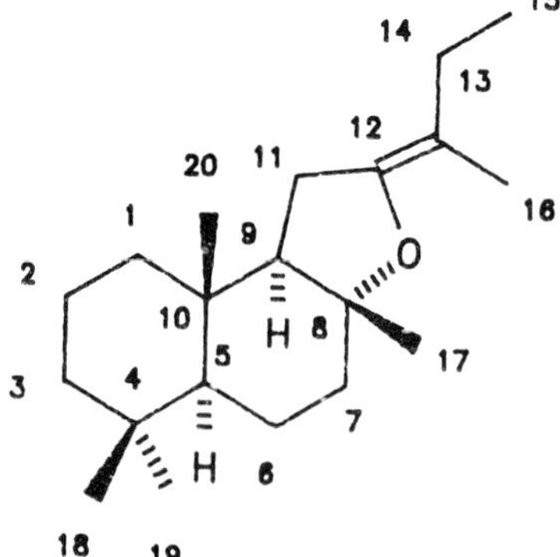

| 212 | 15(=O);16(=O) |
| 213 | 14(=O);15(=O);16(OAc) |

TABLE 2. (*contd.*)

Compound:	
No.	Substituents
208	15(OH);16(OH);13α(-O-)14α;14β(H)
209	15(OH);16(OH);13β(-O-)14β;14α(H)

20. Normal-Labdane-15,16-Lactone

164	7(8);18(=O,OH)
965	7(8);17(OH)
966	7(8);17(=O)

21. Normal-Labdane-20,12-Lactone

214	1(=O);5α(OH);9α(OH);8β(H)

22. Normal-Labdane Spiroketal

215	7(8);13(14);2α(OAng);3α(OAng)

TABLE 2. (*contd.*)

Compound:	
No.	Substituents
216	7(8);13(14);17α(OH);2α(OAng);3α(OAng)
217	7(8);13(14);17β(OH);2α(OAng);3α(OAng)

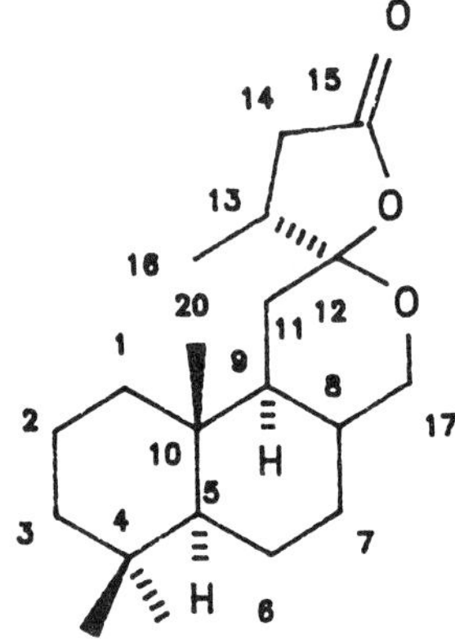

218	7(8);13(14);2α(OH);3α(OAng)

88. Normal-Labdane-15,12-Lactone

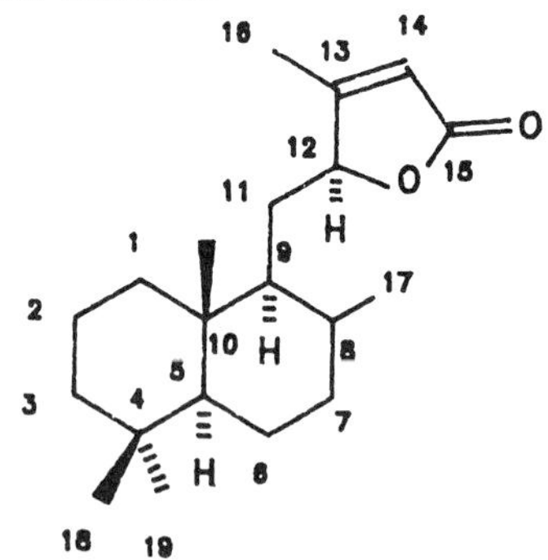

938	3α(OH);7α(OAc);8β(H)

23. Normal-Labdane, 7,8-Seco-

219	6(7);14(15);12(13)E;19(OH)

TABLE 2. (*contd.*)

Compound:
No.   Substituents

24. Normal-Labdane, 3,4-Seco-

220   7(8);15(=O,OH)

25. Grindelane, 8,9-Seco-

223

89. Grindelane, 19-Nor-

221   7(8);15(=O,OH);4α(OH)

222   7(8);15(=O,OH);4α(OCHO)

1154   7(8);15(=O,OMe);6(=O);4α(OH)

TABLE 2. (*contd.*)

Compound:	
No.	Substituents

26. Grindelane, 8,17-Bis-Nor-8,9-Seco-

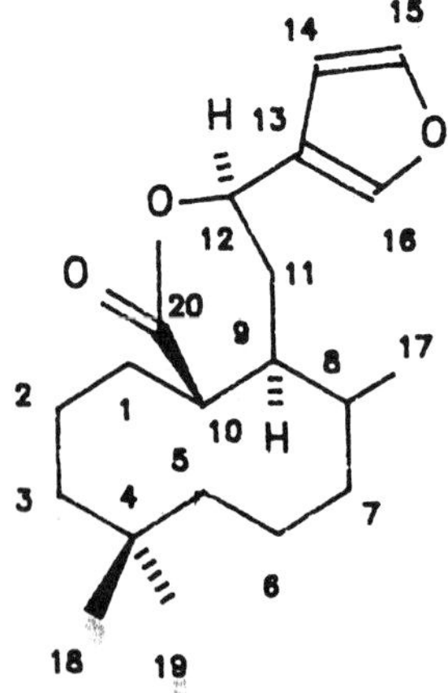

224

27. Normal -Labdane-20,12-Lactone, 5,10-Seco-

225	5(=O);6α(OH);1(-O-)10
226	5(=O);6α(OAng);1(-O-)10
227	5(=O);6α(Oi-But);1(-O-)10
228	5(=O);6α(O2-Mebut);1(-O-)10
229	5(=O);6(Oi-Val);1(-O-)10
230	5(OH);6α(OAng);1(-O-)10

TABLE 2. *(contd.)*

Compound:

No.   Substituents

100. Normal-Labdane, 14,15-Bis-Nor-

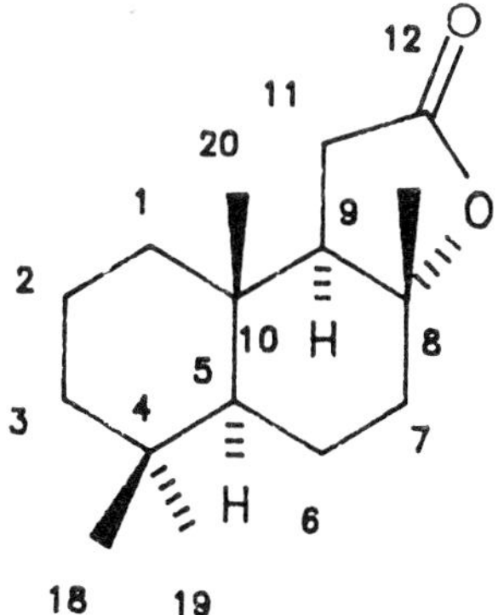

1155   11(12)E;13(=O);8α(OH)

1157   11(12)E;13(=O);6α(OH);7β(OH);8α(OH)

1158   11(12)E;13(=O);7β(OH);8α(OH);6α(OAc)

1159   11(12)E;13(=O);6α(OH);8α(OH);7β(OAc)

1160   11(12)E;13(=O);7β(OH);8α(OH)

101. Normal-Labdane, 13,14,15,16-Tetranor-

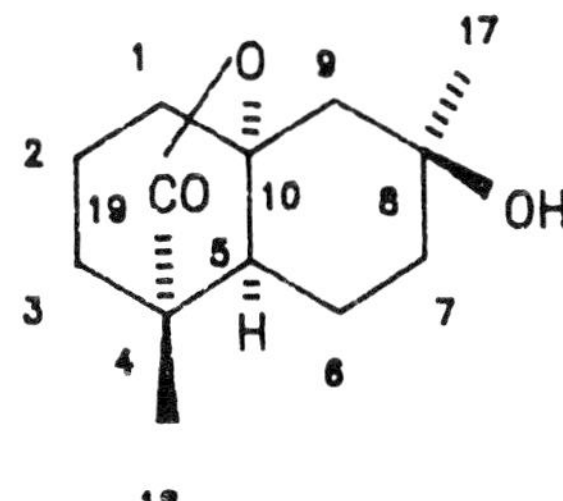

1156

87. Hexanorlabdane

TABLE 2. *(contd.)*

Compound:
No.   Substituents

28. Relhaniane

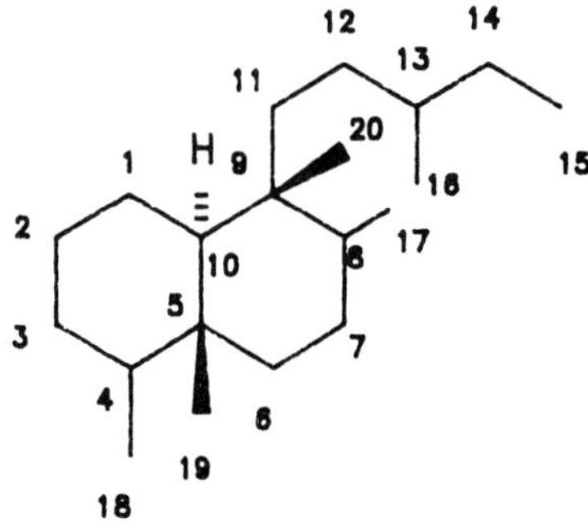

231   1(10);13(14)E;15(=O,OH);2(=O);8β(H)

232   5(6);19(=O,OH);15(OH);10α(H);8α(H)

233   5(6);13(16);14(15);19(=O,OH);8α(H)

234   5(6);13(14);19(=O,OH);16(=O);8α(H)

235   5(6);13(14);19(=O,OH);15(=O);8α(H)

29. <u>trans</u>-Normal-Clerodane

(No proven structures of this type)

TABLE 2. (*contd.*)

Compound:

No. Substituents

30. Clerodane, 5,10-Seco-

255 1(2);3(4);18(=O,OH);8β(H) [Revised to 9α-Me]

256 1(2);3(4);18(=O,OMe);8β(H) [Revised to 9α-Me]

31. <u>cis</u>-Normal-Clerodane

238 3(4);13(16);14(15);19(OH);15(-O-)16;8β(H)

253 3(4);13(14);15(=O);19(OH);8β(H)

254 3(4);13(14);16(=O);19(OH);8β(H)

257 3(4);15(=O,OH);8α(H)

258 3(4);13(14);15(=O);6β(OAng);15(-O-)16;8β(H)

259 3(4);13(14);15(=O);6β(OTig);15(-O-)16;8β(H)

260 13(14);15(=O);3β(-O-)4β;6β(OAng);15(-O-)16;8β(H)

261 13(14);15(=O);3α(-O-)4α;6β(OTig);15(-O-)16;8β(H)

262 3(4);13(14);6(=O);15(=O);15(-O-)16;8β(H)

263 3(4);13(14);15(=O);2α(OH);6β(OAng);15(-O-)16;8β(H)

TABLE 2. (*contd.*)

Compound: No.	Substituents
264	3(4);13(14);15(=O);2α(OH);6β(OTig);15(-O-)16;8β(H)
265	3(4);13(14);6(=O);15(=O);2β(OH);15(-O-)16;8β(H)
266	13(14);15(=O);3β(OH);4α(OH);6β(OAng);15(-O-)16;8β(H)
267	13(14);15(=O);3β(OH);4α(OH);6β(OTig);15(-O-)16;8β(H)
268	13(14);3(=O);15(=O);6β(OAng);15(-O-)16;8β(H)
269	13(14);3(=O);15(=O);6β(OTig);15(-O-)16;8β(H)
545	3(4);13(14);15(=O);6β(OH);15(-O-)16;8β(H)
546	3(4);13(14);15(=O);6β(OAc);15(-O-)16;8β(H)
551	13(14);15(=O);2α(OH);3α(-O-)4α;15(-O-)16;8β(H)
552	13(14);15(=O);6β(OAng);3α(-O-)4α;15(-O-)16;8β(H)
553	13(14);15(=O);3β(-O-)4β;6β(OAc);15(-O-)16;8β(H)
988	13(14);15(=O);3β(-O-)4β;6β(OTig);15(-O-)16;8β(H)
239	3(4);13(16);14(15);19(=O,OH);15(-O-)16;8β(H)
240	3(4);13(16);14(15);19(=O);15(-O-)16;8β(H)
241	3(4);13(16);14(15);19(OH);15(-O-)16;8β(H)
242	3(4);13(16);14(15);18(OH);19(OH);15(-O-)16;8β(H)
243	3(4);13(16);14(15);18(=O);19(=O);15(-O-)16;8β(H)
244	3(4);13(16);14(15);19(=O,OH);18(OAng);15(-O-)16;8β(H)
246	3(4);13(16);14(15);18(-O-)19;19(OH);15(-O-)16;8β(H)
247	18(-O-)19;3(-O-)4;19(OH);15(-O-)16;8β(H)
270	13(14);15(=O);3α(Cl);4β(OH);19(OMe);15(-O-)16; 18(-O-)19
556	13(14);15(=O);3β(-O-)4β;15(-O-)16
557	13(14);15(=O);3(-O-)4;7β(OAng);15(-O-)16
1025	3(4);14(15);2(=O);13β(OH);6β(OAc);8β(H)
1062	3(4);13(16);14(15);19(=O);15(-O-)16;8β(H);
1063	3(4);13(16);14(15);18(=O);19(=O);15(-O-)16;8β(H)
1064	3(4);13(16);14(15);19(OH);15(-O-)16;8β(H)
1065	3(4);13(16);14(15);18(OH);15(-O-)16;8β(H)
1066	3(4);13(16);14(15);18(=O);15(-O-)16;18(-O-)19;8β(H)
1067	3(4);13(16);14(15);19(OH);15(-O-)16;18(-O-)19;8β(H)

TABLE 2. *(contd.)*

Compound:	
No.	Substituents

1068   13(16);14(15);19(OH);3α(-O-)4α;15(-O-)16;
18(-O-)19;8β(H)

1083   3(4);13(14);15(=O);6α(OH);18(OH);15(-O-)16;8α(H)

1084   3(4);13(14);15(=O);18(OH);19(OH);15(-O-)16;8α(H)

1085   3(4);13(14);15(=O);18(=O);15(-O-)16;18(-O-)19;8α(H)

1086   3(4);13(14);15(=O);19α(OH);15(-O-)16;
18(-O-)19;8α(H);19β(H);

1087   13(14);15(=O);18β(OH);19α(OH);3α(-O-)4α;15(-O-)16;
18(-O-)19;8α(H);18α(H);19β(H)

1088   13(14);15(=O);18(=O);19α(OH);3α(-O-)4α;
15(-O-)16;18(-O-)19;8α(H);19β(H)

1089   13(14);15(=O);19α(OH);3α(-O-)4α;15(-O-)16;
18(-O-)19;8α(H);19β(H)

1090   13(14);15(=O);3α(OH);4β(OH);19α(OH);15(-O-)16;
18(-O-)19;8α(H);19β(H)

1091   3(4);13(14);15(=O);19(=O,O-Arab);15(-O-)16;8α(H)

1092   3(4);13(14);15(=O);18(=O);2β(OH);15(-O-)16;
18(-O-)6α;8α(H)

90. Tricycloclerodane

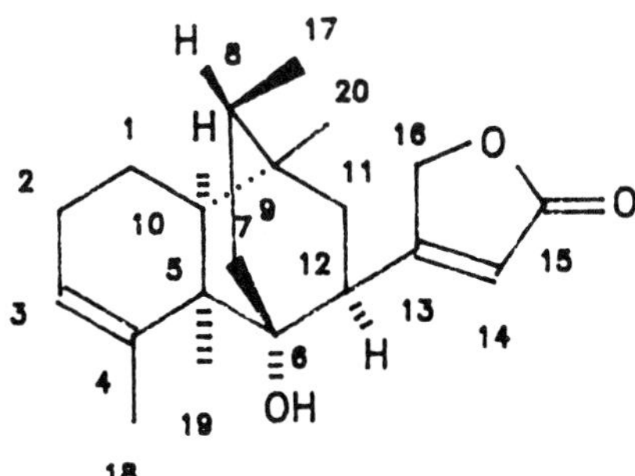

947

72. Chrysolane

271

TABLE 2. *(contd.)*

Compound:

No.   Substituents

III. Normal-Tricarbocyclics and Derivatives

32. Sandaracopimarane (Normal-Pimarane)

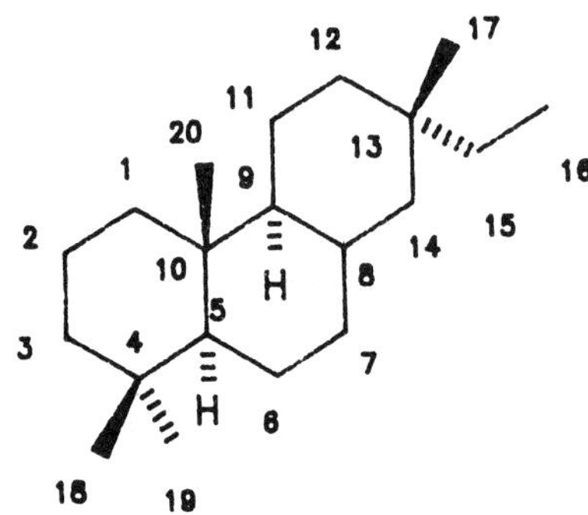

274	8(14);15(16)
927	8(14);15(16);7α(OH)
928	8(14);15(16);1β(OH);7α(OH)
929	8(14);15(16);7α(OH);1β(OAc)
930	8(14);15(16);1β(OH);7α(OAc)
931	8(14);15(16);1β(OH);11α(OH)
932	8(14);15(16);1β(OAc);11α(OAc)
933	8(14);15(16);1β(OH);11α(OH);7α(OAc)
934	8(14);15(16);7α(OH);1β(OAc);11α(OAc)
935	8(14);15(16);1β(OAc);7α(OAc);11α(OAc)
936	8(14);15(16);6(-O-)7;1β(OAc);11α(OAc)
937	8(14);15(16);7(=O);1β(OAc);11α(OAc)
275	7(8);15(16)
276	15(16);8β(OH)
277	15(16);6β(OH);8β(OH)
278	15(16);8β(OH);6β(OAc)
279	15(16);8β(OH);11α(OH)
280	15(16);8β(OH);11α(OSen)

TABLE 2. (*contd.*)

Compound: No.	Substituents
281	15(16);8β(OH);11α(OTig)
282	15(16);8β(OH);11β(OAc)
283	15(16);11(=O);8β(OH)
284	15(16);8β(OH);12β(OAc)
285	15(16);8β(OH);18(OH)
286	15(16);8β(OH);20(OH)
287	15(16);20(OH);8β(-O-)20
288	15(16);8β(OH);11α(OH);12β(OH)
289	15(16);8β(OH);12β(OH);11α(OAc)
290	15(16);8β(OH);12β(OH);11α(OSen)
291	15(16);8β(OH);12β(OH);11α(OTig)
292	15(16);8β(OH);11α(OH);12β(OAc)
293	15(16);8β(OH);11α(OH);12β(OCoum)
294	15(16);8β(OH);11α(OAc);12β(OAc)
295	15(16);8β(OH);12α(OH);11β(OAc)
296	15(16);11(=O);8β(OH);12α(OH)
297	15(16);11(=O);8β(OH);12α(OAc)
298	15(16);6β(OH);8β(OH);11α(OH)
299	15(16);8β(OH);11α(OH);6β(OAc)
300	8(14);15(16);18(=O,OH)
301	7(8);15(16);18(=O,OH)
302	8(14);15(16);18(=O,OH);3β(OAc)
807	8(14);3β(OH);15(OH);16(OH);17(OH)
303	7(8);15(16);18(=O,OH);3β(OAc)
921	8(14);19(=O,OH);6β(OH);15(OH);16(OH)
304	7(8);3α(OH);15(OH);16(OH)
305	7(8);9α(OH);15(OH);16(OH)
922	8(14);15(OH);16(OH);6β(OGlu)
923	8(14);6α(OH);15(OH);16(OH);18(OH)

TABLE 2. (*contd.*)

Compound:

No.   Substituents

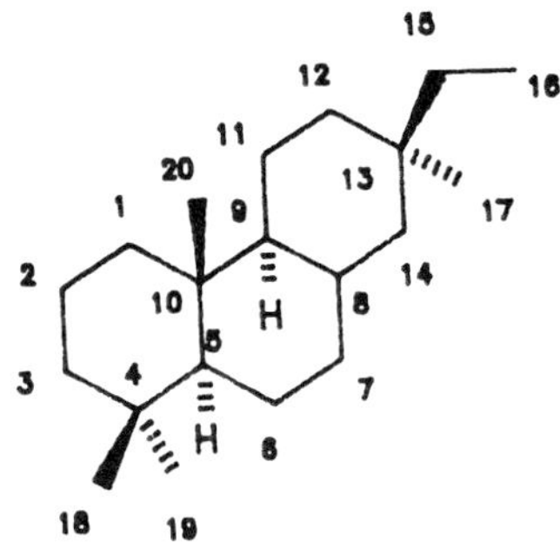

891   15(16);8β(OH);5α(H);5α(H)

306   7(8);3α(OH);15(OH);16(OH);19(OH);5α(H) [Revised to rosane]

307   7(8);3α(OH);9α(OH);15(OH);16(OH);5α(H) [Revised to rosane]

308   5(6);3(OH);15(OH);16(OH);19(OH)

34. Cassane

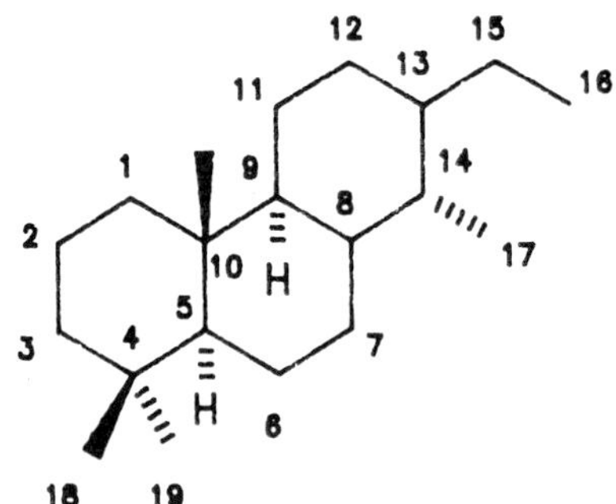

310   12(13);15(16);11(=O);2α(OH)

311   12(13);15(16);11(=O);1β(OH);2α(OH)

312   12(13);11(=O);1β(OH);2α(OH)

35. Rosane

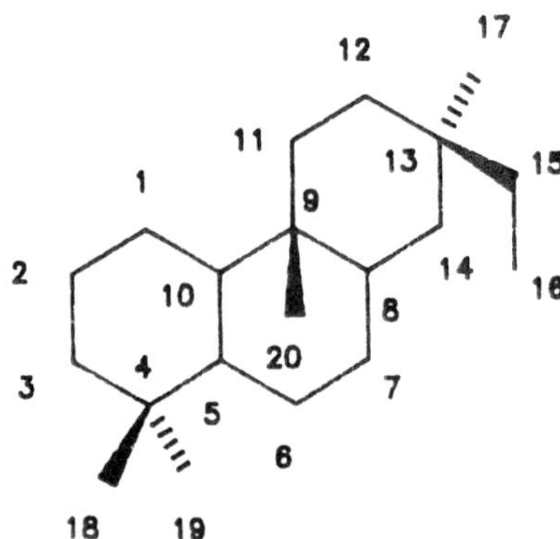

TABLE 2. (*contd.*)

Compound: No.	Substituents
306	5(6);3α(OH);15(OH);16(OH);19(OH);10α(H)
307	5(6);3α(OH);15(OH);16(OH);10α(H)
313	5(6);3β(OH);15(OH);16(OH);10α(H)
314	5(6);3β(OH);15(OH);16(OH);18(OH);10α(H)
315	5(6);3(=O);15(OH);16(OH);10α(H)
881	15(16);5β(OH);10α(H)
882	15(16);5β(OH);18(OH);10α(H)
883	15(16);5β(OH);18(OAc);10α(H)
884	15(16);3α(OH);5β(OH);10α(H)
885	15(16);5β(OH);18(OTig);10α(H)
886	15(16);3α(OH);5β(OH);18(OH);10α(H)
887	5(6);15(16);3α(OH);18(OH);10α(H)
888	5(6);15(16);18(OH);3α(OAc);10α(H)
889	5(6);15(16);3α(OH);18(OAc)
890	1(10);15(16);18(=O,OH)
1161	5(6);14(OH);15(OH);3β(OAc);18(OAc)
1162	5(6);14(OH);15(OH);18(OH);3β(OAc)

36. Sandaracopimarane, 19-Nor-13-Epi-

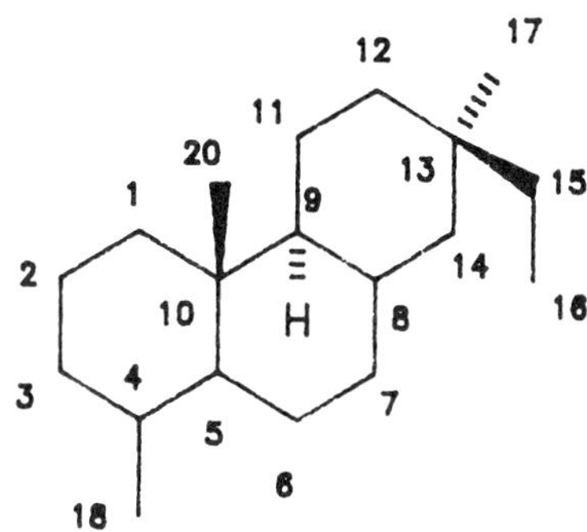

309	4(5);3(=O);15(OH);16(OH)

TABLE 2. (*contd.*)

Compound:

No.   Substituents

37. Normal-Abietane

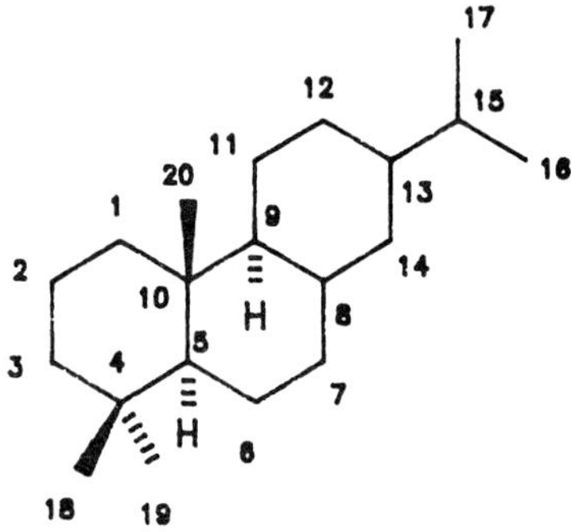

316   7(8);13(14);2α(OAc)

317   7(8);13(14);3β(OH)

318   7(8);13(14);3(=O)

319   8(9);11(12);13(14);12(OH)

320   8(9);11(12);13(14);7(=O);11(OH);14(OH);12(OMe)

321   8(9);12(13);11(=O);14(=O);12(OH)

322   6(7);8(9);12(13);11(=O);14(=O);12(OH)

323   8(9);12(13);11(=O);14(=O);12(OH);7α(OAc)

324   8(9);12(13);7(=O);11(=O);14(=O);12(OH)

1072   8(14);6α(OH);13(OH);7α(OSen)

1073   8(14);7α(OH);13(OH);6α(OSen)

1074   8(14);7α(OH);13(OH)

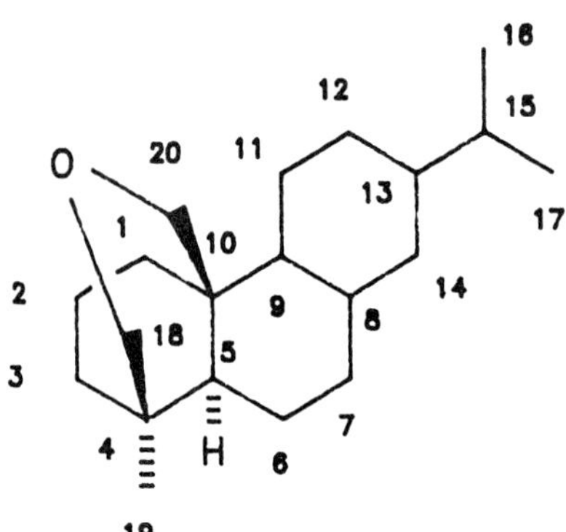

1163   8(9);12(13);11(=O);14(=O);7α(OH);12(OH);20(OH)

TABLE 2. *(contd.)*

Compound:
No.   Substituents

94. Stevisalane

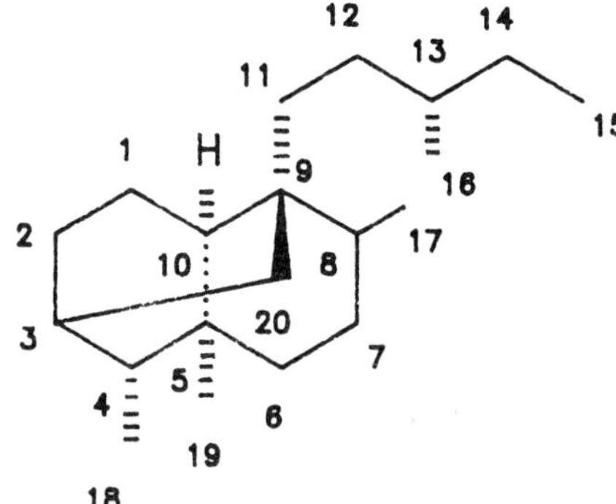

1026   14(15);2(=O);6β(OH);13β(OH);4β(H);8β(H)

IV. Normal-Tetracarbocyclics

73. Normal-Stachane

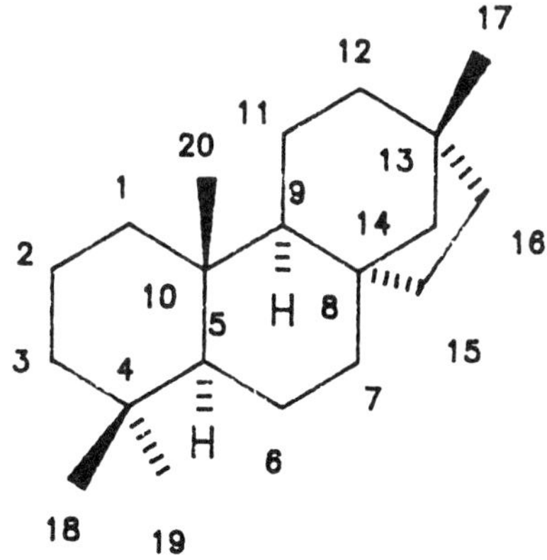

325   15(16);18(=O,OH)

V. <u>ent</u>-Bicarbocyclics and Derivatives

38. <u>ent</u>-Labdane

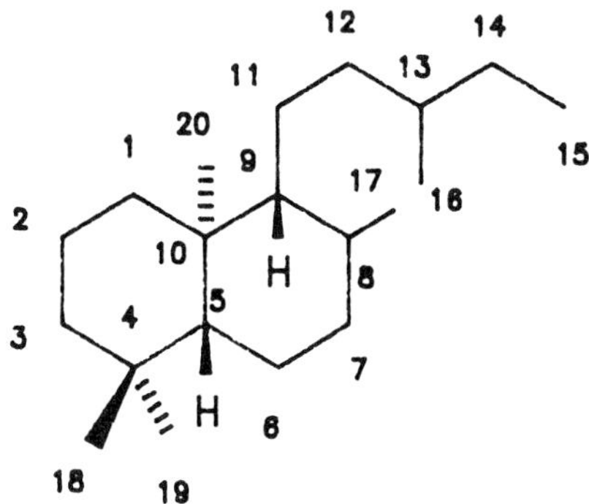

TABLE 2. (*contd.*)

Compound: No.	Substituents
326	8(17);13(14)E;15(OH)
848	8(17);13(14)E;2α(OH);15(OH)
327	8(17);13(14)E;15(Obern)
849	8(17);13(14)E;2(=O);15(OH)
328	8(17);13(14)E;15(OH);7β(OAc)
844	8(17);13(14)E;15(OH);18(OH);3α(OAng)
329	8(17);13(14)E;18(=O,OH);15(OH)
330	8(17);13(14)E;18(=O,OH);15(OAc)
331	8(17);13(14)E;15(=O);2α(OH)
332	8(17);13(14)Z;15(=O);2α(OH)
333	8(17);13(14)E;2(=O);15(=O)
334	8(17);13(14)Z;2(=O);15(=O)
335	8(17);13(14)E;15(=O,OH)
336	8(17);13(14)Z;15(=O,OH)
337	8(17);13(14)Z;15(=O,OH);3α(OH)
338	8(17);13(14)Z;15(=O,OH);3α(OH);7α(OH)
339	13(14)Z;15(=O,OH);17(=O);3α(OH);8β(H)
340	13(14)Z;15(=O,OH);17(=O);3α(OH);8α(H)
341	13(14)Z;15(=O,OH);3(=O);17(=O);8β(H)
342	13(14)Z;15(=O,OH);3(=O);17(=O);8α(H)
343	13(14)Z;15(=O,OH);17(=O);3α(OH);8α(OH)
344	13(14)Z;15(=O,OH);17(=O,OH);8β(H)
345	13(14)E;15(=O,OH);17(=O,OH);8β(H)
346	13(14)Z;15(=O,OH);3(=O);8(-O-)17
347	13(14)E;8β(-O-)17;15(Obern)
348	13(14)E;8(OH);15(OH)
349	13(14)E;15(=O);8(OH)
350	13(14)Z;15(=O);8(OH)
351	7(8);13(14)E;6β(OH);15(OH)
352	7(8);13(14)Z;6(=O);15(OH)

TABLE 2. (*contd.*)

Compound: No.	Substituents
353	7(8);13(14)E;11(OH);3α(OAc);15(OAc)
354	7(8);13(14)E;15(=O,OH)
355	7(8);13(14)E;15(=O,OH);2β(OAng)
356	6(7);8(17);15(=O,OH)
357	8(17);2α(OH);15(OH)
358	13(S);8(17);15(=O,OH);2α(Oi-Val)
857	8(17);15(=O,OH);2α(OH)
858	8(17);15(=O,OH);2α(OAng)
859	8(17);15(=O,OH);2α(OTig)
860	8(17);15(=O,OH);2α(O2-Mebut)
861	8(17);15(=O,OH);2(=O)
862	15(=O,OH);2α(OAng);8β(-O-)17
863	15(=O,OH);2α(O2-Mebut);8β(-O-)17
359	8(17);2(=O);15(OH)
360	7(8);15(=O,OH);17(OMe)
361	7(8);15(=O,OH);2β(OAng)
362	7(8);13(14)Z;15(OH);16(OH);17(OH)
363	7(8);13(14)Z;15(OAc);16(OAc)
364	7(8);13(14)Z;15(OAc);16(OAc);17(OAc)
365	6(7);8(17);13(14)E;15(=O,OH);16(OH)
366	7(8);13(14)Z;15(=O,OH);16(OH)
367	7(8);13(14)Z;15(=O,OH);16(OH);17(OH)
368	7(8);13(14)Z;15(=O,OH);16(=O);12(OH)
369	7(8);13(14)Z;15(=O,OH);17(=O);16(OH)
370	7(8);13(14)Z;15(=O,OH);17(OH);16(OAc)
371	7(8);13(14)E;15(=O,OH);17(=O);16(OAc)
372	7(8);13(14)E;15(=O,OH);17(=O);6α(OH);16(OAc)
373	7(8);13(14)E;15(=O,OH);17(=O);6β(OH);16(OAc)
374	7(8);13(14)Z;15(=O,OH);6(=O);16(OAc)
375	7(8);13(14)Z;15(=O,OH);6(=O);16(=O)

TABLE 2. (*contd.*)

Compound:	
No.	Substituents
376	7(8);13(14)E;15(=O,OH);6(=O);16(OH)
377	7(8);13(14)E;15(=O,OH);6(=O);16(OAc)
378	7(8);13(14)E;15(=O,OH);6(=O);16(OMe)
379	6(7);13(14)E;15(=O,OH);8α(OH);16(OH)
380	6(7);13(14)E;15(=O,OH);17(=O);16(OAc);8α(H)
381	6(7);13(14)E;15(=O,OH);17(=O,OH);16(OAc);8α(H)
382	8(17);13(14)Z;7β(OH);15(OH);16(OH)
383	8(17);15(=O,OH);16(=O)
384	6(7);8(17);13(14);15(=O);15(-O-)16
385	8(17);13(14);15(=O);19(OH);15(-O-)16
386	7(8);13(14);15(=O);15(-O-)16
387	7(8);13(14);15(=O);12(OH);15(-O-)16
388	7(8);13(14);15(=O);16α(OH);15(-O-)16
389	7(8);13(14);15(=O);16β(OH);15(-O-)16
390	7(8);13(14);15(=O);12(OH);16(OH);15(-O-)16
391	7(8);13(14);15(=O);12(OH);16α(OH);15(-O-)16
392	7(8);13(14);15(=O);12(OH);16β(OH);15(-O-)16
393	7(8);13(14);6(=O);15(=O);15(-O-)16
394	7(8);13(14);6(=O);15(=O);16(OH);15(-O-)16
395	7(8);13(14);15(=O);17(=O);16α(OH);15(-O-)16
396	7(8);13(14);15(=O);17(=O);16β(OH);15(-O-)16
397	8(17);13(14);16(=O);7β(OH);15(-O-)16
843	8(17);13(14);18(=O,OH);16(=O);15(-O-)16
398	7(8);13(14);16(=O);15(-O-)16
399	7(8);13(14);16(=O);12(OH);15β(OH);15(-O-)16
400	7(8);13(14);16(=O);12(OH);15α(OH);15(-O-)16
401	8(17);13(16);14(15);7β(OH);15(-O-)16
402	8(17);13(16);14(15);7β(OAc);15(-O-)16
403	8(17);13(16);14(15);19(OH);15(-O-)16
404	8(17);13(16);14(15);6β(OH);19(OH);15(-O-)16

TABLE 2. (*contd.*)

Compound: No.	Substituents
839	8(17);13(16);14(15);19(=O,OH);15(-O-)16
840	8(17);13(16);14(15);19(=O,OH);3α(OH);15(-O-)16
841	8(17);13(16);14(15);18(OH);19(OH);15(-O-)16
842	8(17);13(16);14(15);3α(OH);18(OH);15(-O-)16
974	8(17);13(16);14(15);18(OH);3α(OAng);15(-O-)16
975	8(17);13(16);14(15);3α(OH);18(OAng);15(-O-)16
830	8(17);13(16);14(15);18(=O,OH);15(-O-)16
831	8(17);13(16);14(15);18(=O,OH);3α(OH);15(-O-)16
832	8(17);13(16);14(15);18(=O,OH);3α(OAng);15(-O-)16
833	8(17);13(16);14(15);18(=O,OH);3α(O2-Mebut);15(-O-)16
834	8(17);13(16);14(15);18(=O,OH);3α(Oi-But);15(-O-)16
835	8(17);13(16);14(15);18(=O,OH);19(OH);15(-O-)16
836	13(16);14(15);18(=O,OH);8α(-O-)17;15(-O-)16
837	13(16);14(15);18(=O,OH);8β(-O-)17;15(-O-)16
838	13(16);14(15);18(=O,OH);17(=O);8β(H);15(-O-)16
405	1(2);8(17);13(16);14(15);18(=O,OH);15(-O-)16
406	8(17);13(16);14(15);2(=O);12(OH);15(-O-)16
407	8(17);13(16);14(15);2(=O);12(=O);15(-O-)16
408	8(17);13(16);14(15);18(OH);7β(OAc);15(-O-)16
409	8(17);13(16);14(15);7β(OAc);18(OAc);15(-O-)16
410	8(17);13(16);14(15);18(=O);7β(OH);15(-O-)16
411	8(17);13(16);14(15);18(=O);7β(OAc);15(-O-)16
412	8(17);13(16);14(15);2(=O);12(=O);3α(OH);15(-O-)16
413	8(17);13(16);14(15);2(=O);3α(OH);12(OH);15(-O-)16
414	7(8);13(16);14(15);17(OH);15(-O-)16
415	7(8);13(16);14(15);18(=O,OH);15(-O-)16
416	13(16);14(15);8(OH);3α(OAng);15(-O-)16
417	8(17);13(14);2(=O);12(=O);15α(OH);15(-O-)16
418	8(17);13(14);2(=O);12(=O);15β(OH);15(-O-)16
421	8(17);12(13);14(15);6β(OH);12(-O-)15

TABLE 2. *(contd.)*

Compound: No.	Substituents
422	8(17);12(13)Z;14(15)
423	8(17);12(13)Z;14(15);18(=O)
424	8(17);12(13)Z;14(15);18(=O,OH)
425	8(17);12(13)E;14(15);18(=O,OH)
426	8(17);12(13)Z;14(15);18(=O,OH);2α(OH)
427	8(17);12(13)Z;14(15);18(=O,OH);2(=O)
428	12(13)Z;14(15);6β(OH);7α(OH);8(OH)
429	12(13)Z;14(15);7α(OH);8(OH);6β(OAc)
808	12(13)Z;14(15);8β(OH);6β(OAng);7α(OAc)
430	8(17);12(13)E;14(15)
431	8(17);12(13)E;14(15);19(OH)
432	8(17);12(13)E;14(15);3α(OH);19(OH)
433	8(17);12(13)E;14(15);18(OH)
434	8(17);12(13)E;14(15);18(OH);19(OH)
435	8(17);12(13)E;14(15);18(=O,OH);1β(OAc)
436	12(13)E;14(15);3β(OH);8β(OH)
809	12(13)E;14(15);6β(OH);7α(OH);8β(OH)
810	12(13)E;14(15);7α(OH);8β(OH);6β(OAc)
811	12(13)E;14(15);8β(OH);6β(OAng);7α(OAc)
441	13(16);14(15);8α(OH)
442	8(17);14(15);13α(OH)
443	8(17);14(15);2β(OH);13α(OH)
444	8(17);14(15);2α(OH);13α(OH)
445	8(17);14(15);13α(OH);2β(OSucc)
446	8(17);14(15);2(=O);13α(OH)
447	7(8);14(15);13(OH)
448	7(8);14(15);3(OH);13(OH)
449	7(8);14(15);3(=O);13(OH)
450	14(15);8α(OH);13α(OH)
460	13(14)Z;15(=O,OH);6(=O);16(=O);8(-O-)12

TABLE 2. *(contd.)*

Compound: No.	Substituents
461	13(S);1(OH);6β(OH);7β(-O-)8β;15(OSen);5β(H);9β(-O-)13
1012	7(8);13(14)E;2β(OH);15(OAc);18(OAc)
1013	7(8);13(14)E;2β(OH);15(OH);18(OAc)
1014	8(17);13(14)E;15(OH);3α(OAc);18(OTig)
1015	12(13)E;14(15);8β(OH);18(OH)
1016	8(9);14(15);7(=O);13β(OH);18(OH)
1017	14(15);7(=O);18(OH);9β(-O-)13β;8β(H)
1018	14(15);7(=O);18(=O);9β(-O-)13β;8β(H)
1019	14(15);18(=O,OH);7(=O);9β(-O-)13β;8β(H)
1037	8(17);13(16);14(15);18(OH);3α(OAng);15(-O-)16
1038	8(17);13(16);14(15);2β(OH);18(OH);3α(OAng);15(-O-)16
1039	8(17);13(14)Z;2β(OH);15(OH);16(OH);18(OH);3α(OAng)
1040	8(17);13(14)E;15(=O,OH)
1041	8(17);13(14)E;15(=O,OH);18(OH)
1042	7(8);13(14)E;15(=O,OH);18(=O);6β(OH)
1043	7(8);13(14)E;15(=O,OH);18(=O,OH);6β(O-i-Val)
1044	13(14)E;15(=O,OH);5β(-O-O-)8β
1045	7(8);13(14)E;15(=O,OH);6β(OH);18(OH)
1046	7(8);15(=O,OH);17(OH)
1047	7(8);15(=O,OH);17(OAc)
1048	7(8);15(=O,OH);17(=O)
1049	7(8);15(=O,OH);2α(OH);17(OH)
1050	7(8);15(=O,OH);2α(OH);17(OAc)
1051	7(8);15(=O,OH);17(=O);2α(OH)
1052	7(8);15(=O,OH);17(=O,OH)
1053	7(8);15(=O,OH);19(OH);17(OAc)
1164	8(17);13(14)E;15(=O,OH);18(=O,OH)
1173	8(17);19(=O,OH);15(OH)
1174	13(14)E;2β(OH);8β(OH);15(OH)
1175	7(8);15(=O,OH);3α(OH);17(OH)

TABLE 2. (*contd.*)

Compound: No.	Substituents
1176	7(8);15(=O,OH);3α(OH);17(OAc)
1177	7(8);15(=O,OH);17(=O);3α(OH)
1178	7(8);15(=O,OH);3α(OH);17(OMe)
1179	7(8);15(=O,OH);2(OH);17(OMe)
1180	7(8);13(14)Z;15(=O,OH);16(OH);17(OMe)
1181	8(17);13(14)Z;15(=O);18(OH);2β(OH);15(-O-)16
1182	8(17);13(14)Z;15(=O);18(OH);3α(OAng);15(-O-)16
1183	8(17);13(14)Z;15(=O);18(OGlu);3α(OAng);15(-O-)16
1184	8(17);13(14)Z;15(=O);2β(OH);18(OH);3α(OAng);15(-O-)16
1185	8(17);12(13)E;14(15);18(OAc);19(OAc)
1186	8(17);12(13)E;3α(OH);14(OH);15(OH)
1188	7(8);13(16);14(15);2β(OH);3α(OH);15(-O-)16
1189	8(17);13(14)E;15(=O,OH);3α(OH);18(OH)
1190	8(17);13(14)E;15(=O,OH);18(OH);3α(OAng)
1191	8(17);13(14)Z;15(=O,OH);2β(OH);16(OH);18(OH);3α(OAng)
1192	7(8);13(14)E;2β(OH);15(OH);3α(OAng)
1193	7(8);13(14)E;2β(OH);3β(OH);15(OH)
1194	6(7);8(17);13(14)E;15(=O,OH)
1195	6(7);8(17);13(14)E;15(=O,OH);18(OH)
1196	7(8);15(=O,OH);2β(OH);3β(OH)
1197	7(8);2β(OH);3β(OH);15(OH)

1165	7(8);13(14)E;15(OH);3α(OAng);2'β(OH);3'α(OH);4'α(OH)
1166	7(8);13(14)E;15(OH);3α(OAng);2'β(OH);3'α(OH);4'α(OAc)
1167	7(8);13(14)E;15(OH);3α(OAng);2'β(OH);3'α(OAc);4'α(OH)

TABLE 2. (*contd.*)

Compound:
No.   Substituents

1168   7(8);13(14)Z;15(OH);16(OH);3α(OAng);2'β(OH);
           3'α(OH);4'α(OAc)

1169   7(8);13(16);14(OH);15(OH);3α(OAng);2'β(OH);
           3'α(OH);4'α(OAc)

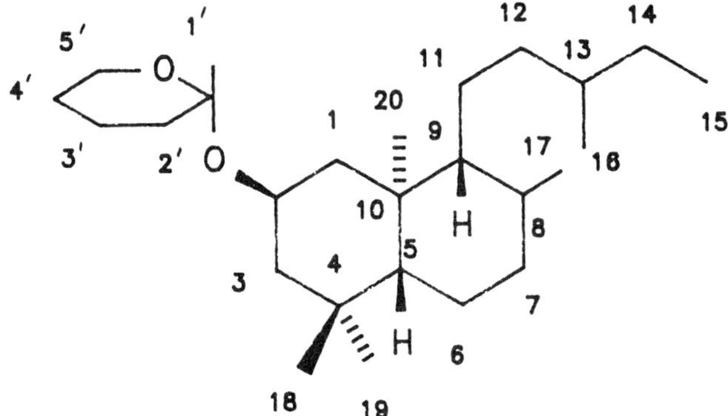

1170   7(8);13(14)E;15(OH);3α(OAng);2'β(OH);3'α(OH);4'β(OAc)

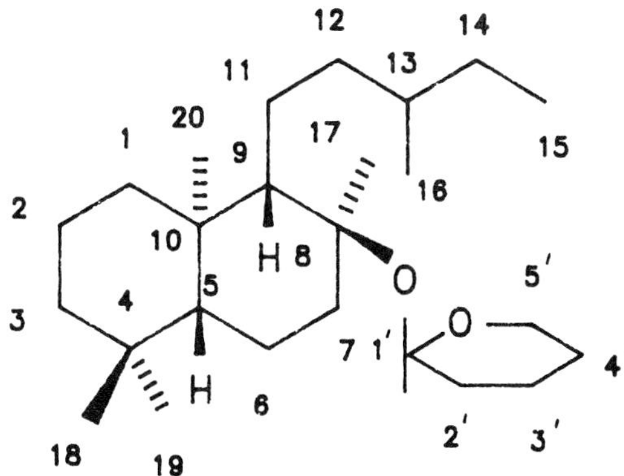

1171   7(8);13(14)E;15(OH);3α(OAng);2'β(OH);3'α(OH);4'β(OH)

1172   13(14)E;15(OH);2'α(OH);3'β(OH);4'α(OH)

TABLE 2. *(contd.)*

Compound:
No.   Substituents

437   8(17);14(15);18(=O,OH)

438   8(17);14(15);18(=O,OH)

439   8(17);14(15);18(=O,OH)

TABLE 2. (*contd.*)

Compound:
No.   Substituents

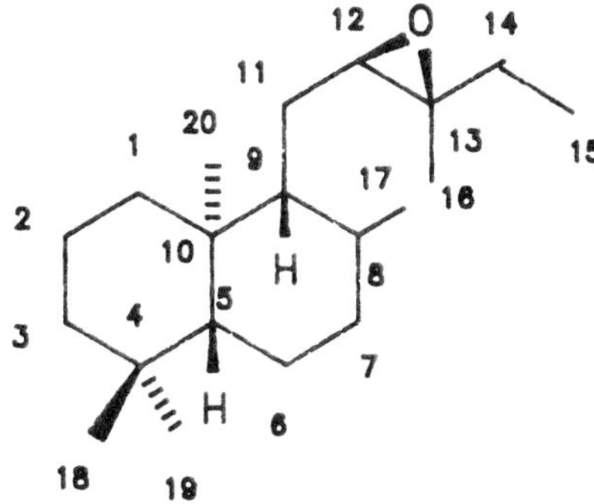

440   8(17);14(15);18(=O,OH)

39. <u>ent</u>-Labdane-19,6-Lactone

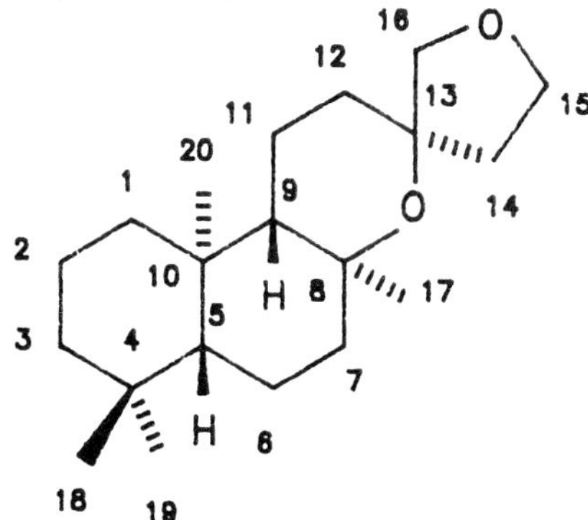

462   7(8);15(=O,OH)

40. <u>ent</u>-Labdane-8,13-Epoxide-15,16-Hemiacetal

463   14(OH);16β(OH)

TABLE 2. *(contd.)*

Compound:

No.	Substituents
464	14(OH);16β(OAc)

97. <u>ent</u>-Labdane Dimer

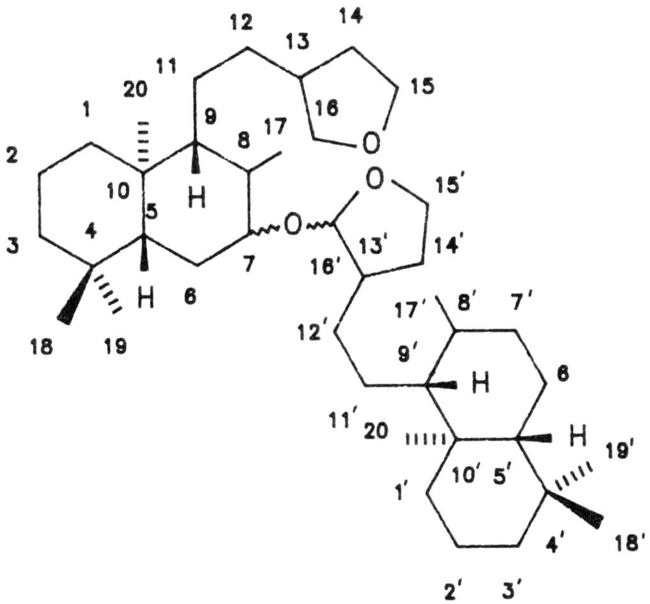

| 419 | 8(17);13(16);14(15);7α(H);8'(17');15'(=O);16'α(H) |
| 420 | 8(17);13(16);14(15);7α(H);8'(17');15'(=O);16'β(H) |

98. <u>ent</u>-Manoyl Oxide, 13-Epi-

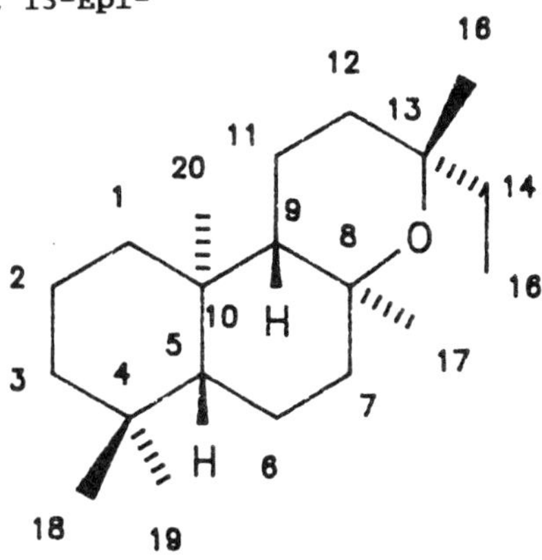

| 451 | 14(15) |

TABLE 2. *(contd.)*

Compound:	
No.	Substituents
452	14(15);3α(OH)
453	14(15);3(=O)
454	14(15);2(=O)
455	14(15);19(OH)
456	14(15);2(=O);19(OH)
457	14(-O-)15;16(OAc)
1198	14(15);3α(OAc)
1199	14(15);3α(OH);19(OCinn)
1200	14(15);3α(OH);19(OH)

99. **ent-Manoyl Oxide**

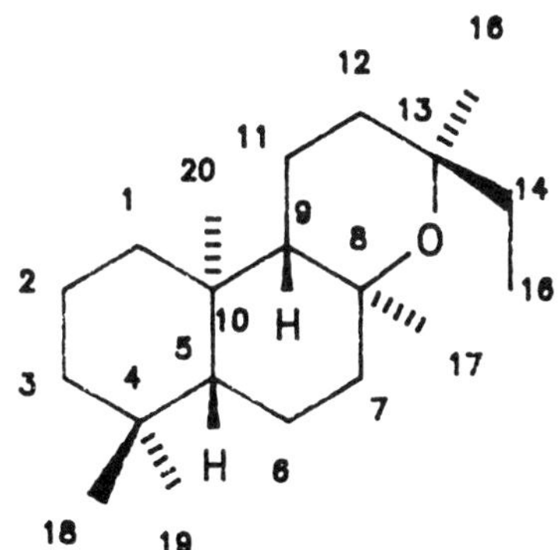

458	14(15)
459	14(15);3β(OH)

41. **ent-Labdane, 17-Nor-**

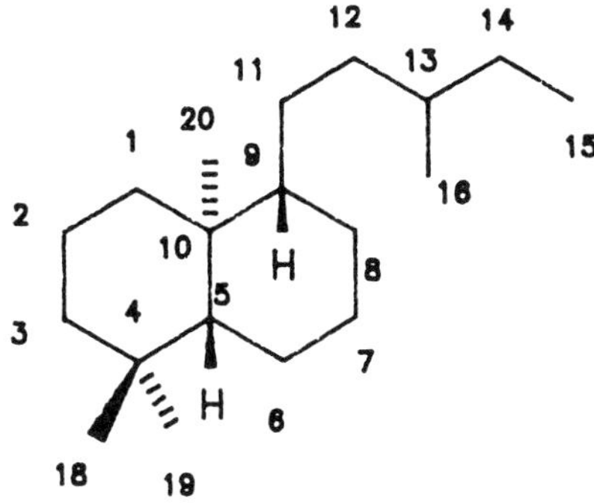

TABLE 2. (*contd.*)

Compound:

No.	Substituents
465	6(7);13(14)E;15(=O,OH);8β(OCHO);16(OAc)
466	6(7);13(14)E;15(=O,OH);8(=O);16(OAc)
467	13(14)Z;15(=O,OH);8(=O);3α(OH)
468	13(14)Z;15(=O,OH);3(=O);8(=O)
1201	13(14)Z;15(=O,OH);8(=O)
1202	13(14)Z;15(=O,OH);3(=O);8(=O)

42. _ent_-Labdane, 15-Nor-

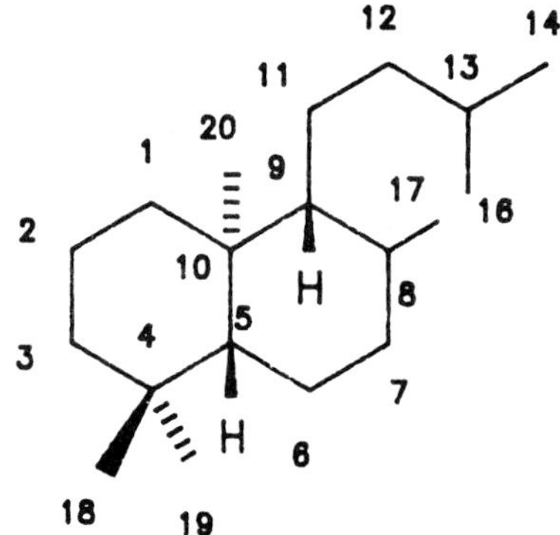

469	8(17);12(13)E;19(=O,OH);14(=O)

43. _ent_-Labdane, 14,15-Bis-Nor-

470	8(17);13(=O)

TABLE 2. *(contd.)*

Compound:

No.   Substituents

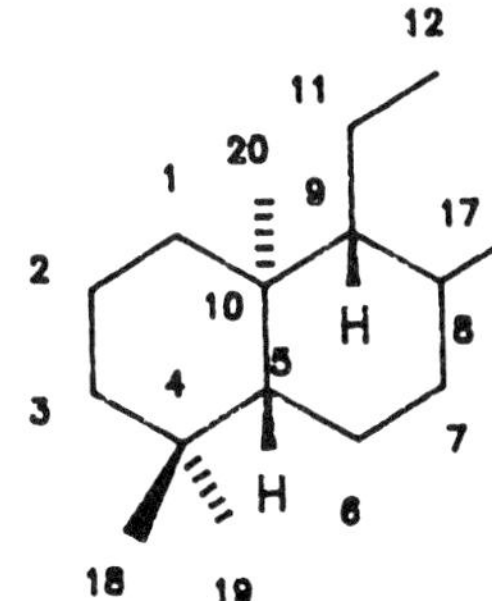

1203   7(8);13(=O);3α(OAng);2'β(OH);3'α(OH);4'α(OAc)

44. <u>ent</u>-Labdane, 13,14,15,16-Tetra-Nor-

471   7(8);12(=O)

472   7(8);12(=O,OH);17(OH)

45. <u>ent</u>-Labdane, 7,8-Seco-

473   12(13)E;14(15);7(=O,OH);8(=O)

TABLE 2. *(contd.)*

Compound:	
No.	Substituents

74. _ent_-Labdane, 18-Nor-

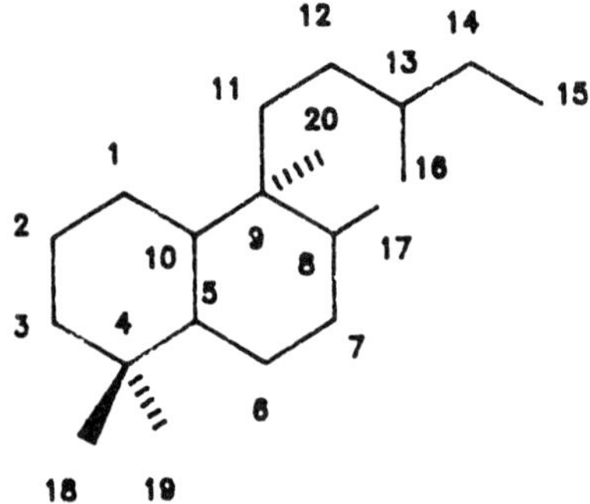

828	7(=O);2α(OH);3α(OH);19(OH);8α(H)
829	7(=O);12(=O);2α(OH);3α(OH);19(OH);8α(H)
1205	7(=O);3α(OH);19(OH);2α(OAc);8α(H)
1206	7(=O);2α(OH);3α(OH);19(OH);8α(H)
1207	7(=O);12(=O);3α(OH);19(OH);2α(OAc);8α(H)

95. _ent_-Friedolabdane

1059	5(6);13(14)E;15(=O,OH);18(OH);10β(H)
1208	5(6);13(14)E;15(=O,OH);18(OH);10β(H)
1209	1(10);13(14)E;2(=O);15(OH);5β(H)
1210	1(10);13(14)E;2α(OH);15(OH);5β(H)
1211	1(10);14(15) ;19(=O,OH);13(OH);5β(H)

TABLE 2. (*contd.*)

Compound:

No.   Substituents

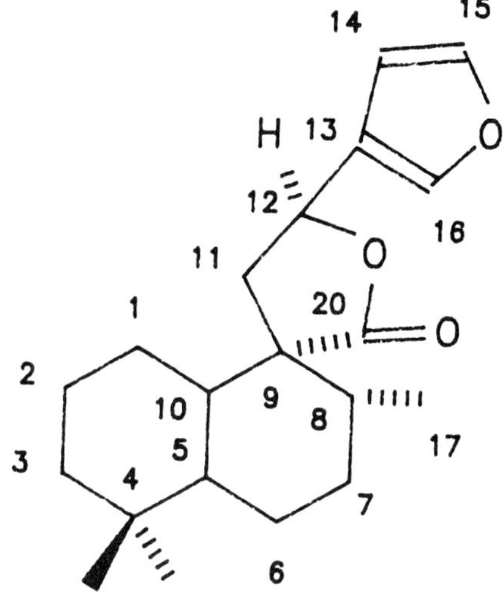

1212	3α(OH);5α(OH)
1213	5(6);3α(OH)
1214	3α(OH);5β(-O-)10β

102. *ent*-Friedolabdane, 3,4-Seco-

| 1215 | 3(=O) |
| 1216 | 3(=O,OH) |

TABLE 2. (*contd.*)

Compound:	
No.	Substituents
1217	3(=O,OH);19(OH)
1218	3(=O,OMe);19(OH)

1219　　　　4(18)

1220　　　　4(5);3(=O,OH)

TABLE 2. *(contd.)*

Compound:
No.   Substituents

103. <u>ent</u>-Friedolabdane, Rearranged 3,4-Seco-

1221

46. <u>trans</u>-<u>ent</u>-Clerodane

236   4(18);13(14)E;15(OH)

248   1(2);3(4);13(16);14(15);18(=O,OH);15(-O-)16

249   3(4);13(16);14(15);18(=O,OH);15(-O-)16

TABLE 2. *(contd.)*

Compound: No.	Substituents
250	1(2);3(4);13(16);14(15);18(=O,OH);15(-O-)16
251	1(2);3(4);13(16);14(15);18(=O,OMe);15(-O-)16
245	3(4);13(16);14(15);18(=O,OH);15(-O-)16
475	3(4);13(14)E;15(OH)
476	3(4);13(14)E;15(OH);17(OH)
477	13(14)E;3β(-O-)4β;2α(OH);15(OAc)
478	13(14)Z;3β(-O-)4β;2α(OH);16(OH);15(OAc)
1010	3(4);13(14)E;17(OH);15(OArac)
1011	3(4);13(14)E;17(OH);15(OBehen)
479	3(4);14(15);13(OH)
480	3(4);14(15);13(OH);6β(OAng)
481	3(4);15(OH);17(OH)
482	3(4);17(OH);15(OArac)
483	3(4);17(OH);15(OBehen)
484	3(4);15(OH);18(OSucc)
485	3α(-O-)4α;2α(OH);15(OH)
486	3(4);17(=O);15(OH)
487	3(4);18(=O);18(-O-)6α;15(OH);17(OAc)
488	3(4);6α(OH);15(OH);18(OH);17(OAc)
489	3(4);6α(OH);15(OH);18(OH);17(OPhe-Ac)
490	3(4);18(=O);6α(OH);15(OH);17(OPhe-Ac)
491	3(4);18(=O);6α(OH);15(OH);17(OAc)
492	3(4);15(=O);17(OH)
493	3(4);13(14)Z;15(=O,OH)
494	3(4);13(14)Z;15(=O,OH);2β(OAc)
495	3(4);13(14)E;15(=O,OMe)
496	3(4);13(14)E;15(=O,OMe);6β(OAc)
497	3(4);13(14)E;15(=O,OMe);6β(OAng)
498	3(4);13(14)E;15(=O,OH);6α(OAng)
499	3(4);13(14)E;15(=O,OH);6α(OTig)

TABLE 2. (*contd.*)

Compound: No.	Substituents
500	3(4);13(14)E;15(=O,OH);7α(OAc)
962	13(14)Z;15(=O,OH);2α(OH);3α(OH);4β(OH)
501	3(4);13(14)E;15(=O,OH);6α(OH);18(OH);17(OAc)
502	3(4);13(14)E;15(=O,OH);6α(OH);18(OH);17(OPhe-Ac)
503	3(4);15(=O,OH)
504	3(4);15(=O,OH);18(OAc)
961	3(4);13(14)E;15(=O,OH);2(=O)
505	13(S);3(4);15(=O,OH);18(=O,OH)
506	15(=O,OH);2(=O);3β(-O-)4β
507	3(4);13(16);14(15);15(-O-)16
508	3(4);13(16);14(15);18(OMal);15(-O-)16
509	3(4);13(16);14(15);7α(OH);18(OH);15(-O-)16
846	3(4);13(16);14(15);18(OH);19(OH);15(-O-)16
510	3(4);13(16);14(15);19(OAc);18(OMal);15(-O-)16
967	3(4);13(16);14(15);17(=O);15(-O-)16
511	3(4);13(16);14(15);18(=O,OH);15(-O-)16
953	3(4);13(16);14(15);18(=O,OH);12α(O2-Mebut);15(-O-)16
512	1(2);3(4);13(16);14(15);18(=O,OH);15(-O-)16
911	1(2);3(4);13(16);14(15);18(=O,OH);19(OAc);15(-O-)16
513	3(4);13(16);14(15);18(=O,OH);2α(OH);15(-O-)16
514	7(8);13(16);14(15);3α(OH);18(Oi-Val);4β(H);15(-O-)16
515	7(8);13(16);14(15);3α(OH);18(O2-Mebut);4β(H);15(-O-)16
516	7(8);13(16);14(15);3α(OH);18(OAng);4β(H);15(-O-)16
518	3(4);13(16);14(15);18(=O,OH);19(OH);15(-O-)16
519	3(4);13(16);14(15);18(=O,OH);2β(OH);19(OH);15(-O-)16
900	3(4);13(16);14(15);18(=O,OH);19(OAc);15(-O-)16
901	3(4);13(16);14(15);18(=O,OH);19(OMe);15(-O-)16
902	3(4);13(16);14(15);18(=O,OH);19(OAng);15(-O-)16
903	3(4);13(16);14(15);18(=O,OH);19(Oi-Val);15(-O-)16
904	3(4);13(16);14(15);18(=O,OH);19(=O);15(-O-)16

TABLE 2. (*contd.*)

Compound: No.	Substituents
524	3(4);13(16);14(15);20(=O,OH);15(-O-)16
520	3(4);13(16);14(15);18(=O);15(-O-)16;18(-O-)19
950	3(4);13(16);14(15);18(=O);12α(OH);15(-O-)16;18(-O-)19
951	3(4);13(16);14(15);18(=O);7α(OH);12α(OH); 15(-O-)16;18(-O-)19
521	3(4);13(16);14(15);18(=O);1α(OH);15(-O-)16;18(-O-)19
522	3(4);15(=O,OH);18(=O);17(OPhe-Ac);6α(-O-)18
523	3(4);13(16);14(15);7(=O);18(=O);6α(-O-)18;15(-O-)16
534	3(4);13(14)Z;15(=O);17(OH)
535	3(4);13(14)E;15(=O);17(OH)
536	3(4);13(14)Z;17(=O,OH);15(=O)
537	3(4);13(14)E;17(=O,OH);15(=O)
865	13(14)E;3(4);15(=O,OH);16(OH)
962	13(14)E;3(4);15(=O,OH);6α(OAc)
538	3(4);13(14)Z;15(=O,OH);16(OH);18(OH)
539	3(4);13(14)Z;15(=O,OH);18(OH);16(OAc)
540	3(4);13(14)Z;15(=O,OH);18(=O);16(OH)
541	3(4);13(14)Z;15(=O,OH);18(=O);16(OAc)
542	3(4);15(=O,OH);16(=O)
543	3(4);13(14);15(=O);15(-O-)16
544	3(4);13(14);15(=O);2α(OH);15(-O-)16
864	3(4);13(14);15(=O);16(OH);15(-O-)16
550	3(4);13(14);15(=O);18(OH);15(-O-)16
558	3(4);13(14);18(=O,OH);15(=O);15(-O-)16
909	3(4);13(14);18(=O,OH);15(=O);16α(OAc);19(OAc); 15(-O-)16
910	3(4);13(14);18(=O,OH);15(=O);16β(OAc);19(OAc); 15(-O-)16
559	3(4);13(14);15(=O);18(=O);7α(OH);15(-O-)16;18(-O-)19
560	3(4);13(14);15(=O);18(=O);12(OH);15(-O-)16;18(-O-)19
561	3(4);7(=O);15(=O);18(=O);15(-O-)16;18(-O-)19

TABLE 2. *(contd.)*

Compound:
No.   Substituents

562   3(4);15(=O);18(=O);7α(OH);15(-O-)16;18(-O-)19

563   3(4);15(=O);18(=O);15(-O-)16;18(-O-)19

564   3(4);15(=O);15(-O-)16

565   3(4);15(=O);16(OH);15(-O-)16

963   3(4);15(=O);18(OAc);16(OMe);15(-O-)16

566   3(4);13(14);16(=O);20(=O);15(-O-)16

905   3(4);13(14);18(=O,OH);16(=O);15α(OH);19(OH);15(-O-)16

906   3(4);13(14);18(=O,OH);16(=O);15β(OH);19(OH);15(-O-)16

907   3(4);18(=O,OH);16(=O);19(OH);14α(-O-)15α;15(-O-)16

908   3(4);18(=O,OH);16(=O);19(OH);14β(-O-)15β;15(-O-)16

567   3(4);13(14);16(=O);2β(OH);15(-O-)16

568   3(4);13(14);2(=O);16(=O);15(-O-)16

569   3(4);15(OH);18(OMal);15(-O-)16

1027   3(4);13(14)Z;15(=O,OH);6α(O-i-But);7β(OAc)

1028   3(4);13(14)Z;15(=O,OH);6α(OAng);7β(OAc)

1029   3(4);13(14)Z;15(=O,OH);6α(OAc);7β(OAc)

1030   3(4);13(14)E;15(=O,OH);2(=O);6α(OH);7β(OH)

1031   3(4);15(=O,OH);6α(O-i-But);7β(OAng)

1033   3(4);13(16);14(15);18(=O,OH);19(OH);15(-O-)16;8β(H)

1034   3(4);13(16);14(15);18(=O,OH);2β(OH);19(OH);
           15(-O-)16;8β(H)

1070   1(2);3(4);13(16);14(15);18(=O,OH);19(OH);15(-O-)16

1071   13(16);14(15);20(=O,OH);3α(-O-)4α;15(-O-)16;8β(H)

1075   13(14)Z;15(=O);3α(-O-)4α;8β(H)

1076   13(16);14(15);3α(-O-)4α;15(-O-)16;8β(H)

1077   13(14)E;15(=O);3α(-O-)4α;8β(H)

1078   13(14)Z;3(=O);15(=O);4β(H);8β(H)

1079   13(14)E;3(=O);15(=O);4β(H);8β(H)

1080   13(14);15(=O);3α(-O-)4α;15(-O-)16;8β(H)

1081   13(14);3(=O);15(=O);15(-O-)16;4β(H);8β(H)

TABLE 2. (*contd.*)

Compound:	
No.	Substituents
1082	13(14);15(=O);3α(OH);4β(OH);15(-O-)16;8β(H)
1222	3(4);13(14)Z;15(OH);16(OH);18(OH)
1223	3(4);13(14)E;18(=O,OH);6α(OH);15(OH)
1224	3(4);13(14)E;17(=O,OH);15(OH)
1225	3(4);13(14)Z;17(=O,OH);16(OH);15(OAc)
1227	3(4);13(14)E;15(OAc);18(OAc)
1228	13(14)Z;16(OH);3α(OAc);15(OAc);3β(-O-)4β
1229	3(4);15(OAc);18(OAc)
1230	3(4);15(OAc);16(OAc);18(OAc)
1231	3(4);13(16);14(15);18(OH);19(OMal);15(-O-)16
1232	3(4);13(16);14(15);18(OH);15(-O-)16
1233	3(4);13(16);14(15);18(OAc);19(OAc);15(-O-)16
1234	3(4);13(16);14(15);18(=O);15(-O-)16;18(-O-)6α
1235	3(4);13(16);14(15);18(=O);7α(OH);15(-O-)16;18(-O-)6α
1236	3(4);13(14)Z;15(=O);16(OMe);18(OAc);15(-O-)16
1237	3(4);13(14)Z;15(=O);18(=O);2(OH);15(-O-)16;18(-O-)19
1238	3(4);13(14)Z;15(=O);18(=O);2(OAc);15(-O-)16;18(-O-)19
1239	3(4);18(OAc);15(OMe);15(-O-)16
1240	3(4);13(14)Z;16(=O);18(OAc);15(-O-)16
1241	3(4);18(OH);15(OMe);15(-O-)16
1242	3(4);18(OH);19(OH);15(OMe);15(-O-)16
1243	3(4);18(OAc);19(OAc);15(OMe);15(-O-)16

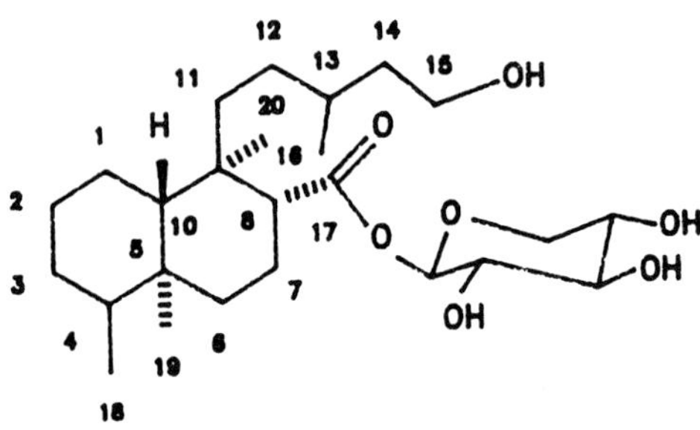

1226	3(4);13(14)E

TABLE 2. (*contd.*)

Compound:

No.   Substituents

570   3(4);13(14)Z;15(=O,OH);20(OH)

525   3(4);8(17);7(=O);12α(H)

526   3(4);7(8);17(=O);7(OH);12α(H)

527   3(4);17(=O);8β(H);12β(H)

528   3(4);17(-O-)20;8β(H);12β(H);17β(H)

969   3(4);8(17);7(=O);12β(H)

970   3(4);7(=O);8β(H);12β(H)

971   1(=O);17(=O);6α(OAng);8α(H);12α(H)

972   3(4);7(8);17(=O);6β(OH);12β(H)

973   2(3);17(=O);8α(H);12α(H)

1069   3(4);17(=O);1α(OH);12β(H)

TABLE 2. (*contd.*)

Compound:	
No.	Substituents
1257	3(4);17(=O);1α(OAc);8β(H);12β(H)
1258	3(4);17(=O);7α(OH);8β(H);12β(H)

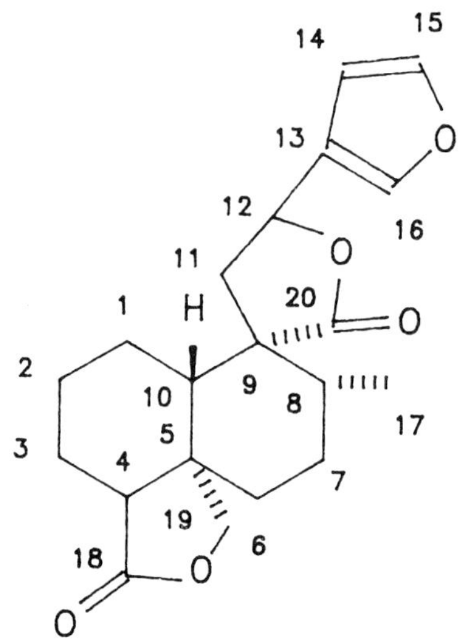

529	3(4);12β(H)
968	3(4);2α(OAng);12β(H)
530	3(4);7β(OAng);12β(H)
531	2α(OAng);3α(OSen);12β(H)
532	3(4);2α(OSen);12β(H)
533	3(4);2α(O2-Mebut);12β(H)

1244	3(4);3'(4');18(OMal);18'(OMal)

TABLE 2. *(contd.)*

Compound:	
No.	Substituents
1245	3(4);3'(4');18(OSucc);18'(OSucc)

47. Printziane

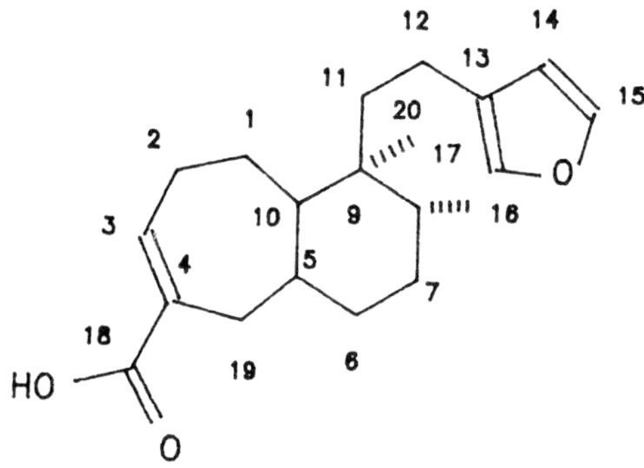

571	5(10)
572	5(6);10β(H)
913	5α(OH);10β(H)
914	5α(OMe);10β(H)
915	1(2);5α(OH);10β(H)

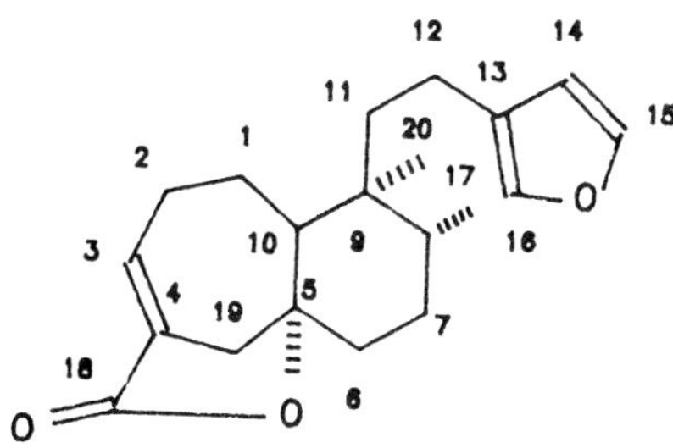

916	10β(H)

76. Conyscabrane

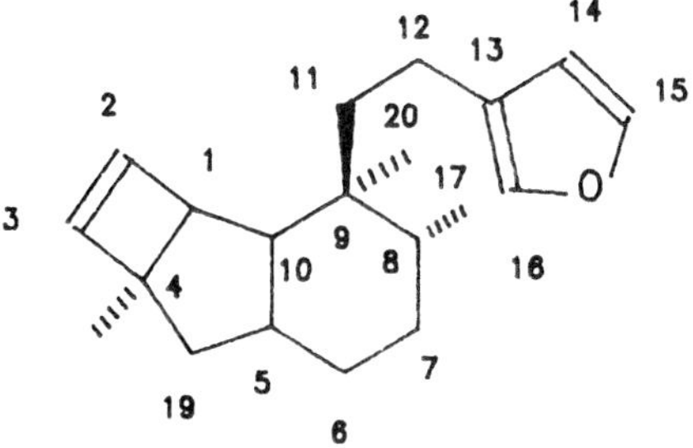

TABLE 2. (*contd.*)

Compound:	
No.	Substituents
917	5(6);18(=O,OH);1α(H);10β(H)
919	18(=O,OH);5α(OH);1α(H);10β(H)

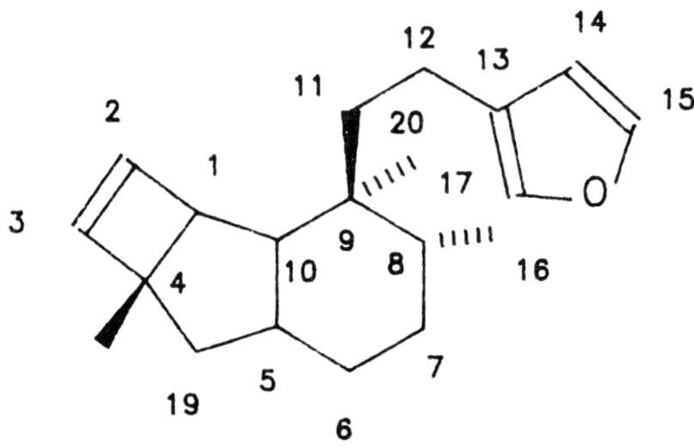

918	5(6);18(=O,OH);1β(H);10β(H)
920	18(=O,OH);5α(OH);1β(H);10β(H)

48. <u>ent</u>-**Clerodane, 5,10-Seco-**

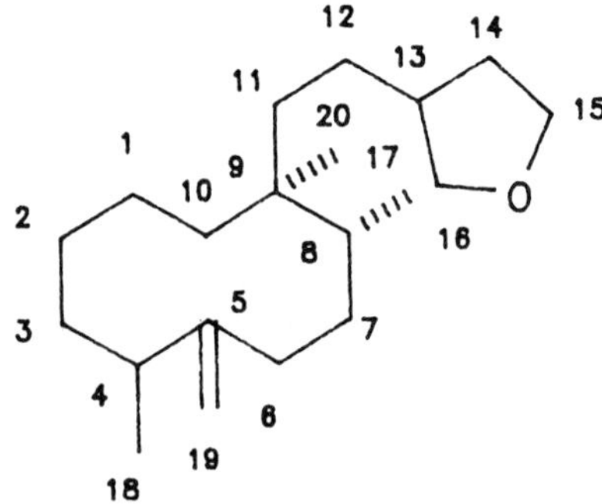

573	1(2);3(4);13(16);14(15);18(=O,OH)
912	1(2);3(4);13(16);14(15);18(=O,OH);19(OAc)Z
985	1(2);3(4);13(14);18(=O,OH);16(=O)
986	1(2);3(4);13(14);18(=O,OH);15(=O)
987	1(2);3(4);18(=O,OH);16(=O);14(-O-)15;
952	1(2);3(4);13(16);14(15);18(=O,OH);12α(O2-Mebut)

TABLE 2. (*contd.*)

Compound:

No.   Substituents

96. <u>trans-ent</u>-Clerodane, 17-Nor-

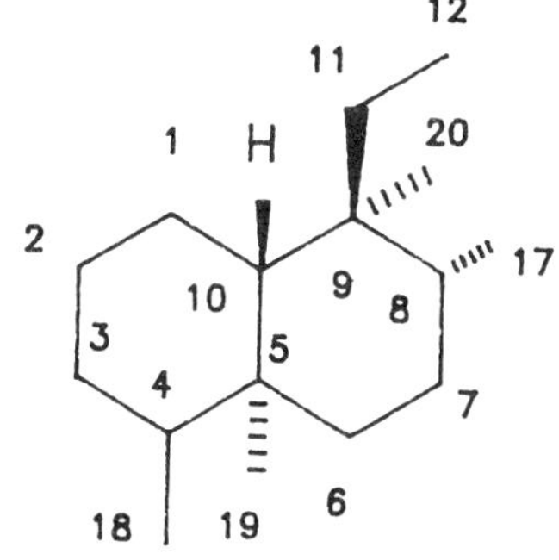

1061   3(4);13(16);14(15);7(=O);18(=O);12(OH);18(-O-)19

49. <u>trans-ent</u>-Clerodane, 13,14,15,16-Tetra-Nor-

574   3(4);12(OH);18(OH)

575   3(4);12(=O,OH);18(OH)

1032   3(4);12(=O,OH);6α(OH);7β(OH)

1093   1(2);3(4);12(=O,OH);18(=O,OH);19(OH);8β(H)

Undetermined

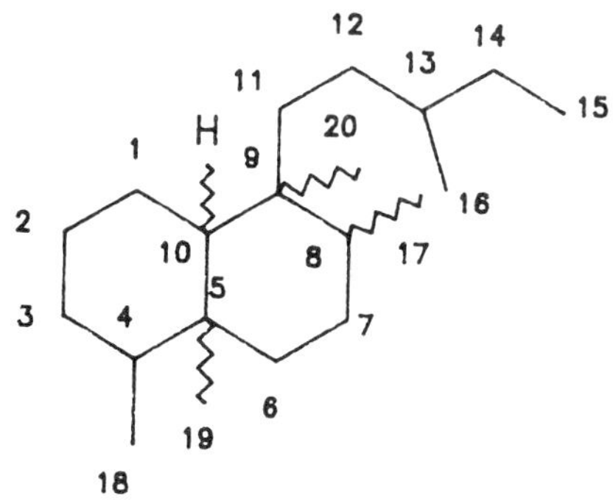

TABLE 2. *(contd.)*

Compound:	
No.	Substituents
237	8(S);3(4);13(14)E;15(OH)

50. <u>cis</u>-<u>ent</u>-Clerodane

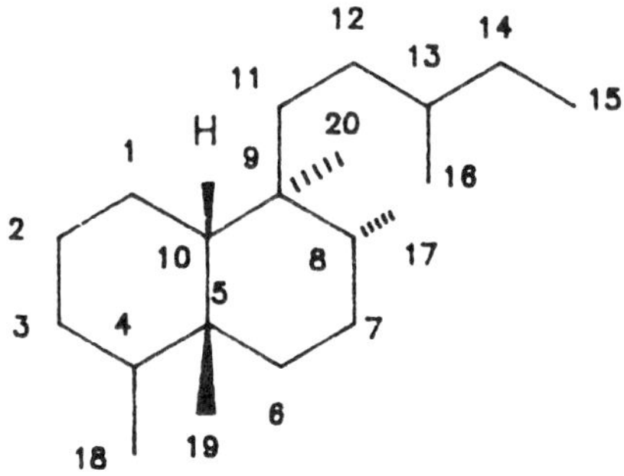

578	3(4);13(16);14(15);15(-O-)16
579	3(4);13(16);14(15);18(OH);15(-O-)16
580	3(4);13(16);14(15);18(OAc);15(-O-)16
576	3(4);13(14);15(=O,OH);18(=O,OH)
577	3(4);13(4);15(=O,OH);18(=O,OMe)
581	3(4);13(16);14(15);6α(OH);18(OH);15(-O-)16
582	3(4);13(16);14(15);6α(OH);18(OAc);15(-O-)16
583	3(4);13(16);14(15);18(=O);15(-O-)16;6α(-O-)18
584	3(4);15(=O);18(=O);15(-O-)16;6α(-O-)18
585	3(4);15(=O);18(=O);2β(OH);15(-O-)16;6α(-O-)18
1055	3(4);13(14)E;15(=O,OH);18(OAc)
1056	3(4);13(14)E;15(=O,OH);18(OH)
1057	3(4);15(=O,OH);18(OH)
1058	3(4);15(=O,OH);19(OH)
1246	13(16);14(15);20(=O);3β(-O-)4β;20(-O-)12; 15(-O-)16;12α(H)

TABLE 2. *(contd.)*

Compound:
No.   Substituents

82. Solidagonane

960

VI. ent-Tricarbocyclics and Derivatives

51. ent-Labdane, Cyclopropyl-

586   14(15);8α(OH);13α(OH)

77. Gutierreziane

845   17(=O);18(=O,OH)

Table 2. (*contd.*)

Compound:

No.   Substituents

52. <u>ent</u>-Pimarane

587	15(16);8β(OH);9β(H)
588	15(16);8β(OH);11α(OH);12β(OAc);9β(H)
589	8(14);2β(OH);15(OH);16(OH);19(OH);9β(H)
590	8(14);15(16);19(=O,OH);9β(H)
820	7(8);15(16);19(=O,OH);9β(H)
821	8(9);15(16);19(=O,OH);7(=O)
823	9(11);19(=O,OH);12β(OAc)
591	8(14);3α(OH);15β(OH);16(OH);9β(H)
592	8(14);15β(OH);16(OH);3α(OGlu);9β(H)
593	8(14);15β(OH);16(OH);6β(OGlu);9β(H)
594	8(14);6β(OH);15β(OH);16(OH);9β(H)
595	8(14);2β(OH);15β(OH);16(OH);19(OH);9β(H)
944	8(14);6β(OH);15(OH);16(OH);18(OH);9β(H)
1247	8(14);15(16);3α(OH);19(OH);9β(H)
1248	8(14);15(16);3α(OH);18(OH);9β(H)
1249	8(9);15(16);7(=O);19(OH)
1250	9(11);15(16);19(OH)

TABLE 2. (*contd.*)

Compound:
No.   Substituents

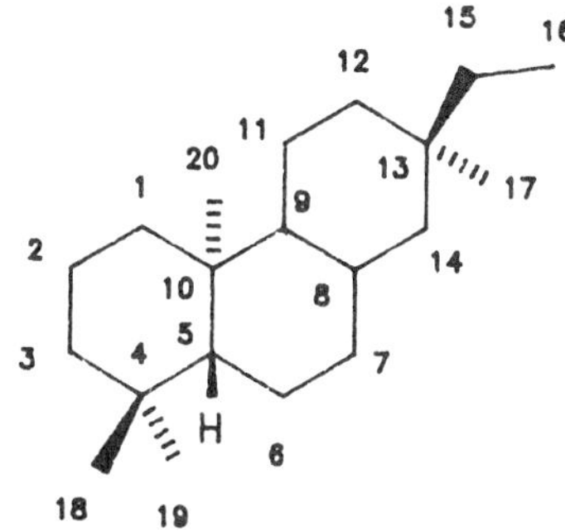

599	9(11);15(16);19(OH)
600	9(11);15(16);19(=O,OH)
1251	9(11);15(16);18(OH)
1252	9(11);15(16);18(OMal)
1253	9(11);15(16);19(OMal)
1254	9(11);15(16);3α(OH);19(OH)

53. <u>ent</u>-Pimarane-8,15-Tetrahydrofuran

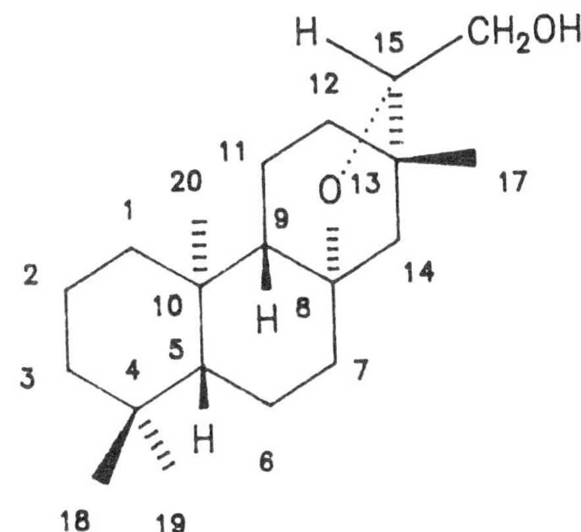

| 596 | 3α(OH);12β(OH);16(OH) |
| 598 | 3(=O);12β(OH);16(OH) |

TABLE 2. (*contd.*)

Compound:
No.   Substituents

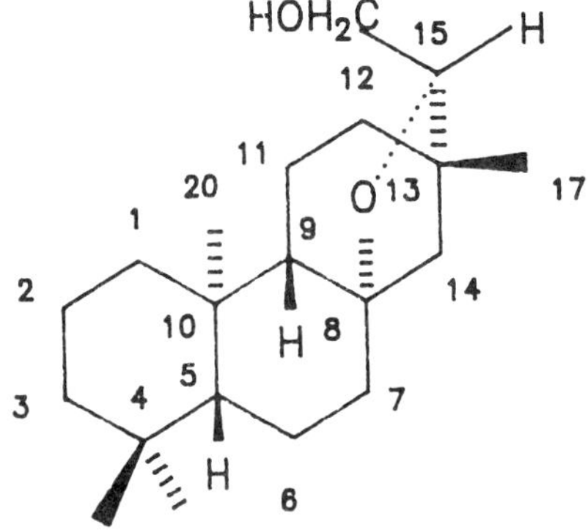

824   9(11);19(=O,OH);12β(OAc)

597   3α(OH);12β(OH)

54. Cleistanthane (Cleistanthol)

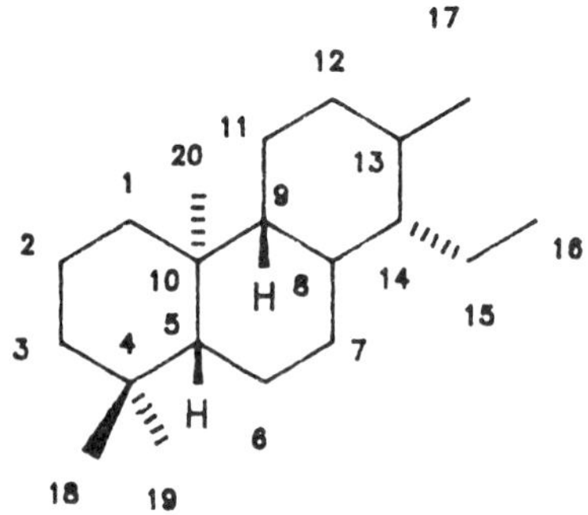

601   12(13);11(=O);15(-O-)16

602   12(13);11(=O);17(=O);15(-O-)16

603   12(13);11(=O);15(-O-)16;17(OAc)

604   12(13);11(=O);15(-O-)16;3α(OAng)

605   12(13);11(=O);14β(OH);15(-O-)16;17(OAc)

606   12(13);11(=O);15(-O-)16;3α(OAng);17(OAc)

607   12(13);11(=O);17(=O);8(-O-)9;15(-O-)16

608   12(13);11(=O);14β(OH);15β(OH);16(-O-)17;15α(H)

TABLE 2. *(contd.)*

Compound:

No.   Substituents

85. Cleistanthane, Iso-

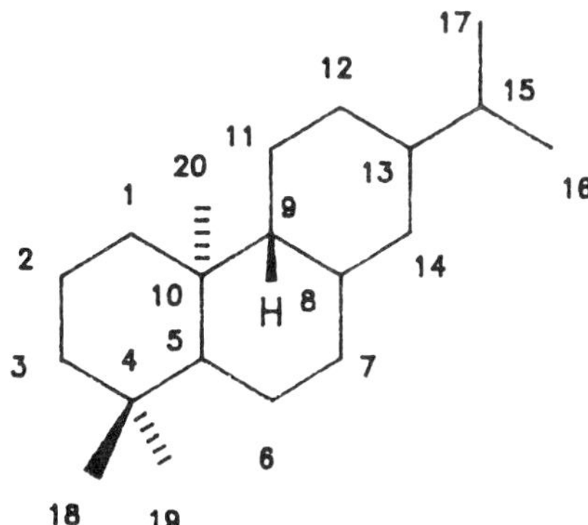

609   3(4);12(13);11(=O);15(-O-)16;17(OAc)

610   12(13);11(=O);3(-O-)4;15(-O-)16;17(OAc)

55. <u>ent</u>-Abietane

611   8(14);3(=O);13(OH);5β(H)

612   8(14);3β(OH);13(OH);5β(H)

613   8(14);13α(OH);2β(OAc);5β(H)

614   7(8);13(14);5β(H)

615   7(8);13(14);5(OAc)

616   7(8);13(14);19(=O,OH);5β(H)

1054   7(8);13(14);19(O-Succ);5β(H)

Compound:

No.   Substituents

56. Acritoconfertane

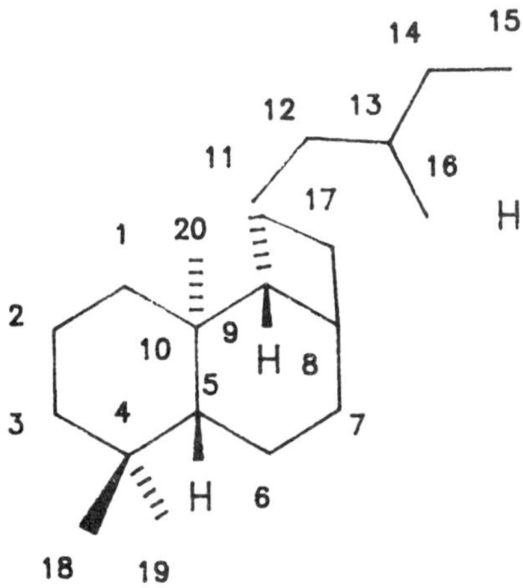

617   7(8);13(14)E;15(=O,OH);6(=O);17(OH);16(OAc)

57. Acritoconfertane-Acetal

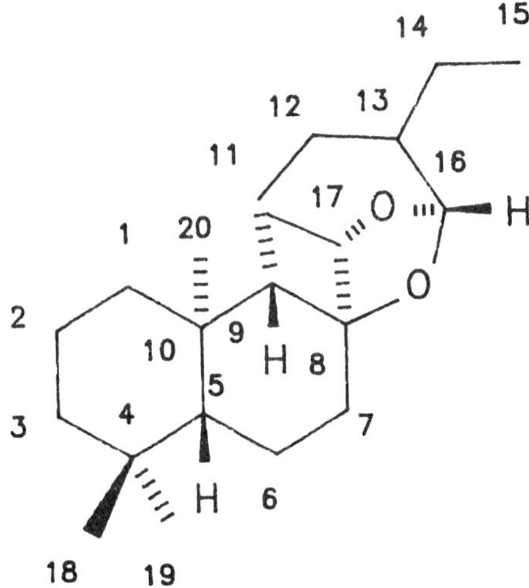

1100   13(14)Z;15(=O,OH);6(=O)

TABLE 2. (*contd.*)

Compound:	
No.	Substituents

83. Erythroxane

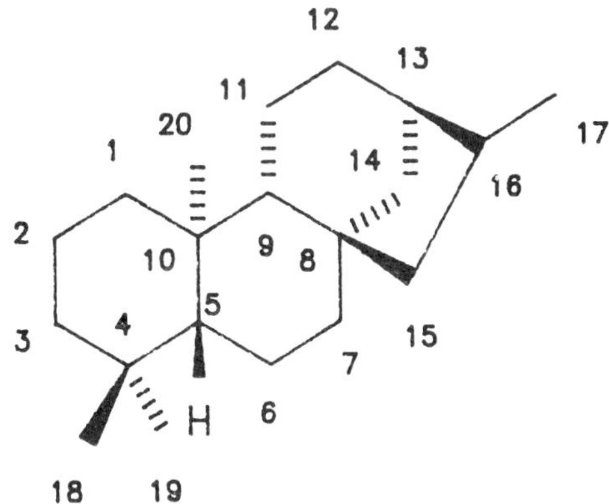

| 976 | 3(4);15(16);18(=O,OH) |

VII. ent-Tetracarbocyclics and Derivatives

58. ent-Kaurane

618	16(17)
619	16(17);19(OH)
997	16α(OH)
620	16(17);19(OAc)
621	16(17);19(OMeform)
622	16(17);19(O-p-OH-H-Cinn)

TABLE 2. (*contd.*)

Compound: No.	Substituents
623	16(17);3α(OH);19(OH)
624	16(17);19(=O)
625	16(17);19(=O,OH)
626	16(17);18(OAc)
926	16(17);18(=O,OH)
939	16(17);18(=O.OH);19(OAng)
627	16(17);19(=O,OGly)
628	16(17);19(=O,OH);3α(OH)
629	16(17);19(=O,OH);3α(OAc)
630	16(17);19(=O,OH);3α(OAng)
631	16(17);19(=O,OH);3α(OTig)
632	16(17);19(=O,OH);3α(OCinn)
633	16(17);19(=O,OH);3α(OSen}
634	16(17);19(=O,OH);3α(Oi-Val)
635	16(17);18(=O)
636	16(17);19(=O,OH);9β(OH);13β(OH);15α(OAng)
637	16(17);19(=O,OH);11β(OH);15α(OH)
638	18(OH);16α(-O-)17
1000	16(17);19(=O,OH);7β(OH)
639	16(17);19(=O,OMe);9β(OH)
640	16(17);19(=O,OH);11β(OAc)
641	16(17);19(=O,OH);12α(OH)
642	16(17);19(=O,OH);12α(OAc) [=1108]
796	16(17);19(=O,OH);12β(OAc)
643	16(17);19(=O,OH);18(OAng)
644	16(17);19(=O,OH);18(OSen)
645	16(17);19(=O,OH);18(Oi-Val)
646	16(17);19(=O,OH);15α(OH)
647	16(17);19(=O,OH);15β(OH)
648	16(17);19(=O,OH);15α(OAc)

Table 2. (*contd.*)

Compound: No.	Substituents
649	16(17);19(=O,OH);15α(OAng)
1007	16(17);19(=O,OMe);9β(OH);15α(OAng)
650	16(17);19(=O,OH);15α(OTig)
651	16(17);19(=O,OH);15α(OSen)
652	16(17);19(=O,OH);15α(Oi-Val)
653	16(17);19(=O,OH);15α(O2-Mac)
654	16(17);19(=O,OH);15α(Oi-But)
655	16(17);19(=O,OH);15α(OCinn)
656	16(17);19(=O,OH);15α(OEpoxyang)
657	16(17);19(=O,OH);15α(OBenz)
818	16(17);19(=O,OH);15α(O2-Mebut-2,3-OH)
892	16(17);19(=O,OH);15α(O2-Mebut)
658	16(17);19(=O,OH);15β(OAng)
659	16(17);19(=O,OH);15β(OTig)
660	16(17);19(=O,OH);15β(OSen)
661	16(17);19(=O,OMe);15α(OH)
662	16(17);19(=O,OH);9β(OH);3α(OAng)
663	16(17);19(=O,OH);9β(OH);3α(OCinn)
664	16(17);19(=O,OH);9β(OH);15α(OH)
665	16(17);19(=O,OH);9β(OH);15α(OAng)
666	16(17);19(=O,OH);9β(OH);15α(OTig)
667	16(17);19(=O,OH);9β(OH);15α(OAc)
668	16(17);19(=O,OH);9β(OH);15α(OSen)
669	16(17);19(=O,OH);9β(OH);15α(Oi-Val)
670	16(17);19(=O,OH);9β(OH);15α(OCinn)
819	16(17);19(=O,OH);9β(OH);15α(O2-Mebut-3-OH)
671	16(17);19(=O,OH);11β(OH);15β(OH)
812	19(=O,OH);17(=O,OMe);16β(H)
672	16(17);19(=O,OH);11β(OH);15β(OAc)
673	16(17);19(=O,OH);11(=O);15β(OH)

TABLE 2. (*contd.*)

Compound:	
No.	Substituents
674	16(17);19(=O,OH);15(=O);11β(OH)
675	16(17);19(=O,OH);3β(OH);9β(OH);15α(OAng)
869	16(17);19(=O,OH);9β(OH);11β(OH);15α(OAng)
676	16(17);19(=O,OH);3β(OH);9β(OH);15α(OSen)
677	16(17);19(=O,OH);3β(OH);9β(OH);15α(OTig)
690	9(11);16(17);19(=O,OH)
691	9(11);16(17);19(=O,OH);2β(OH)
692	9(11);16(17);19(=O,OH);3β(OH)
693	9(11);16(17);19(=O,OH);7β(OH)
694	9(11);16(17);19(=O,OH);12β(OH)
998	9(11);16(17);19(=O,OH);12β(OEt)
695	9(11);16(17);19(=O,OH);12(=O)
696	9(11);16(17);19(=O,OH);15α(OH)
697	15(16);17(OH)
698	15(16);17(OH);19(OH)
699	15(16);17(OH);19(Op-OH-H-Cinn)
946	15(16);19(=O)
700	15(16);19(=O,OH)
701	15(16);19(=O,OH);17(OH)
702	15(16);19(=O,OH);17(=O)
703	15(16);17(=O,OH);19(=O,OH)
704	9(11);15(16);19(=O,OH);3α(Oi-Val)
705	9(11);15(16);19(=O,OH);3α(Oi-But)
706	9(11);15(16);19(=O,OH);3α(OTig)
707	15(16);19(=O,OH);9β(OH);3α(Oi-Val)
708	15(16);19(=O,OH);9β(OH);3α(Oi-But)
733	16α(-O-)17
709	16α(-O-)17;3α(OAng)
734	19(OH);16α(-O-)17
710	16α(-O-)17;19(OAng)
711	19(=O);16α(-O-)17;3α(OAng)

TABLE 2. *(contd.)*

Compound: No.	Substituents
735	19(=O);16α(-O-)17
712	19(=O);16β(-O-)17
713	19(=O,OH);16α(-O-)17
714	19(=O,OH);16β(-O-)17
715	19(=O,OH);16(-O-)17;15α(OAng)
716	19(=O,OH);16(-O-)17;15α(OAc)
717	19(=O,OH);16(-O-)17;15β(OAng)
718	19(=O,OH);16(-O-)17;15β(OTig)
719	9(11);19(=O,OH);12β(OH);16α(-O-)17
720	16α(OH)
1001	16β(OH)
721	16α(OH);3β(OAc)
722	16α(OH);19(OAc)
723	19(=O,OH);16α(OH)
724	11(12);19(=O,OH);16α(OH)
725	16α(OH);17(OH)
726	16α(OH);17(OH);19(OH)
727	3(=O);16α(OH);17(OH)
851	16β(OH)
852	16β(OH);19(OH)
728	19(=O,OH);16α(OH);17(OH)
924	19(=O,OH);17(OH);16α(H)
925	17(=O,OH);19(=O,OH);16α(H)
822	19(=O,OH);16β(OH);17(OH)
729	17(=O);16β(H)
730	17(=O);19(OH);16β(H)
731	17(=O,OH);16β(H)
732	17(=O,OH);19(OH);16β(H)
736	17(=O,Me);19(OH);16β(H)
737	17(=O);19(=O);16β(H)
738	19(=O,OH);17(OH);16β(H)

TABLE 2. *(contd.)*

Compound:

No.	Substituents
739	9(11);19(=O,OH);16α(OH);17(OH)
989	17(=O,OH);19(=O);16β(H)
990	17(=O,OH);19(=O,OH);16β(H)
740	17(=O);19(Op-OH-H-Cinn)
741	19(=O,OH);17(Oi-But);16β(H)
742	19(=O,OH);17(ODodec);16β(H)
743	19(=O,OH);17(OTetradec);16β(H)
744	19(=O,OH);17(OHexadec);16β(H)
745	19(=O,OH);17(OOctadec);16β(H)
746	19(=O,OH);15α(-O-)16α;17(OH)
747	19(=O,OH);15(=O);11β(OH);16α(H)
748	19(=O,OH);15(=O);11β(OH);16α(H)
815	19(=O,OH);15(=O);11β(OH);16α(H)
942	9(11);16(17);19(=O)
943	9(11);16(17);19(=O,OH);15α(OCinn)
1255	19(=O);16α(OH)
1256	9(11);19(=O,OH);16α(OH)
1260	16(17);19(=O,OH);11β(OH);12β(OH);15β(OH)

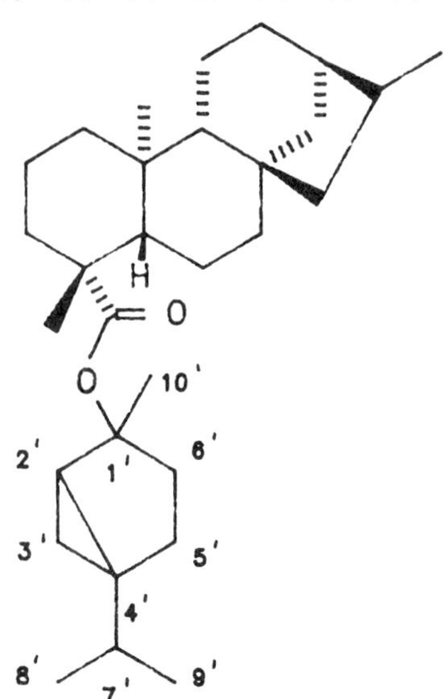

850	

TABLE 2. *(contd.)*

Compound:
No.   Substituents

945

678   R1=R2=COOH;R3=OCOCH2CH(CH3)2;R4=R5=HO3SO;15α(OH)

679   R1=R2=COOH;R3=NHCOCH2CH(CH3)2;R4=CO(CH2)2C6H5;
           R5=OH;13(OH);15(OH)

680   R1=R2=COOH;R3=NHCOCH2CH(CH3)2;R4=CO(CH2)2C6H5;
           R5=OL-Rha;13(OH);15(OH)

681   R1=H;R2=COOH;R3=OCOCH2CH(CH3)2;R4=R5=HO3SO;15α(OH)

TABLE 2. (*contd.*)

Compound:
No.   Substituents

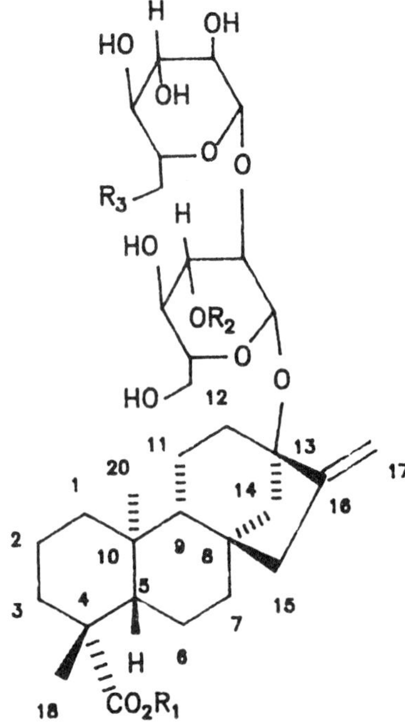

895	16(17);15β(OH)
896	16(17);11β(OH);15β(OH)
897	16(17);15(=O);11β(OH)
898	16α(OH);17(OH)
899	16(17);15β(OGlu)

| 682 | R1=R2=H;R3=OH |
| 683 | R1=βGlu;R2=H;R3=OH |

TABLE 2. (*contd.*)

Compound:

No.	Substituents
684	R1=H;R2=βGlu;R3=OH
685	R1=R2=βGlu;R3=OH
686	R1=βGlu;R2=R3=H
687	R1=R2=βGlu;R3=H
688	R1=βGlu[2]-[1]βGlu;R2=βGlu;R3=H
689	R1=βGlu[2]-[1]βGlu;R2=H;R3=OH

59. ent-Kaurane, 17-Nor-

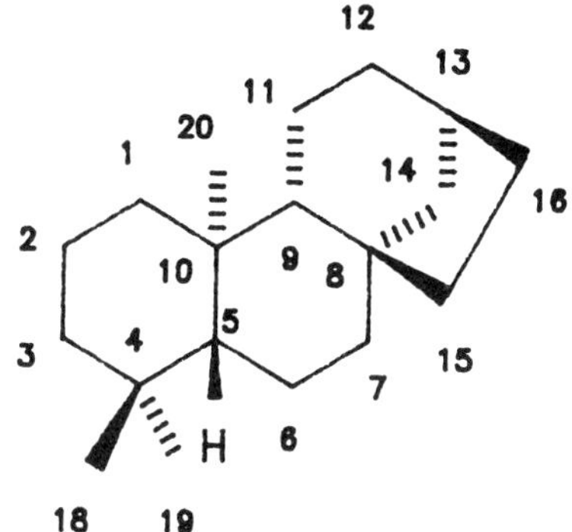

749	19(=O,OH);16(=O)

60. ent-Kaurane, 18-Nor-

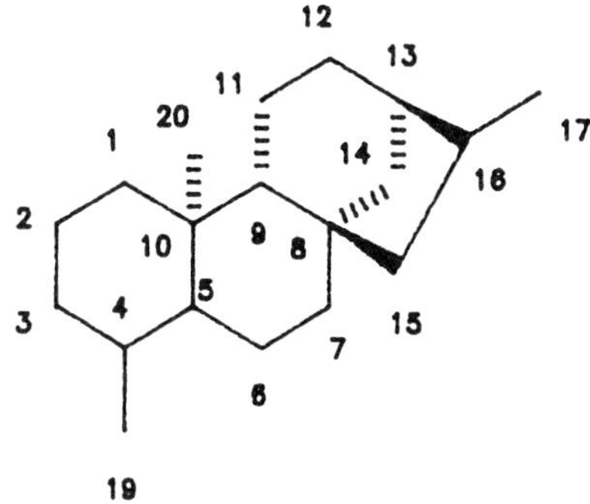

752	16(17);4β(H);5β(H)
753	16(17);4β(OH);5β(H)
754	4(5);6(7);16(17);3(=O);19(OH)

TABLE 2. *(contd.)*

Compound:	
No.	Substituents
755	4(5);6(7);16(17);3(=O);19(=O)
756	4(5);6(7);16(17);3(=O);19(=O,OMe)
757	4(5);6(7);3(=O);16(-O-)17
758	16(17);19(=O,OH);2β(OH);15α(OH);4β(H);5β(H)
813	16(17);4β(-O-)19;5β(H)

61. _ent_-Kaurane, 19-Nor-

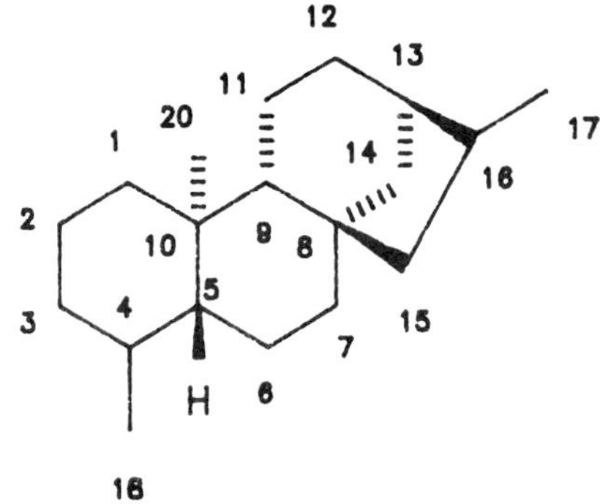

750	16(17);4α(H)
751	16(17);4α(OH)

62. _ent_-Kaurane, 9,10-Seco-

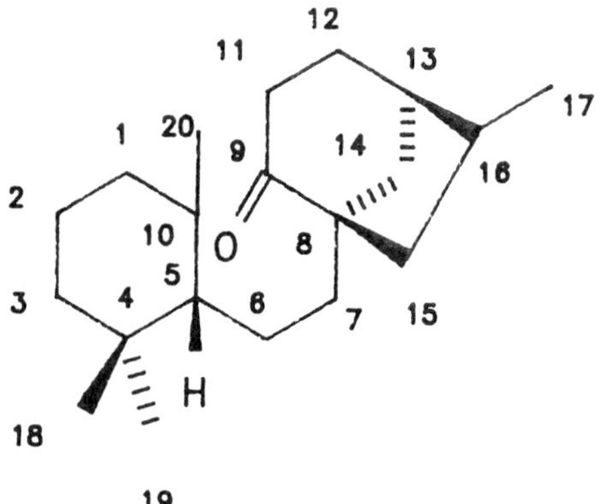

759	1(10);16(17);19(=O,OH)

TABLE 2. (*contd.*)

Compound:

No.   Substituents

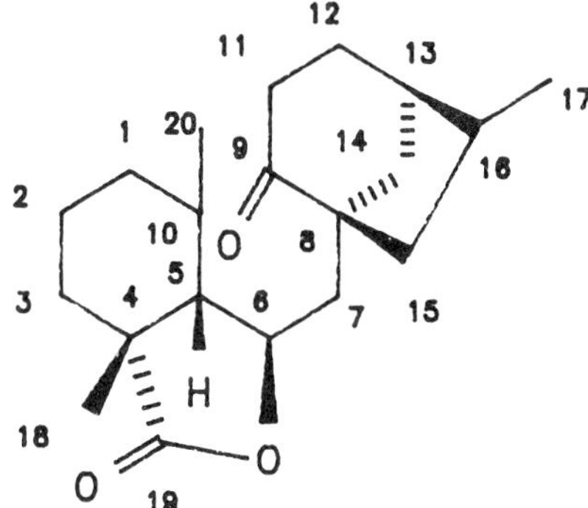

866	16(17);10β(OH);15α(OH)
867	16(17);10β(OH);15β(OH)
868	16(17);10β(OH);15β(OAc)

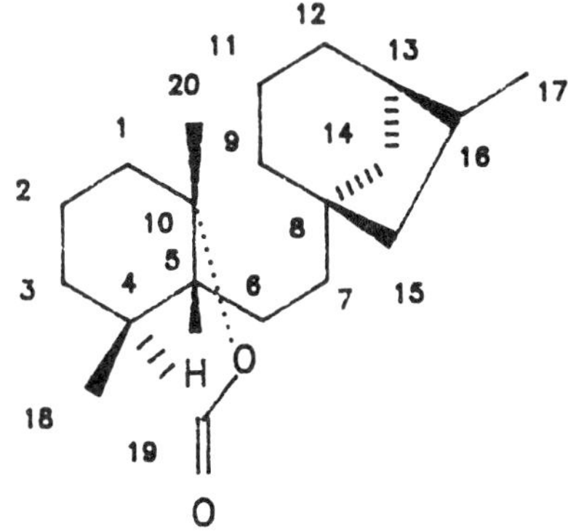

760	16(17);9(=O)
761	16(17);9α(OH)
762	16(17);9β(OH)
763	16(17);9(=O);15β(OH)

TABLE 2. (*contd.*)

Compound:

No.   Substituents

64. _ent_-Stachane (_ent_-Beyerane)

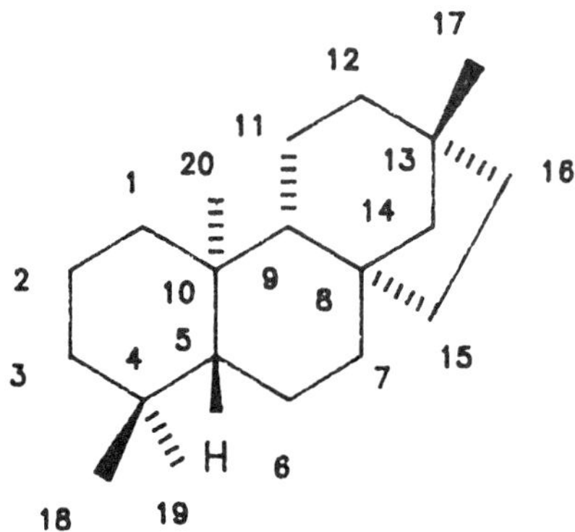

764	15(16);3α(OH)
765	15(16);19(OH)
766	15(16);18(OH)
767	15(16);3α(OH);19(OH)
768	15(16);3α(OH);12β(OH)
769	15(16);19(OMal)
770	15(16);3α(OH);17(OH)
817	15(16);12β(OH);19(OH)
771	15(16);17(OH);19(OH)
772	15(16);18(OH);19(OH)
774	15(16);19(=O,OH)
773	15(16);19(=O)
775	15(16);19(=O,OH);3α(OH)
776	15(16);19(=O,OH);3α(OTig)
777	9(11);15(16);19(=O,OH)
778	9(11);15(16);19(=O,OH);3α(OTig)
779	18(OH);15β(-O-)16β
780	19(OH);15β(-O-)16β

TABLE 2. *(contd.)*

Compound:

No.   Substituents

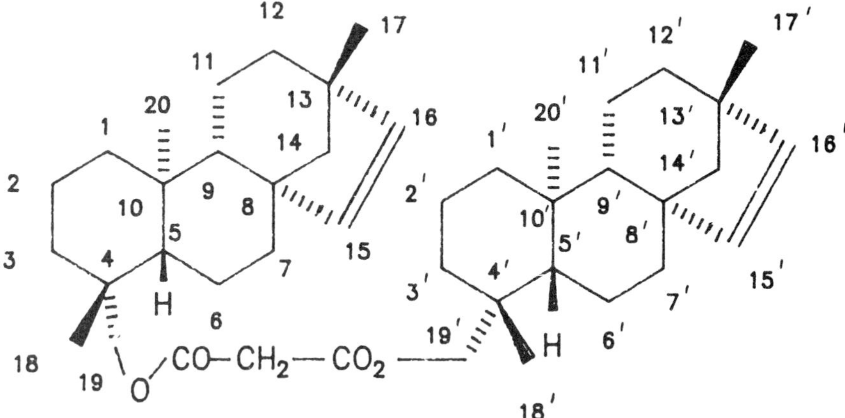

781

93. <u>ent</u>-Stachane, 19-Nor-

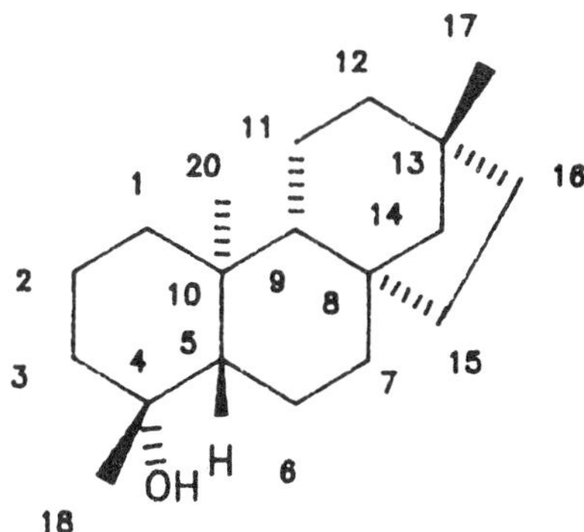

782   15(16)

65. Atisirane (Atisane)

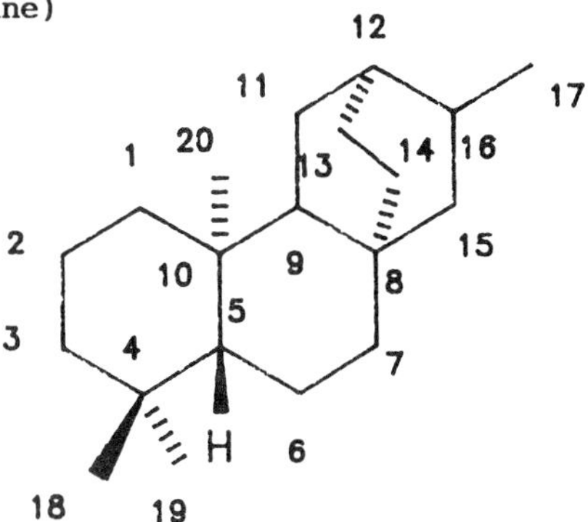

TABLE 2. (*contd.*)

Compound: No.	Substituents
783	16(17);19(=O,OH);9β(H)
853	16α(OH);9β(H)
854	16β(OH);9β(H)
816	13(14);3α(OH);16β(OH);9β(H)
784	16(17);19(=O,OH);7α(OH);9β(H)
785	16(17);19(=O,OH);11α(OAc);9β(H)
786	16(17);19(=O,OH);13α(OAng);9β(H)
787	16(17);19(=O,OH);13α(Oi-Val);9β(H)
788	16(17);19(=O,OH);13α(Oi-But);9β(H)
789	9(11);16(17);19(=O,OH);7α(OAc)
790	9(11);16(17);19(=O,OH);14β(OAc)
814	16(17);19(=O,OH);13(OH);9β(H)

66. Trachylobane

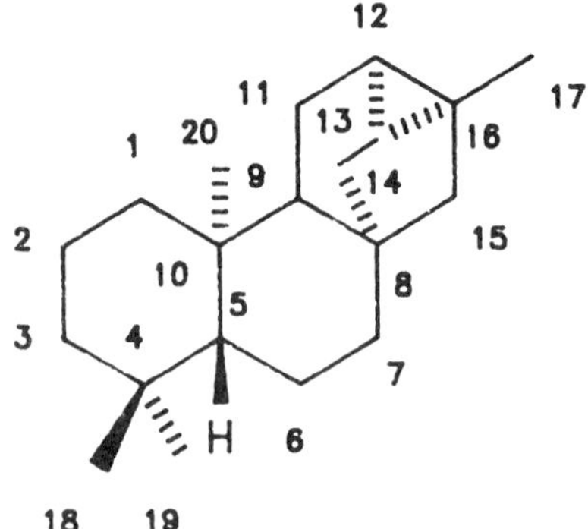

855	19(=O);9βH
791	19(=O,OH);9β(H)
825	19(=O,OH);15α(OAc);9β(H)
826	19(=O,OH);15α(OAng);9β(H)
827	19(=O,OH);15α(Oi-Val);9β(H)
893	19(=O,OH);15α(Oi-But);9β(H)
792	19(=O,OH);7α(OH);9β(H)
793	19(=O,OH);11(=O);9β(H)

TABLE 2. (*contd.*)

Compound:	
No.	Substituents
794	9(11);19(=O,OH)
1204	19(=O,OMe);9β(H)
1259	19(=O,OH);7β(OH);9β(H)

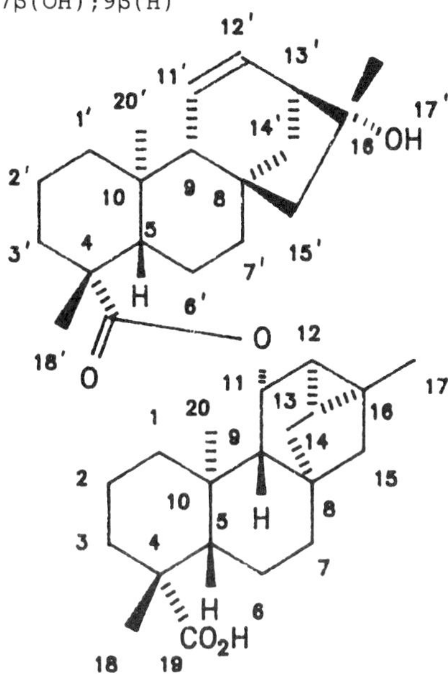

795

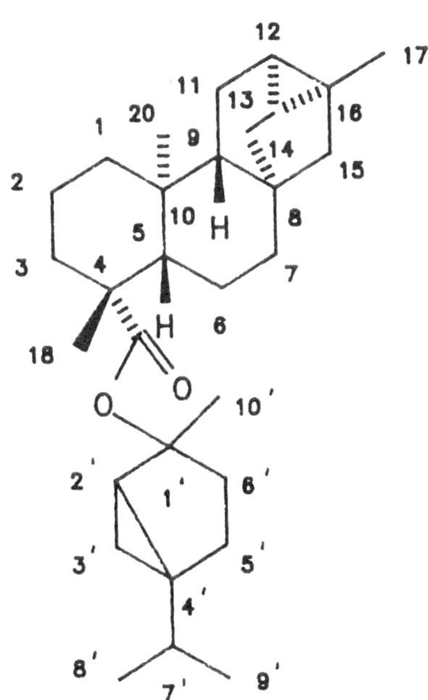

856

TABLE 2. (*contd.*)

Compound:
No.   Substituents

67. Helifulvane

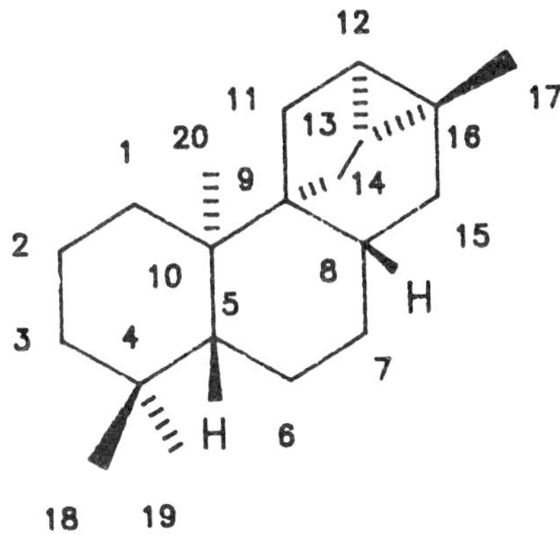

796	19(OH)
797	19(=O,OH)
798	19(=O,OH);11α(OH)
799	19(=O,OH);11α(OAc)

68. Tetrachyrane

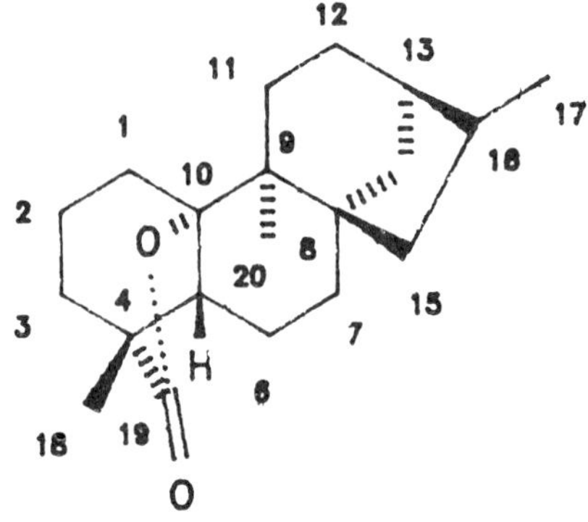

800	16(17)
847	16(17);15(=O)
805	7α(OH);16α(OH)
806	6α(OH);16α(OH)

TABLE 2. (*contd.*)

Compound:
No.   Substituents

78. Villanovane

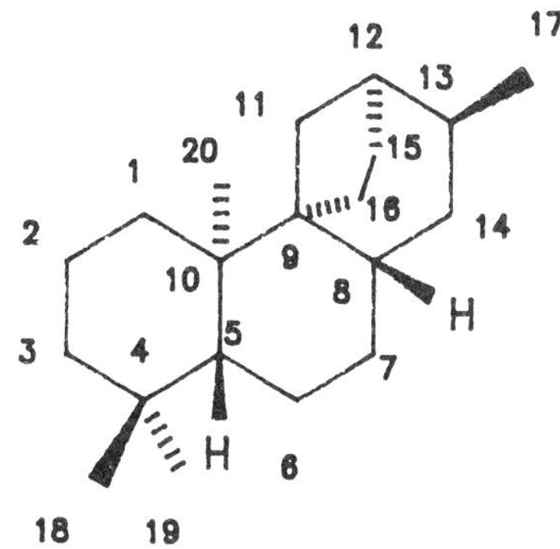

875	13α(OH);17(OH);19(Oi-Val)
876	13α(OH);17(OAc);19(Oi-Val)
877	13α(OH);17(OH);19(OMeVal)
878	13α(OH);17(OAc);19(OMeVal)
879	13α(OH);17(OAc);3α(Oi-Val)
880	13α(OH);17(OAc);3α(OMeVal)
894	13α(OH);15α(OH)

69. Lycoctonine

TABLE 2. (*contd.*)

Compound:	
No.	Substituents
801	

802	R=H
803	R=Anth

VIII. Undetermined Stereochemistry

79. Inulaefane

804	2(OH);3(OH);19(OH)

TABLE 2. (*contd.*)

Compound:

No.   Substituents

84. Hebeclinane (9,10-Seco-Labdane)

940

999

TABLE 3. Diterpene compound numbers, the species from which these compounds were reported and references (an asterisk indicates unpublished result)

Compound:

No.	Species	Reference
1.	Helichrysum mimetes	351
	Mikania goyazensis	46
2.	Pegolettia senegalensis	104
3.	Bejaranoa semistriata	44
	Disynaphia halimifolia	*
	Kingianthus paradoxus	216
	Stevia myriadenia	191
4.	Bejaranoa semistriata	44
	Trichogoniopsis morii	192
5.	Mikania sessilifolia	48
6.	Wyethia helenioides	202
7.	Stylotrichium rotundifolium	148
8.	Mikania officinalis	48
9.	Lasiolaena morii	106
10.	Lasiolaena morii	106
11.	Grazielia intermedia	189
	Lasiolaena morii	
	Lasiolaena santosii	99
12.	Grazielia intermedia	189
13.	Mikania goyazensis	46
14.	Disynaphia halimifolia	67
15.	Lasiolaena morii	106
16.	Zexmenia gnaphaloides	125
17.	Mikania sessilifolia	48
18.	Mikania sessilifolia	48
19.	Koanophyllon admantium	42
20.	Koanophyllon conglobatum	42
21.	Lasiolaena santosii	99
22.	Zinnia tenuiflora	215
	Zinnia verticillata	215
23.	Geigeria burkei subsp. burkei var. burkei	172
	Geigeria burkei subsp. fruticulosa	172
	Helichrysum mundii	171

TABLE 3. (*contd.*)

Compound: No.	Species	Reference
	Helichrysum nanum	*
	Helichrysum subfacultatum	171
	Inula cuspidata	140
	Mikania officinalis	48
	Stevia myriadenia	191
24.	Geigeria burkei subsp. burkei var. burkei	172
	Geigeria burkei subsp. diffusa	172
25.	Geigeria burkei subsp. burkei var. burkei	172
26.	Geigeria burkei subsp. burkei var. elata	172
27.	Geigeria burkei subsp. burkei var. elata	172
	Geigeria burkei subsp. burkei var. intermedia	172
28.	Geigeria burkei subsp. burkei var. elata	172
	Geigeria burkei subsp. burkei var. intermedia	172
	Geigeria burkei subsp. burkei var. zeyheri	172
29.	Geigeria burkei subsp. burkei var. burkei	172
	Geigeria burkei subsp. burkei var. elata	172
	Geigeria burkei subsp. burkei var. intermedia	172
30.	Geigeria burkei subsp. burkei var. burkei	172
31.	Geigeria burkei subsp. burkei var. burkei	172
32.	Helichrysum Krebsianum	171
	Helichrysum oreophilum	351
	Helichrysum oxyphyllum	171
33.	Geigeria burkei subsp. diffusa	172
34.	Geigeria burkei subsp. diffusa	172
35.	Geigeria burkei subsp. burkei var. burkei	172
	Geigeria burkei subsp. diffusa	172
36.	Geigeria burkei subsp. burkei var. burkei	172
37.	Geigeria burkei subsp. fruticulosa	172

TABLE 3. (*contd.*)

Compound: No.	Species	Reference
38.	Geigeria burkei subsp. burkei var. zeyheri	172
39.	Geigeria burkei subsp. burkei var. burkei	172
	Geigeria burkei subsp. burkei var. zeyheri	172
40.	Acanthospermum australe	96
41.	Acanthospermum australe	96
42.	Melampodium longipilum	434
43.	Melampodium longipilum	434
44.	Melampodium diffusum	435
45.	Melampodium diffusum	435
46.	Melampodium longipilum	434
47.	Melampodium longipilum	434
48.	Acanthospermum australe	96
49.	Ichthyothere ulei	107
50.	Ichthyothere ulei	107
51.	Ichthyothere ulei	107
52.	Ichthyothere ulei	107
53.	Ichthyothere ulei	107
54.	Ichthyothere ulei	107
55.	Montanoa leucantha subsp. arborescens	382
	Montanoa leucantha subsp. leucantha	382
	Montanoa mollissima	382
	Montanoa tomentosa subsp. microcephala	382
	Montanoa tomentosa subsp. tomentosa	356
	Montanoa tomentosa subsp. xanthiifolia	382
56.	Montanoa tomentosa subsp. xanthiifolia	459
57.	Montanoa tomentosa subsp. tomentosa	356
58.	Mikania sessilifolia	48

TABLE 3. (*contd.*)

Compound:

No.	Species	Reference
59.	Baccharis trinervis	117
	Centaurea cineraria	*
	Conyza canadensis	*
	Conyza ulmifolia	*
	Felicia erigeroides	*
	Heteromma simplicifolium	*
	Nidorella auriculata subsp. polycephala	*
	Osteospermum barberia	151
	Plagiocheilus prostratus	133
60.	Centipeda orbicularis	128,133
	Plagiocheilus prostratus	133
61.	Conyza podocephala	154
62.	Dimerostemma asperatum	214
	Dimerostemma brasilianum	141
63.	Dimerostemma asperatum	214
64.	Dimerostemma brasilianum	141
65.	Helichrysum calliconum	39
66.	Helichrysum calliconum	39
67.	Helichrysum calliconum	39
68.	Achyrocline vargasiana	*
	Gnaphalium microcephalum	*
	Helichrysum Krebsianum	*
	Helichrysum odoratissimum	298
	Helichrysum nudifolium	39
	Senecio pseudoorientalis	*
	Senecio trichopterygius	*
69.	Helichrysum acutatum	40
70.	Greenmaniella resinosa	*
	Helichrysum acutatum	
	Helichrysum Krebsianum	171
	Helichrysun odoratissimum	*
71.	Helichrysum acutatum	40
72.	Helichrysum acutatum	40
73.	Helichrysum subfalcatum	171
74.	Liatris elegans	335
75.	Liatris elegans	335
76.	Hemizonia lutescens	93
	Zexmenia gnaphaloides	*

TABLE 3. (*contd.*)

Compound: No.	Species	Reference
77.	Acanthospermum hispidum	398
78.	Brickellia eupatoriedes	59
79.	Baccharis scoparia	57
	Oxylobus arbutifolius	87
	Oxylobus glanduliferus	6
80.	Gymnosperma glutinosa	393
81.	Oxylobus glanduliferus	6
82.	Carterothamnus anomalochaeta	185
83.	Carterothamnus anomalochaeta	185
84.	Carterothamnus anomalochaeta	185
85.	Hemizonia lutescens	93
86.	Madia sativa	100
87.	Madia sativa	100
88.	Hemizonia congesta	93
	Hemizonia lutescens	93
89.	Hemizonia congesta	93
	Hemizonia lutescens	93
	Madia sativa	100
90.	Hemizonia lutescens	93
91.	Hemizonia lutescens	93
92.	Hemizonia lutescens	93
93.	Hemizonia lutescens	93
94.	Carterothamnus anomalochaeta	185
95.	Gutierrezia sphaerocephala	272
96.	Gutierrezia sphaerocephala	272
97.	Brickellia diffusa	50
98.	Brickellia eupatoriedes	59
99.	Brickellia eupatoriedes	59
100.	Pleurocoronis pluriseta	147
101.	Brickellia argyrolepis	147
102.	Brickellia argyrolepis	147
103.	Brickellia eupatoriedes	59

TABLE 3. (*contd.*)

Compound: No.	Species	Reference
104.	Brickellia eupatoriedes	59
105.	Brickellia eupatoriedes	59
106.	Brickellia corymbosa Brickellia eupatoriedes Brickellia squarrosa	158 59 158
107.	Brickellia squarrosa	158
108.	Brickellia corymbosa Brickellia squarrosa	158 158
109.	Brickellia squarrosa	158
110.	Baccharis tucumanensis	477
111.	Senecio erosus	209
112.	Trichogoniopsis morii	192
113.	Solidago sempervirens	430
114.	Chrysothamnus nauseusus	70
115.	Chrysothamnus nauseusus	70
116.	Chrysothamnus nauseusus	70
117.	Chrysothamnus nauseusus	70
118.	Stevia jaliscensis	206
119.	Brickellia vernicosa Chrysothamnus nauseusus	29 70
120.	Chrysothamnus nauseusus Grindelia discoidea	70 471
121.	Macowania glandulosa Stevia salicifolia	163 170
122.	Brickellia argyrolepis	147
123.	Brickellia argyrolepis	147
124.	Baccharis salicifolia Brickellia veronicaefolia	503 222
125.	Brickellia veronicaefolia	222
126.	Brickellia eupatoriedes Brickellia paniculata Brickellia sp.	59 287 59,286
127.	Brickellia eupatoriedes	59
128.	Brickellia eupatoriedes	59

TABLE 3. (*contd.*)

Compound: No.	Species	Reference
129.	Brickellia veronicaefolia	158
	Brickellia veronicaefolia var. typica	158
	Chrysothamnus nauseusus	70
130.	Gutierrezia lucida	188
131.	Gutierrezia lucida	188
132.	Denekia capensis	180
	Helichrysum confertum	180
	Helichrysum sutherlandia	171
	Palafoxia rosea	66
133.	Palafoxia rosea	66
	Palafoxia texana var. ambigua	*
134.	Palafoxia rosea	66
	Palafoxia texana var. ambigua	*
135.	Palafoxia rosea	66
	Palafoxia texana var. ambigua	*
136.	Palafoxia rosea	66
137.	Chromolaena collina	175
	Fleischmannia deborabellae	68
	Helichrysum confertum	180
138.	Chromolaena collina	175
139.	Fleischmannia pycnocephaloides	68
	Geigeria burkei Harv. subsp. burkei var. intermedia	172
	Geigeria burkei Harv. subsp. burkei var. zeyheri	172
	Inula crithmoides	
	Koanophyllon conglobatum	*
	Podachaenium eminens	121
	Silphium connatum	*
140.	Hofmeisteria fasciculata	135
	Stevia andina	*
141.	Nidorella auriculata DC. subsp. polycephala	72
	Stevia andina	*
142.	Nidorella auriculata DC. subsp. polycephala	72
	Stevia andina	*
143.	Stevia monardaefolia	433
144.	Koanophyllon conglobatum	135
145.	Koanophyllon conglobatum	135
	Stevia berlandiera	170

TABLE 3. (*contd.*)

Compound:		
No.	Species	Reference
146.	Koanophyllon conglobatum	135
147.	Koanophyllon conglobatum	135
148.	Koanophyllon conglobatum	135
149.	Aristeguietia buddleaefolia	134
150.	Aristeguietia buddleaefolia	134
151.	Aristeguietia buddleaefolia	134
152.	Bishovia boliviensis	186
153.	Stevia monardaefolia	433
154.	Stevia salicifolia	415
155.	Smallanthus fruticosus	211
	Helichrysum coriaceum	*
	Helichrysum nudifolium var. nudifolium	*
	Helichrysum pinifolium	171
	Helichrysum vernum	171
156.	Koanophyllon conglobatum	135
157.	Koanophyllon conglobatum	135
158.	Solidago canadensis	16,17,73
	Solidago gigantea	
	Solidago sempervirens	249
159.	Gutierrezia dracunculoides	80
160.	Gutierrezia mandonii	188
161.	Gutierrezia mandonii	188
162.	Gutierrezia mandonii	188
163.	Gutierrezia dracunculoides	80
164.	Gutierrezia lucida	188
165.	Chrysothamnus paniculatus	475
	Grindelia boliviana	348
	Grindelia camporum	49,473,475
	Grindelia chiloensis	294
	Grindelia paludosa	49
	Grindelia pulchella	294
	Grindelia squarrosa	181
	Grindelia stricta	49
166.	Chrysothamnus paniculatus	475
	Grindelia camporum	473,475
	Grindelia chiloensis	294
	Grindelia pulchella	475
	Grindelia squarrosa	473,475

TABLE 3. (*contd.*)

Compound: No.	Species	Reference
	Heterotheca subaxillaris	*
167.	Chrysothamnus paniculatus	475
	Grindelia aphanactis	348
	Grindelia acutifolia	*
	Grindelia boliviana	348
	Grindelia camporum	49,473
	Grindelia chiloensis	294,348
	Grindelia nauseosus	440
	Grindelia paludosa	49
	Grindelia perennis	348
	Grindelia pulchella	294
	Grindelia robusta	440
	Grindelia squarrosa	68,220,473
	Grindelia stricta	49
	Haplopappus venetus	74
	Heterotheca subaxillaris	*
	Isocoma coronopifolia	253
	Solidago petradoria	*
168.	Grindelia pulchella	293
169.	Grindelia stricta	49
170.	Grindelia camporum	49
	Grindelia humilis	441
	Grindelia paludosa	49
	Grindelia squarrosa	473
	Grindelia stricta	49
171.	Grindelia camporum	49
	Grindelia stricta	49
172.	Chrysothamnus paniculatus	475
	Grindelia camporum	473,475
	Grindelia humilis	441
	Grindelia stricta	475
173.	Grindelia aphanactis	348
	Grindelia boliviana	348
	Grindelia chiloensis	294
	Grindelia perennis	348
	Grindelia pulchella	294
	Grindelia squarrosa	220,473
	Grindelia stricta	49
	Haplopappus tenuisecta	74
	Heterotheca subaxillaris	*
	Isocoma coronopifolia	253
174.	Chrysothamnus paniculatus	475
	Grindelia camporum	473,475
	Grindelia chiloensis	294
	Grindelia perennis	348
	Grindelia pulchella	294
	Grindelia squarrosa	473
	Grindelia stricta	49
	Solidago petradoria	*

TABLE 3. (*contd.*)

Compound: No.	Species	Reference
175.	Chrysothamnus paniculatus	475
	Grindelia camporum	473,475
	Grindelia squarrosa	473
	Grindelia stricta	49
176.	Chrysothamnus paniculatus	475
	Grindelia camporum	475
	Grindelia perennis	348
	Grindelia squarrosa	473
	Grindelia stricta	49
177.	Chrysothamnus paniculatus	475
	Grindelia camporum	475
	Grindelia squarrosa	473
	Grindelia stricta	49
178.	Grindelia perennis	348
	Grindelia stricta	49
179.	Chrysothamnus paniculatus	475
	Grindelia camporum	475
	Grindelia squarrosa	473
180.	Chrysothamnus nauseusus	440
	Grindelia camporum	49
	Grindelia chiloensis	294
	Grindelia paludosa	49
	Grindelia pulchella	294
	Grindelia stricta	49
	Isocoma coronopifolia	253
181.	Grindelia camporum	49
	Grindelia paludosa	49
	Grindelia stricta	49
182.	Grindelia stricta	49
183.	Grindelia stricta	49
184.	Grindelia stricta	49
185.	Chrysothamnus nauseusus	440
186.	Grindelia stricta	49
187.	Grindelia stricta	49
188.	Grindelia camporum	49
	Grindelia chiloensis	294
	Grindelia paludosa	49
	Grindelia pulchella	294
	Grindelia stricta	49
189.	Chrysothamnus paniculatus	341,475
	Grindelia acutifolia	
	Grindelia aphanactis	*
	Grindelia camporum	49,473,475
	Grindelia paludosa	49
	Grindelia robusta	380

TABLE 3. (*contd.*)

Compound: No.	Species	Reference
	Grindelia squarrosa	473
	Grindelia stricta	49,475
	Heterotheca subaxillaris	*
190.	Grindelia stricta	49
191.	Chrysothamnus paniculatus	475
	Grindelia camporum	475
	Grindelia chiloensis	475
	Grindelia pulchella	475
	Grindelia stricta	49
192.	Grindelia camporum	473
	Grindelia pulchella	294
	Grindelia robusta	381
	Grindelia squarrosa	473
193.	Solidago canadensis	17,73
194.	Solidago canadensis	73
195.	Eupatorium jhanii	289
	Stevia rebaudiana	461
196.	Eupatorium jhanii	289
197.	Eupatorium jhanii	289
198.	Aristeguietia buddleaefolia	134
	Eupatorium jhanii	289
199.	Eupatorium jhanii	289
200.	Palafoxia rosea	66,250
201.	Palafoxia rosea	250
202.	Baccharis tola	450
203.	Lindheimera texana	337
	Silphium asteriscus	91
	Silphium compositum	*
	Silphium integrifolium	91
	Silphium perfoliatum	91
	Silphium terebinthinaceum	91
204.	Echinacea purpurea	88
	Lindheimera connata	
	Lindheimera texana	337
	Silphium asteriscus	91
	Silphium compositum	*
	SIlphium connatum	*
	Silphium integrifolium	91
	Silphium perfoliatum	91
	Silphium terebinthinaceum	91
	Trichogoniopsis morii	
205.	Silphium perfoliatum	91

TABLE 3. (*contd.*)

Compound: No.	Species	Reference
206.	Silphium asteriscus	91
	Silphium integrifolium	91
	Silphium perfoliatum	91
	Silphium terebinthinaceum	91
207.	Silphium asteriscus	91
	Silphium integrifolium	91
	Silphium perfoliatum	91
	Silphium terebinthinaceum	91
208.	Silphium perfoliatum	91
209.	Silphium perfoliatum	91
210.	Carterothamnus anomalochaeta	185
211.	Carterothamnus anomalochaeta	185
212.	Silphium perfoliatum	91
	Silphium terebinthinaceum	91
213.	Silphium perfoliatum	91
214.	Nidorella hottentotica	155
215.	Brickellia diffusa	50
216.	Brickellia diffusa	50
217.	Brickellia diffusa	50
218.	Brickellia diffusa	50
219.	Koanophyllon conglobatum	135
220.	Brickellia veronicaefolia	158
221.	Grindelia paludosa	49
	Grindelia stricta	49
222.	Grindelia stricta	49
223.	Chrysothamnus paniculatus	341
	Grindelia acutifolia	
	Grindelia stricta	49
224.	Grindelia stricta	49
225.	Nidorella hottentotica	155
226.	Nidorella hottentotica	155
227.	Nidorella hottentotica	155
228.	Nidorella hottentotica	155
229.	Nidorella hottentotica	155
231.	Relhania acerosa	90

TABLE 3. (*contd.*)

Compound: No.	Species	Reference
232.	Koanophyllon conglobatum	42
233.	Koanophyllon conglobatum	42
234.	Koanophyllon conglobatum	42
235.	Koanophyllon conglobatum	42
236.	Bedfordia salicinia	122
237.	Bedfordia salicinia	122
238.	Nidorella agria	72
239.	Solidago gigantea var. serotina Solidago serotina	15,401 14
240.	Solidago serotina	14
241.	Solidago serotina	14
242.	Solidago serotina	14
243.	Solidago serotina	14
244.	Solidago gigantea var. serotina Solidago serotina	15,401 14
245.	Koanophyllon admantium	42
246.	Solidago serotina	14
247.	Solidago serotina	14
248.	Centipeda orbicularis	128
249.	Centipeda orbicularis	128
250.	Koanophyllon admantium	42
251.	Nidorella resedifolia	72
253.	Nidorella agria	72
254.	Nidorella agria	72
255.	Conyza stricta Grangea maderaspatana Koanophyllon admantium	376 42
256.	Nidorella resedifolia	42
257.	Macowania glandulosa	163
258.	Solidago altissima Solidago virgaurea	143 291

TABLE 3. (*contd.*)

Compound:		
No.	Species	Reference
259.	Solidago altissima	143
	Solidago virgaurea	291
260.	Solidago altissima	143
	Solidago rugosa	73
	Solidago virgaurea	291
261.	Solidago altissima	143
	Solidago virgaurea	291
262.	Solidago altissima	143
	Solidago virgaurea	291
263.	Solidago virgaurea	291
264.	Solidago virgaurea	291
265.	Solidago virgaurea	291
266.	Solidago virgaurea	291
267.	Solidago virgaurea	291
268.	Solidago virgaurea	291
269.	Solidago virgaurea	291
270.	Gutierrezia dracunculoides	235
271.	Chrysothamnus paniculatus	474
272.	Othonna sedifolia	*
	Palafoxia rosea	66
	Senecio variabilis	*
	Stevia berlandiera	170
273.	Palafoxia rosea	66
	Stevia salicifolia	170
274.	Senecio subrubriflorus	169
275.	Senecio subrubriflorus	169
276.	Chrysanthemoides monolifera subsp. canescens	79
	Garuleum bipinnatum	156
	Garuleum pinnatifidum	156
	Garuleum sonchifolium	
	Osteospermum auriculatum	151
	Osteospermum barberiae	151
	Osteospermum ciliatum	151
	Osteospermum corymbosum	157
	Osteospermum fruticosum	156
	Osteospermum jucundum	151
	Osteospermum junceum	156
	Osteospermum muricatum subsp. muricatum	151
	Osteospermum rotundifolium	157
	Senecio subrubriflorus	169

TABLE 3. (*contd.*)

Compound: No.	Species	Reference
277.	Garuleum bipinnatum	156
	Garuleum pinnatifidum	156
	Osteospermum auriculatum	151
278.	Garuleum bipinnatum	151
	Garuleum pinnatifidum	151
	Osteospermum auriculatum	151
279.	Chrysanthemoides monolifera	79
	Chrysanthemoides monolifera subsp. canescens	79
	Garuleum bipinnatum	156
	Garuleum pinnatifidum	156
280.	Senecio subrubriflorus	79
281.	Senecio subrubriflorus	79
282.	Senecio subrubriflorus	79
283.	Senecio subrubriflorus	169
284.	Garuleum bipinnatum	156
	Garuleum pinnatifidum	156
	Senecio subrubriflorus	79
285.	Osteospermum fruticosum	156
286.	Osteospermum barberiae	151
287.	Osteospermum barberiae	151
288.	Chrysanthemoides monolifera subsp. canescens	79
	Senecio subrubriflorus	169
289.	Chrysanthemoides monolifera subsp. canescens	79
	Garuleum bipinnatum	156
	Garuleum pinnatifidum	156
	Osteospermum auriculatum	151
	Osteospermum ciliatum	151
	Osteospermum corymbosum	157
	Osteospermum incanum	*
	Osteospermum jucundum	151
	Osteospermum junceum	156
	Osteospermum muricatum subsp. muricatum	151
	Osteospermum oppositifolium	156
	Osteospermum polygaloides	*
	Osteospermum rotundifolium	157
	Osteospermum scariosum subsp. scariosum	151
	Osteospermum thodei	151
290.	Senecio subrubriflorus	169
291.	Senecio subrubriflorus	169

TABLE 3. (*contd.*)

Compound: No.	Species	Reference
292.	Chrysanthemoides incana	79
	Chrysanthemoides monolifera	79
	Garuleum bipinnatum	156
	Garuleum pinnatifidum	156
	Garuleum sonchifolium	
	Osteospermum corymbosum	157
	Osteospermum junceum	156
	Osteospermum polygalioides	157
	Osteospermum rotundifolium	157
	Senecio sandersonii	173
	Senecio subrubriflorus	*
293.	Garuleum bipinnatum	156
	Garuleum pinnatifidum	156
294.	Chrysanthemoides monolifera subsp. canescens	79
	Helichrysum diosmifolium	*
	Osteospermum subulatum	157
295.	Chrysanthemoides monolifera subsp. canescens	79
	Osteospermum auriculatum	151
	Osteospermum ciliatum	151
	Osteospermum jucundum	151
	Osteospermum thodei	151
	Senecio sandersonii	173
296.	Garuleum bipinnatum	156
	Garuleum pinnatifidum	156
297.	Garuleum pinnatifidum	156
298.	Garuleum bipinnatum	156
299.	Garuleum bipinnatum	156
300.	Achillea filipendulina	25
	Dimorphotheca pluvialis	118
	Mikania pyramidata	131
301.	Dimorphotheca pluvialis	118
302.	Dimorphotheca pluvialis	118
303.	Dimorphotheca pluvialis	118
304.	Palafoxia rosea	66
305.	Palafoxia rosea	66
306.	Palafoxia rosea	66
307.	Palafoxia rosea	66
308.	Croptilon divaricatum	252
309.	Croptilon divaricatum	252
	Palafoxia rosea	66

TABLE 3. (*contd.*)

Compound: No.	Species	Reference
310.	Osteospermum muricatum subsp. muricatum	151
311.	Osteospermum muricatum subsp. muricatum	151
312.	Osteospermum muricatum subsp. muricatum	151
313.	Palafoxia rosea	66,168
314.	Palafoxia rosea	66,168,251
315.	Palafoxia rosea	66,168
316.	Solidago missouriensis	10
317.	Solidago missouriensis	10
318.	Solidago missouriensis	10
319.	Inula royleana	34
320.	Inula royleana	34
321.	Inula royleana	34,254
322.	Inula royleana	34,254
323.	Inula royleana	34,254
324.	Inula royleana	34
325.	Dimorphotheca pluvialis	118
326.	Zexmenia gnaphaloides	125
327.	Relhania acerosa	90
328.	Austroeupatorium chaparense	146
329.	Chrysothamnus viscidiflorus	370
330.	Chrysothamnus viscidiflorus Grindelia aegialitis	370 414
331.	Baccharis sternbergiana	136
332.	Baccharis sternbergiana	136
333.	Baccharis sternbergiana	136
334.	Baccharis sternbergiana	136
335.	Morithamnus crassus Relhania acerosa	102 90
336.	Morithamnus crassus	102

TABLE 3. (*contd.*)

Compound: No.	Species	Reference
337.	Ayapana amygdalina	113
338.	Ayapana amygdalina	113
339.	Ayapana amygdalina	113
340.	Ayapana amygdalina	113
341	Ayapana amygdalina	113
342.	Ayapana amygdalina	113
343.	Ayapana amygdalina	113
344.	Ayapana amygdalina	113
345.	Ayapana amygdalina	113
346.	Ayapana amygdalina	113
347.	Relhania acerosa	90
348.	Acritopappus longifolius Helichrysum confertum	85 180
349.	Acritopappus longifolius	85
350.	Acritopappus longifolius	85
351.	Solidago canadensis	73
352.	Solidago canadensis	73
353.	Lasiolaena santosii	99
354.	Morithamnus crassus	102
355.	Ageratina dendroides Pleurocoronis pluriseta	77 62
356.	Hartwrightia floridana	95
357.	Baccharis sternbergiana	136
358.	Eupatorium petiolare	223
359.	Baccharis sternbergiana	136
360.	Haplopappus pectinatus	348
361.	Pleurocoronis pluriseta	62
362.	Achyrocline alata	43
363.	Baccharis intermixta	178
364.	Baccharis intermixta Lasiolaena santosii	178 99

TABLE 3. (*contd.*)

Compound: No.	Species	Reference
365.	Acritopappus confertus	184
366.	Acritopappus confertus	179,184
367.	Acritopappus confertus	179
368.	Acritopappus confertus	184
369.	Acritopappus confertus	179
370.	Acritopappus confertus	179
371.	Acritopappus confertus	184
372.	Acritopappus confertus	184
373.	Acritopappus confertus	184
374.	Acritopappus confertus	184
375.	Acritopappus confertus	184
376.	Acritopappus confertus	184
377.	Acritopappus confertus	184
378.	Acritopappus confertus	184
379.	Acritopappus confertus	184
380.	Acritopappus confertus	184
381.	Acritopappus confertus	184
382.	Acritopappus morii	179
383.	Acritopappus morii	179
384.	Acritopappus longifolius	85
385.	Xanthocephalum linearifolium	110
386.	Acritopappus longifolius	85
387.	Acritopappus confertus	184
388.	Ageratum fastigiatum	53,126
389.	Ageratum fastigiatum	53,126
390.	Ageratum fastigiatum	126
391.	Ageratum fastigiatum	53
392.	Ageratum fastigiatum	53
393.	Acritopappus confertus	184
394.	Acritopappus confertus	184

TABLE 3. (*contd.*)

Compound: No.	Species	Reference
395.	Acritopappus longifolius	85
396.	Acritopappus longifolius	85
397.	Acritopappus morii	179
398.	Ageratum fastigiatum	126
399.	Acritopappus hagei	179
400.	Acritopappus hagei	179
401.	Acritopappus morii	179
402.	Austroeupatorium chaparense	146
403.	Xanthocephalum linearifolium	110
404.	Psiadia altissima	224
405.	Conyza stricta	460
406.	Austroeupatorium inulaefolium	177
407.	Austroeupatorium inulaefolium	177
408.	Austroeupatorium chaparense	146
409.	Austroeupatorium chaparense	146
410.	Austroeupatorium chaparense	146
411.	Austroeupatorium chaparense	146
412.	Planaltoa lychnophoroides	203
413.	Planaltoa lychnophoroides	203
414.	Acritopappus morii	179
415.	Desmanthodium fruticosum Stevia lucida Lag. var. Bipontini	* 448
416.	Xanthocephalum linearifolium	110
417.	Austroeupatorium inulaefolium	177
418.	Austroeupatorium inulaefolium	177
419.	Acritopappus morii	179
420.	Acritopappus morii	179
421.	Stevia myriadenia	191
422.	Ichthyothere connata Viguiera bishopii	201 92

TABLE 3. (*contd.*)

Compound: No.	Species	Reference
423.	Helianthus decapetalus	98
424.	Helianthus angustifolius	98,409
	Helianthus decapetalus	98
	Helianthus maximiliani	280
	Helianthus occidentalis	
	Helianthus tuberosus	98
	Ichthyothere connata	201
	Ichthyothere latifolia	201
	Ichthyothere terminalis	107
	Mikania banisteriae	228
	Viguiera stenoloba	
425.	Helianthus angustifolius	98
	Helianthus maximiliani	*
	Helianthus occidentalis	
	Helianthus tuberosus	98
	Mikania alvimii	47
	Mikania pyramidata	131
426.	Ichthyothere terminalis	107
427.	Ichthyothere terminalis	107
428.	Austroeupatorium inulaefolium	177
	Stevia andina	350
	Stevia aristata	*
	Stevia origanoides	221
	Stevia rebaudiana	461
429.	Stevia berlandiera	170
	Stevia origanoides	221
	Stevia rebaudiana	461
	Stevia salicifolia	221
430.	Helichrysum confertum	180
	Helichrysum suterlandii	171
	Ichthyothere connata	201
431.	Mikania alvimii	47
432.	Mikania alvimii	47
433.	Mikania alvimii	47
434.	Mikania alvimii	47
435.	Mikania pyramidata	131
436.	Grazielia dimorpholepsis	189
437.	Mikania alvimii	47
438.	Mikania alvimii	47
439.	Mikania alvimii	47
	Mikania pyramidata	131
440.	Mikania alvimii	47

TABLE 3. (*contd.*)

Compound: No.	Species	Reference
	Mikania pyramidata	131
441.	Helichrysum sp.	171
442.	Aspilia parvifolia	213
	Baccharis oxydonta	116
	Denekia capensis	180
	Viguiera bishopii	92
443.	Baccharis oxydonta	116
444.	Baccharis oxydonta	116
445.	Baccharis oxydonta	116
446.	Baccharis oxydonta	116
447.	Helichrysum albirosulatum	171
	Helichrysum pinifolium	171
	Helichrysum vernum	171
448.	Helichrysum albirosulatum	171
449.	Helichrysum albirosulatum	171
450.	Gnaphalium gaudichaudianum	
	Gnaphalium undulatum	210
451.	Coespeletia moritziana	145
	Espeletia hartwegiana	*
	Espeletiopsis guacharaca	145
	Olearia paniculata	426
	Othonna sedifolia	*
	Solidago missouriensis	10,12
452.	Coespeletia moritziana	145
453.	Solidago missouriensis	10,12
454.	Printzia laxa	165
455.	Baccharis tola	451
456.	Printzia laxa	165
457.	Schkuhria multiflora	101
458.	Libanothamnus spectabilis	174
	Libanothamnus wurdackii	174
459.	Coespeletia lutescens	144
	Coespeletia moritziana	145
	Libanothamnus occultus	174
	Libanothamnus spectabilis	174
460.	Acritopappus confertus	184
461.	Erigeron philadelphicus	487

TABLE 3. (*contd.*)

Compound: No.	Species	Reference
462.	Hartwrightia floridana	95
463.	Schkuhria multiflora	101
464.	Schkuhria multiflora	101
465.	Acritopappus confertus	184
466.	Acritopappus confertus	184
467.	Ayapana amygdalina	113
468.	Ayapana amygdalina	113
469.	Mikania alvimii	47
470.	Morithamnus crassus	102
471.	Acritopappus confertus	184
472.	Acritopappus confertus	184
473.	Athrixia elata	150
475.	Melampodium divaricatum Solidago elongata	343 18
476.	Symphyopappus reticulatus	190
477.	Goyazianthus tetrastichus	60
478.	Goyazianthus tetrastichus	60
479.	Solidago elongata	18
480.	Solidago elongata	18
481.	Symphyopappus reticulatus	190
482.	Symphyopappus reticulatus	190
483.	Symphyopappus reticulatus	190
484.	Baccharis alaternoides	117
485.	Symphyopappus compressus	190
486.	Symphyopappus reticulatus	190
487.	Gochnatia paniculata	51
488.	Gochnatia paniculata	51
489.	Gochnatia paniculata	51
490.	Gochnatia paniculata	51
491.	Gochnatia paniculata	51

TABLE 3. (*contd.*)

Compound: No.	Species	Reference
492.	Symphyopappus reticulatus	190
493.	Fleischmannia sinclairii	217
	Stevia myriadenia	191
494.	Stevia myriadenia	191
495.	Solidago elongata	18
496.	Solidago elongata	18
497.	Solidago elongata	18
498.	Solidago altissima	412
499.	Solidago altissima	412
500.	Solidago altissima	143,364,365,412
501.	Gochnatia paniculata	51
502.	Gochnatia paniculata	51
503.	Fleischmannia sinclairii	217
	Hartwrightia floridana	95
	Neomiranda angularis	*
504.	Liatris scariosa	54
505.	Haplopappus ciliatus	35,36
506.	Hartwrightia floridana	95
507.	Baccharis hutchisonii	57
	Solidago arguta	7,386
508.	Baccharis incanum	284
	Baccharis tricuneata var. lineata	200
509.	Baccharis tricuneata (L. f.)Pers. var. tricuneata	490
510.	Baccharis articulata	283
	Baccharis rhetinoides	503
511.	Baccharis macraei	
	Brickellia annulosa	158
	Grangea maderaspatana	423
	Heteropappus altaicus	181
	Koanophyllon adamanticum	42
	Printzia laxa	165
	Solidago juncea	318
512.	Grangea maderaspatana	423
513.	Conyza japonica	423
514.	Hinterhubera imbricata	82

TABLE 3. (*contd.*)

Compound: No.	Species	Reference
515.	Hinterhubera imbricata	82
516.	Hinterhubera imbricata	82
518.	Baccharis calvescens	178
	Baccharis chilco	57
	Baccharis crispa	
	Baccharis macraei	
	Baccharis salicifolia	*
	Baccharis sarothroides	
	Baccharis tricuneata var. lineata	200
	Conyza ivaefolia	76
	Conyza scabrida	83
	Heteropappus althaicus	181
519.	Olearia muelleri	352
520.	Baccharis tricuneata var. lineata	200
	Conyza ivaefolia	76
	Conyza scabrida	83
	Heteropappus altaicus	181
	Printzia laxa	165
521.	Baccharis conferta	295
522.	Gochnatia paniculata	51
523.	Pulicaria gnaphaloides	443
524.	Solidago canadensis	73
	Solidago juncea	318
	Solidago rugosa	73
525.	Conyza podocephala	154
526.	Conyza podocephala	154
527.	Baccharis articulata	283
	Baccharis gilliesii	503
	Baccharis magellanica	503
	Baccharis rhetinoides	503
	Baccharis tricuneata var. tricuneata	490
	Liatris spicata	
528.	Baccharis tricuneata var. tricuneata	283
529.	Baccharis chilco	57
	Baccharis gilliesii	503
	Baccharis subdentata	116
	Baccharis tricuneata var. tricuneata	489,490
530.	Baccharis tricuneata var. tricuneata	489,490
531.	Baccharis cassinaefolia	178

TABLE 3. (*contd.*)

Compound: No.	Species	Reference
532.	Baccharis cassinaefolia	178
533.	Baccharis cassinaefolia	178
534.	Symphyopappus reticulatus	190
535.	Symphyopappus reticulatus	190
536.	Symphyopappus reticulatus	190
537.	Synphyopappus reticulatus	190
538.	Acritopappus confertus Acritopappus hagei	179 179
539.	Acritopappus hagei Morithamnus crassus	179 102
540.	Acritopappus hagei	179
541.	Acritopappus hagei	179
542.	Acritopappus teixeirae	179
543.	Acritopappus longifolius Eupatorium turbinatum Solidago altissima	52 350 143,413
544.	Symphyopappus compressus	190
545.	Solidago altissima Solidago elongata Solidago rugosa	401,413 18,402 73
546.	Solidago elongata	18,402
550.	Acritopappus hagei	179
551.	Symphyopappus compressus	190
552.	Solidago elongata Solidago rugosa	402 73
553.	Solidago elongata	18,402
556.	Solidago Shortii	8,11
557.	Solidago Shortii	8,11
558.	Printzia laxa	165
559.	Baccharis trimera	334
560.	Olearia heterocarpa	426
561.	Baccharis genistelloides Baccharis microcephala	111 57
562.	Baccharis trimera	334

TABLE 3. (*contd.*)

Compound: No.	Species	Reference
563.	Baccharis trimera	334
564.	Bahianthus viscidus	84
565.	Acritopappus teixeirae	179
	Baccharis truncata	116
	Bahianthus viscidus	84
566.	Solidago rugosa	73
567.	Acritopappus hagei	179
568.	Acritopappus hagei	179
569.	Baccharis alaternoides	117
	Baccharis polyphylla	117
570.	Stevia polycephala	8
571.	Printzia laxa	165
572.	Printzia laxa	165
573.	Conyza japonica	423
	Conyza scabrida	83
	Conyza stricta	470
574.	Acritopappus hagei	179
575.	Acritopappus hagei	179
576.	Haplopappus angustifolius	339,462
	Haplopappus foliosus	339,462
577.	Haplopappus angustifolius	462
578.	Solidago arguta	7,386
579.	Solidago arguta	7,386
580.	Solidago arguta	7,386
581.	Solidago arguta	7,386
582.	Solidago arguta	7,386
583.	Solidago arguta	7,263,386
	Solidago itatiayensis	483
584.	Solidago itatiayensis	483
585.	Solidago itatiayensis	483
586.	Gnaphalium undulatum	210
587.	Garuleum sonchifolium	78
588.	Garuleum sonchifolium	78

TABLE 3. (*contd.*)

Compound: No.	Species	Reference
589.	Sigesbeckia pubescens	397
590.	Achillea filipendula	25
	Aspilia ovalifolia	*
	Helianthus strumosus	327
	Mikania alvimii	47
	Mikania pyramidata	131
591.	Palafoxia arida	320
	Sigesbeckia orientalis	28
592.	Sigesbeckia orientalis	358
	Sigesbeckia pubescens	358
593.	Sigesbeckia pubescens	358
594.	Sigesbeckia pubescens	358
595.	Sigesbeckia pubescens	358
596.	Liatris laevigata	325
597.	Liatris laevigata	325
598.	Liatris laevigata	325
599.	Othonna floribunda	112
600.	Othonna cylindrica	112
	Othonna floribunda	112
601.	Brickellia eupatoriedes	59
602.	Brickellia eupatoriedes	59
603.	Brickellia eupatoriedes	59
604.	Brickellia eupatoriedes	59
605.	Brickellia eupatoriedes	59
606.	Brickellia eupatoriedes	59
607.	Brickellia eupatoriedes	59
608.	Brickellia eupatoriedes	59
609.	Brickellia eupatoriedes	59
610.	Brickellia eupatoriedes	59
611.	Solidago missouriensis	12
612.	Solidago missouriensis	12
613.	Solidago missouriensis	12
614.	Helichrysum chionosphaerum	45

TABLE 3. (*contd.*)

Compound: No.	Species	Reference
615.	Solidago nemoralis	73
616.	Mikania banisteriae	228
	Solidago rugosa	73
617.	Acritopappus confertus	184
618.	Arctotis revoluta	119
	Aspilia foliacea	*
	Athrixia pinifolia	164
	Baccharis minutiflora	117
	Balsamorhiza sagittata	130
	Caramboa pithieri	*
	Coespeletia marcana	145
	Espeletia garciae	
	Espeletia uribei	144
	Espeletiopsis guacharaca	145
	Espeletiopsis purpurascens	145
	Espeletiopsis tachirensis	145
	Gnaphalium undulatum	210
	Helichrysum davyi	*
	Libanothamnus granatesianus	174
	Libanothamnus tamanus	174
	Libanothamnus wurdackii	174
	Lychnophora sellowii	204
	Montanoa pteropoda	123
	Montanoa tomentosa	*
	Ruilopezia jahnii	144
	Ruilopezia lindenii	144
	Ruilopezia marcescens	*
	Ruilopezia palanoides	*
	Ruilopezia ruizii	*
	Tamania chardonii	*
619.	Abrotanella nivigena	13
	Arctotis revoluta	119
	Aspilia foliacea	*
	Aspilia jolyana	*
	Aspilia parvifolia	213
	Austroeupatorium inulaefolium	177
	Baccharis minutiflora	117
	Baccharis quitensis	*
	Cacalia bulbifera	257,373
	Campovassouria bupleurifolia	*
	Coespeletia lutescens	144
	Coespeletia marcana	145
	Coespeletia moritziana	145
	Critonia daleoides	217
	Disynaphia multicrenulata	*
	Espeletia garciae	
	Espeletia grandiflora	*
	Espeletia schultzii	132
	Espeletia wedellii	144
	Espeletiopsis garciae	145
	Espeletiopsis glandulosa	145
	Espeletiopsis guacharaca	145
	Espeletiopsis purpurascens	145
	Espeletiopsis tachirensis	145
	Goyazianthus tetrastichus	60

TABLE 3. (*contd.*)

Compound: No.	Species	Reference
	Helianthus giganteus	98
	Helichrysum davyi	*
	Helichrysum dendroideum	373
	Helichrysum heterolasium	39
	Helichrysum miconiifolium	171
	Helichrysum pilosellum	351
	Helichrysum vernum	*
	Ichthyothere rufa	201
	Ichthyothere terminalis	201
	Libanothamnus granatesianus	174
	Libanothamnus neriifolia	144
	Mikania arrojadoi	142
	Mikania belemii	48
	Mikania luetzelburgii	48
	Mikania sessilifolia	48
	Montanoa pteropoda	123
	Oyedaea lanceolata	166
	Oyedaea rusbyi	*
	Oyedaea verbesinoides	*
	Perymenium ecuadoricum	162
	Perymenium featherstoni	*
	Perymenium serratum	*
	Polymnia fruticosum	161
	Polymnia pyramidalis	161
	Ruilopezia bromeloides	*
	Ruilopezia figerasii	*
	Ruilopezia jahnii	144
	Ruilopezia paltonioides	*
	Ruilopezia lindenii	144
	Smallanthus fruticosus	211
	Smallanthus uvedalia	211
	Steiractinia mollis	115
	Steiractinia sodiroi	124
	Tamania chardonii	*
	Viguiera grammatoglossa	*
	Wedelia grandiflora	120
	Wedelia hookeriana	193
620.	Campovassouria bupleurifolia	146
	Espeletiopsis guacharaca	145
	Helichrysum miconiifolium	171
621.	Baccharis minutiflora	117
622.	Baccharis quitensis	117
	Polymnia fruticosa	116
623.	Helichrysum dendroideum	374
624.	Arctotis revoluta	119
	Aspilia jolyana	*
	Aspilia ovalifolia	*
	Aspilia parvifolia	213
	Aspilia pluriseta subsp. pluriseta	*
	Aster bakeranus	478
	Austroeupatorium inulaefolium	139
	Baccharis minutiflora	117
	Baccharis ramosissima	116
	Baccharis truncata	116

　　2. Compound Names, Structures, and Sources

TABLE 3. (*contd.*)

Compound: No.	Species	Reference
	Cacalia bulbifera	257
	Campovassouria bupleurifolia	146
	Coespeletia lutescens	144
	Coespeletia marcana	145
	Coespeletia moritziana	145
	Disynaphia multicrenulata	*
	Espeletia garciae	
	Espeletia grandiflora	
	Espeletia hartwegiana	*
	Espeletia neriifolia	
	Espeletia schultzii	132
	Espeletia uribei	144
	Espeletia wedellii	144
	Espeletiopsis guacharaca	145
	Gnaphalium undulatum	219
	Goyazianthus tetrastichus	60
	Grazielia intermedia	*
	Helianthus annuus	446
	Helichrysum davyi	*
	Helichrysum heterolasium	39
	Helichrysum miconiifolium	171
	Helichrysum pallidum	171
	Helichrysum pilosellum	351
	Ichthyothere terminalis	201
	Ichthyothere ulei	201
	Libanothamnus neriifolia	144
	Mikania arrojadoi	142
	Mikania belemii	48
	Mikania luetzelburgii	48
	Mikania micrantha	*
	Mikania sessilifolia	48
	Montanoa tomentosa subsp. tomentosa	227
	Montanoa pteropoda	123
	Oyedaea buphthalmoides	
	Oyedaea lanceolata	166
	Oyedaea rusbyi	*
	Perymenium ecuadoricum	162
	Perymenium klattianum	*
	Ruilopezia figueirasii	*
	Ruilopezia lindenii	144
	Ruilopezia marcescens	*
	Ruilopezia paltonioides	*
	Smallanthus fruticosus	211
	Smallanthus glabratus	108
	Smallanthus uvedalia	211
	Solidago multiradiata var. multiradiata	*
	Solidago rugosa	73
	Steiractinia mollis	115
	Viguiera grammatoglossa	*
	Wedelia grandiflora	120
	Zaluzania angusta	*
	Zaluzania subcordata	*
625.	Abrotanella nivigena	13
	Arctotheca prostata	*
	Arctotis revoluta	119

TABLE 3. (*contd.*)

Compound:

No.	Species	Reference
	Aspilia floribunda	
	Aspilia foliacea	*
	Aspilia jolyana	*
	Aspilia mossambicensis	*
	Aspilia ovalifolia	*
	Aspilia parvifolia	213
	Aspilia pluriseta	*
	Aspilia pluriseta subsp. pluriseta	94
	Aspilia riddellii	*
	Aster bakeranus	478
	Aster pleiocephalus	*
	Athrixia elata	150
	Austroeupatorium inulaefolium	177
	Baccharis concinnia	116
	Baccharis elliptica	*
	Baccharis intermixta	178
	Baccharis minutiflora	117
	Baccharis polyphylla	178
	Baccharis pylicoides	57
	Baccharis quintensis	116
	Baccharis ramosissima	116
	Baccharis salzmannii	178
	Baccharis truncata	116
	Cacalia bulbifera	257
	Campovassouria bupleurifolia	146
	Caramboa pithieri	*
	Caramboa trugillensis	*
	Chromalaena cryptantha	*
	Coespeletia lutescens	144
	Coespeletia marcana	145
	Coespeletia moritziana	145
	Critonia daleoides	217
	Critonia hebebotrya	*
	Disynaphia multicrenulata	137
	Enhydra fluctuans	422
	Espeletia figueirasii	480
	Espeletia floccosa	480
	Espeletia garciae	
	Espeletia grandiflora	428
	Espeletia moritziana	480
	Espeletia schultzii	132
	Espeletia uribei	144
	Espeletiopsis garciae	145
	Espeletiopsis glandulosa	145
	Espeletiopsis guacharaca	145
	Espeletiopsis purpurascens	145
	Espeletiopsis tachirensis	145
	Eupatorium petiolare	223
	Gnaphalium oligandrum	210
	Gnaphalium undulatum	210
	Goyazianthus tetrastichus	60
	Grazielia dimorpholepsis	189
	Grazielia intermedia	*
	Grazielia serrata	189
	Helianthopsis bishopii	198
	Helianthopsis microphylla	198
	Helianthopsis utcubambensis	198

TABLE 3. (*contd.*)

Compound: No.	Species	Reference
	Helianthus angustifolius	98
	Helianthus annuus	264,358,491
	Helianthus debilis subsp. cucumerifolius	332
	Helianthus debilis subsp. debilis	411
	Helianthus decapetalus	31,98
	Helianthus occidentalis var. dowellianus	105
	Helianthus giganteus	31,98
	Helianthus heterophyllus	321
	Helianthus hirsutus	31
	Helianthus niveus subsp. canescens	410
	Helianthus nuttallii	31
	Helianthus occidentalis	332
	Helianthus petiolaris	327
	Helianthus radula	326
	Helianthus rigidus	
	Helianthus simulans	332
	Helianthus tuberosus	98
	Helichrysum argentissimum	171
	Helichrysum argyrolepis	*
	Helichrysum aureum var. monocephalum	
	Helichrysum bellum	167
	Helichrysum chionosphaerum	45
	Helichrysum confertum	
	Helichrysum cooperi	
	Helichrysum davyi	*
	Helichrysum fulvum	208
	Helichrysum heterolasium	39
	Helichrysum miconiifolium	171
	Helichrysum pallidum	171
	Helichrysum pilosellum	351
	Helichrysum platypterum	167
	Helichrysum refluxum	86
	Helichrysum subulifolium	351
	Helichrysum vernum	171
	Heterocondylus grandis	*
	Ichthyothere connata	201
	Ichthyothere cunabi	
	Ichthyothere latifolia	201
	Ichthyothere rufa	201
	Ichthyothere terminalis	201
	Ichthyothere ulei	201
	Iostephane heterophylla	*
	Lagascea rigida	89
	Lasianthaea podocephala	124
	Libanothamnus granatesianus	174
	Libanothamnus neriifolia	144
	Libanothamnus occultus	174
	Libanothamnus spectabilis	174
	Libanothamnus tamanus	174
	Libanothamnus wurdackii	174
	Loxothysanus sinuatus	129
	Melampodium paniculatum	*
	Melampodium perfoliatum	*
	Mikania arrojadoi	142
	Mikania belemii	48

TABLE 3. (*contd.*)

Compound: No.	Species	Reference
	Mikania cordata	131
	Mikania hirsutissima var. hirsutissima	396
	Mikania luetzelburgii	48
	Mikania micrantha	*
	Mikania monagasensis	383,384
	Mikania schenkii	*
	Mikania sessilifolia	48
	Montanoa atriplicifolia	64
	Montanoa frutescens	*
	Montanoa tomentosa subsp. xanthiifolia	227
	Montanoa pteropoda	123
	Othonna cylindrica	112
	Othonna floribunda	112
	Othonna sedifolia	*
	Oyedaea boliviana	166
	Oyedaea buphthalmoides	*
	Oyedaea lanceolata	166
	Oyedaea rusbyi	*
	Oyedaea verbesinoides	*
	Perymeniopsis ovalifolia	103
	Perymenium ecuadoricum	162
	Perymenium discolor	
	Perymenium featherstoni	*
	Perymenium klattianum	103
	Perymenium serratum	*
	Polymnia fruticosum	161
	Polymnia maculata var. maculata	371
	Polymnia pyramidalis	161
	Robinsonia thurifera	339
	Ruilopezia bromeloides	*
	Ruilopezia jahnii	144
	Ruilopezia lindenii	144
	Ruilopezia paltonioides	*
	Ruilopezia ruizii	*
	Simsia dombeyana	*
	Simsia holwayi	*
	Smallanthus fruticosus	211
	Smallanthus glabratus	108
	Smallanthus reparius	109
	Smallanthus siegesbeckia	*
	Smallanthus uvedalia	114,211
	Solidago juncea	
	Solidago missouriensis	369
	Solidago multiradiata	*
	Solidago rigida	11,369
	Solidago rugosa	73
	Steiractinia mollis	115
	Steiractinia sodiroi	124
	Stevia andina	*
	Stevia monardaefolia	433
	Stevia myriadenia	191
	Stevia setifera	69
	Tetrachyron manicata	*
	Tetrachyron orizabaensis var. websteri	411
	Tithonia longiradiata	*
	Verbesina angustifolia	160

TABLE 3. (*contd.*)

Compound: No.	Species	Reference
	Verbesina oncophora	160
	Vernonia divaricata	*
	Vernonia venosissima	*
	Viguiera bishopii	92
	Viguiera cordata	*
	Viguiera cordifolia	199
	Viguiera dentata	275
	Viguiera excelsa	242
	Viguiera grammatoglossa	199
	Viguiera hypargyrea	5
	Viguiera incana	*
	Viguiera insignis	245
	Viguiera linearifolia	*
	Viguiera linearis	454
	Viguiera maculata	241
	Viguiera oaxacana	242
	Viguiera pazensis	*
	Viguiera porteri	331
	Viguiera potosina	276
	Viguiera procumbens	*
	Viguiera trichophylla	*
	Wedelia buphthalmiflora	455
	Wedelia calycina	328
	Wedelia glauca	405
	Wedelia grandiflora	120
	Wedelia helianthoides	120
	Wedelia hispida	328
	Wedelia hookeriana	193
	Wedelia scaberrima	476
	Wedelia speciosa	182
	Wedelia trilobata	120,213
	Wedelia villosa	*
	Zaluzania angusta	*
	Zaluzania cinarescens	*
	Zaluzania discoidea	*
	Zaluzania subcordata	*
	Zaluzania triloba	*
	Zexmenia hispida	124
626.	Mikania banisteriae	228
627.	Mikania sessilifolia	48
	Ruilopezia jahnii	144
628.	Bedfordia salicinia	122
	Helichrysum dendroideum	374
629.	Helichrysum aureum var. monocephalum	180
	Helichrysum cooperi	180
	Helichrysum heterolasium	39
	Mikania arrojadoi	142
630.	Smallanthus uvedalia	114
	Wedelia trilobata	120,213
631.	Wedelia trilobata	114,120
	Wulffia maculata	*

TABLE 3. (*contd.*)

Compound: No.	Species	Reference
632.	Wedelia trilobata	114,120
	Wulffia maculata	*
633.	Smallanthus uvedalia	114
	Wedelia grandiflora	120
	Wedelia trilobata	120
634.	Smallanthus uvedalia	114
635.	Mikania banisteriae	228
636.	Aspilia pluriseta subsp. pluriseta	94
637.	Sciadocephala schultze-rhonhofiae	217
638.	Mikania banisteriae	228
639.	Viguiera stenoloba	237,243
640.	Helichrysum aureum var. monocephalum	180
	Helichrysum cooperi	180
	Helichrysum davyi	*
	Helichrysum heterolasium	39
	Helichrysum pallidum	171
	Helichrysum pilosellum	351
	Helichrysum ruderale	171
641.	Aster tanacetifolius	*
	Ruilopezia figueirasii	*
	Smallanthus uvedalia	114
	Stevia eupatoria	416
642.	Helianthus decapetalus	98
	Helianthus radula	326
	Helianthus simulans	332
	Helianthus tuberosus	98
	Mikania arrojadoi	142
643.	Smallanthus fruticosus	211
	Smallanthus reparius	109
	Smallanthus siegesbeckia	*
	Smallanthus uvedalia	114
644.	Smallanthus fruticosus	211
	Smallanthus glabratus	108
	Smallanthus uvedalia	114
645.	Smallanthus uvedalia	114
646.	Austroeupatorium inulaefolium	406
	Baccharis concinna	116
	Baccharis intermixta	178
	Baccharis polyphylla	178
	Coespeletia lutescens	144
	Coespeletia marcana	145
	Disynaphia multicrenulata	137
	Enhydra fluctuans	421
	Espeletia grandiflora	*

TABLE 3. (*contd.*)

Compound: No.	Species	Reference
	Espeletia humbertii	479
	Espeletia littlei	479
	Espeletia schultzii	219
	Espeletia tenor	394
	Espeletia uribei	144
	Espeletia weddellii	394
	Espeletiopsis glandulosa	145
	Espeletiopsis guacharaca	145
	Espeletiopsis tachirensis	145
	Gnaphalium undulatum	210
	Helianthus annuus	387
	Helianthus debilis subsp. cucumerifolius	332
	Helianthus grosseserratus	332
	Helianthus niveus subsp. canescens	410
	Helianthus nuttallii subsp. nuttallii	367
	Helianthus occidentalis	332
	Helianthus petiolaris	327
	Helianthus simulans	332
	Ichthyothere terminalis	*
	Mikania arrojadoi	142
	Mikania luetzelburgii	48
	Oyedaea verbesinoides	*
	Ruilopezia jahnii	144
	Ruilopezia lindenii	144
	Ruilopezia ruizii	*
	Viguiera dentata	275
	Viguiera maculata	241
	Viguiera porteri	331
	Viguiera potosina	276
	Wedelia calycina	328
	Wedelia scaberrima	476
647.	Mikania arrojadoi	142
648.	Arctotis revoluta	119
	Baccharis truncata	116
	Coespeletia lutescens	144
	Coespeletia marcana	145
	Coespeletia moritziana	145
	Espeletia curialensis	*
	Espeletia hartwegiana	*
	Espeletia humbertii	479
	Espeletia schultzii	132
	Espeletia uribei	144
	Espeletiopsis garciae	145
	Espeletiopsis glandulosa	145
	Espeletiopsis guacharaca	145
	Espeletiopsis purpurascens	145
	Espeletiopsis tachirensis	145
	Helianthus annuus	264
	Helichrysum aureum var. monocephalum	180
	Helichrysum heterolasium	39
	Helichrysum miconiifolium	171
	Helichrysum pallidum	171
	Libanothamnus granatesianus	174

TABLE 3. (*contd.*)

Compound: No.	Species	Reference
	Libanothamnus neriifolia	144
	Libanothamnus occultus	174
	Libanothamnus spectabilis	174
	Libanothamnus tamanus	174
	Libanothamnus wurdackii	174
	Mikania arrojadoi	142
	Ruilopezia figuerasii	*
	Ruilopezia jahnii	144
	Ruilopezia lindenii	144
	Ruilopezia marcescens	*
	Tamania chardonii	*
	Tetrachyron manicata	*
	Wedelia hookeriana	193
649.	Aster tanacetifolius	*
	Austroeupatorium inulaefolium	139
	Baccharis concinna	116
	Baccharis polyphylla	178
	Caramboa pithieri	*
	Disynaphia multicrenulata	*
	Enhydra fluctuans	285
	Goyazianthus tetrastichus	60
	Helianthus angustifolius	98
	Helianthus annuus	264
	Helianthus debilis subsp. cucumerifolius	332
	Helianthus debilis subsp. debilis	411
	Helianthus giganteus	98
	Helianthus occidentalis	332
	Helianthus radula	326
	Helianthus simulans	332
	Ichthyothere terminalis	201
	Ichthyothere ulei	107
	Montanoa pteropoda	123
	Montanoa tomentosa subsp. xanthiifolia	351
	Oyedaea buphthalmoides	
	Oyedaea rusbyi	*
	Perymenium ecuadoricum	162
	Perymenium featherstoni	*
	Perymenium ovalifolia	*
	Polymnia fruticosum	161
	Polymnia pyramidalis	161
	Smallanthus fruticosus	211
	Smallanthus reparius	109
	Smallanthus siegesbeckia	*
	Smallanthus uvedalia	211
	Steiractinia mollis	115
	Stevia monardaefolia	433
	Tamania chardonii	*
	Tetrachyron orizabaensis var. websteri	411
	Viguiera dentata	275
	Viguiera pazensis	205
	Viguiera quinqueradiata	240
	Wedelia buphthalmiflora	455
	Wedelia calycina	328
	Wedelia grandiflora	120

TABLE 3. (*contd.*)

Compound: No.	Species	Reference
	Wedelia helianthoides	120
	Wedelia hookeriana	193
	Wedelia pinetorum	
	Zexmenia pinetorum	
650.	Aspilia jolyana	*
	Aspilia ovalifolia	*
	Aspilia parvifolia	213
	Aspilia pluriseta subsp. pluriseta	94
	Aster pleiocephalus	*
	Baccharis salzmannii	178
	Baccharis truncata	116
	Grazielia dimorpholepsis	189
	Grazielia intermedia	189
	Grazielia serrata	189
	Helianthus occidentalis subsp. dowellianus	105
	Helianthus tuberosus	98
	Ichthyothere connata	201
	Ichthyothere latifolia	201
	Ichthyothere rufa	201
	Ichthyothere terminalis	107,201
	Ichthyothere ulei	107
	Lagascea rigida	89
	Oyedaea verbesinoides	*
	Stevia bupthalmiflora	
	Stevia myriadenia	191
	Viguiera pazensis	205
	Viguiera quinqueradiata	240
	Wedelia buphthalmiflora	455
	Wedelia helianthoides	120
	Wedelia hookeriana	193
	Wedelia scaberrima	476
	Wedelia villosa	*
651.	Aspilia jolyana	*
	Baccharis elliptica	*
	Caramboa pithieri	*
	Coespeletia moritziana	145
	Espeletiopsis garciae	145
	Espeletiopsis glandulosa	145
	Espeletiopsis purpurascens	145
	Espeletiopsis tachirensis	145
	Ichthyothere connata	201
	Ichthyothere latifolia	201
	Ichthyothere rufa	201
	Ichthyothere terminalis	107,201
	Ichthyothere ulei	107
	Ruilopezia bromeloides	*
	Ruilopezia lindenii	144
	Ruilopezia marcescens	*
	Ruilopezia paltonioides	*
	Ruilopezia ruizii	*
	Stevia bupthalmiflora	
	Stevia myriadenia	191
	Tamania chardonii	*
	Wedelia grandiflora	120
	Wedelia helianthoides	*

TABLE 3. (*contd.*)

Compound: No.	Species	Reference
	Wedelia hookeriana	193
652.	Aspilia pluriseta subsp. pluriseta	94
	Austroeupatorium inulaefolium	139
	Campovassouria bupleurifolia	146
	Coespeletia moritziana	145
	Disynaphia multicrenulata	*
	Enhydra fluctuans	421
	Espeletiopsis garciae	145
	Espeletiopsis glandulosa	145
	Espeletiopsis guacharaca	145
	Grazielia intermedia	*
	Helianthus annuus	264
	Ichthyothere terminalis	107
	Libanothamnus granatesianus	174
	Mikania sessilifolia	48
	Montanoa frutescens	
	Montanoa pteropoda	123
	Ruilopezia jahnii	144
	Stevia bupthalmiflora	
	Tamania chardonii	*
	Wedelia buphthalmiflora	455
653.	Ichthyothere connata	201
	Ichthyothere latifolia	201
	Ichthyothere rufa	201
	Ichthyothere terminalis	201
	Mikania arrojadoi	142
	Mikania luetzelburgii	48
	Viguiera pazensis	205
654.	Aspilia pluriseta subsp. pluriseta	94
	Espeletiopsis purpurascens	145
	Loxothysanus sinuatus	129
	Mikania cordata	131
	Wedelia buphthalmiflora	455
655.	Helianthus debilis subsp. curcumerifolius	
	Mikania arrojadoi	142
	Mikania belemii	48
	Mikania luetzelburgii	48
	Mikania oblongifolia	482
	Mikania sessilifolia	48
	Montanoa pteropoda	123
	Montanoa tomentosa subsp. xanthiifolia	227
	Perymeniopsis ovalifolia	103
	Wedelia glauca	405
	Wedelia trilobata	120
656.	Viguiera porteri	331
	Wedelia calycina	328
	Wedelia grandiflora	120
657.	Mikania cordata	131

TABLE 3. (*contd.*)

Compound:		
No.	Species	Reference
658.	Aspilia foliacea	*
	Aspilia parvifolia	213
	Solidago rugosa	73
659.	Aspilia parvifolia	213
	Aspilia pluriseta subsp. pluriseta	*
	Baccharis ramoisissima	*
	Solidago rugosa	73
660.	Aspilia parvifolia	*
	Baccharis ramoisissima	*
	Solidago rugosa	73
661.	Baccharis concinna	117
	Baccharis polyphylla	178
662.	Wedelia trilobata	213
663.	Wedelia trilobata	213
664.	Grazielia intermedia	*
	Ichthyothere terminalis	107
	Ichthyothere ulei	107
	Wedelia calycina	328
665.	Aspilia floribunda	
	Aspilia foliacea	*
	Campovassouria bupleurifolia	146
	Goyazianthus tetrastichus	60
	Helianthus occidentalis	332
	Oyedaea boliviana	166
	Oyedaea lanceolata	166
	Oyedaea rusbyi	*
	Steiractinia mollis	115
	Viguiera porteri	331
	Viguiera trichophylla	*
	Wedelia grandiflora	*
	Wedelia helianthoides	120
	Wedelia hookeriana	193
	Wedelia pinetorum	*
	Wedelia trilobata	120
666.	Ichthyothere terminalis	*
	Ichthyothere ulei	107
	Oyedaea boliviana	166
	Viguiera hypargyrea	5
667.	Wedelia hookeriana	193
668.	Espeletiopsis tachirensis	145
	Ichthyothere terminalis	107
	Ichthyothere ulei	107
	Ruilopezia lindenii	144
	Ruilopezia marcescens	*
	Ruilopezia ruizii	*
	Wedelia helianthoides	
	Wedelia hookeriana	193

TABLE 3. (*contd.*)

Compound: No.	Species	Reference
669.	Campovassouria bupleurifolia	146
	Espeletiopsis guacharaca	145
	Oyedaea rusbyii	*
670.	Mikania sessilifolia	48
671.	Adenostemma caffrum	127
	Adenostemma lavenia	230
	Amphiachyris dracunculoides	*
	Eupatorium album	127
	Grazielia serrata	189
672.	Adenostemma lavenia	230
673.	Adenostemma caffrum	127
	Eupatorium album	127
674.	Adenostemma caffrum	127
	Adenostemma lavenia	230
	Grazielia serrata	189
675.	Ichthyothere terminalis	107
676.	Ichthyothere terminalis	107
677.	Ichthyothere terminalis	107
678.	Atractylis gummifera	239
	Xanthium strumarium	232
	Xanthium strumarium var. strumarium	238
679.	Wedelia asperrima	408
680.	Wedelia asperrima	408
681.	Atractylis gummifera	239
	Wedelia glauca	456
682.	Stevia rebaudiana	362
683.	Stevia rebaudiana	361,362
684.	Stevia rebaudiana	362
685.	Stevia rebaudiana	361,362
686.	Stevia rebaudiana	361
687.	Stevia rebaudiana	361
688.	Stevia rebaudiana	447
689.	Stevia rebaudiana	447
690.	Athrixia elata	150
	Coespeletia lutescens	144
	Coespeletia marcana	145

TABLE 3. (*contd.*)

Compound: No.	Species	Reference
	Coespeletia moritziana	145
	Doronicum hungaricum	*
	Espeletia hartwegiana	*
	Espeletia littlei	
	Espeletia neriifolia	
	Espeletia timotensis	
	Espeletia schultzii	132,218
	Espeletia uribei	144
	Espeletia wedellii	144
	Espeletiopsis garciae	145
	Espeletiopsis glandulosa	145
	Espeletiopsis guacharaca	145
	Espeletiopsis purpurascens	145
	Espeletiopsis tachirensis	145
	Eupatorium deltoideum	
	Gnaphalium oligandrum	210
	Goyazianthus tetrastichus	60
	Helianthus giganteus	98
	Helianthus grosseserratus	31
	Helianthus heterophyllus	321
	Helianthus maximiliani	
	Helichrysum aureum var. monocephalum	
	Helichrysum pilosellum	351
	Helichrysum platypterum	171
	Helichrysum refluxum	*
	Helichrysum ruderale	171
	Ichthyothere terminalis	107
	Lagascea rigida	89
	Lasianthaea podocephala	124
	Libanothamnus neriifolia	144
	Libanothamnus occultus	174
	Libanothamnus spectabilis	174
	Libanothamnus tamanus	174
	Libanothamnus wurdackii	174
	Montanoa pteropoda	123
	Montanoa tomentosa subsp. xanthiifolia	227
	Othonna cylindrica	*
	Oyedaea boliviana	166
	Oyedaea buphthalmoides	
	Oyedaea lanceolata	166
	Oyedaea rusbyi	*
	Perymenium klattianum	103
	Polymnia maculata var. maculata	371
	Riencourtia oblongifolia	*
	Ruilopezia bromeloides	*
	Ruilopezia figueirasii	*
	Ruilopezia jahnii	*
	Ruilopezia lindenii	*
	Ruilopezia marcescens	*
	Ruilopezia margarita	481
	Ruilopezia paltonioides	*
	Ruilopezia ruizii	*
	Smallanthus fruticosus	211
	Steiractinia mollis	115
	Steiractinia sodiroi	124
	Tetrachyron manicata	*

TABLE 3. (*contd.*)

Compound: No.	Species	Reference
	Tithonia longiradiata	*
	Verbesina angustifolia	160
	Verbesina oncophora	160
	Viguiera bishopii	92
	Viguiera dentata	92
	Viguiera excelsa	242
	Viguiera grammatoglossa	
	Viguiera linearis	454
	Viguiera insignis	245
	Viguiera oaxacana	242
	Viguiera procumbens	92
	Viguiera stenoloba var. chihuahuensis	
	Wedelia buphthalmiflora	455
	Wedelia calycina	328
	Wedelia grandiflora	120
	Wedelia helianthoides	120
	Wedelia hispida	328
	Wedelia hookeriana	193
	Wedelia trilobata	120
	Zaluzania discoidea	*
	Zaluzania subcordata	*
	Zaluzania triloba	*
	Zexmenia hispida	124
691.	Helianthus maximiliani	
	Ichthyothere terminalis	107
692.	Ichthyothere terminalis	107
693.	Ichthyothere terminalis	107
694.	Coespeletia lutescens	144
	Coespeletia marcana	145
	Espeletia uribei	144
	Espeletiopsis guacharaca	145
	Espeletiopsis purpurascens	145
	Espeletiopsis tachirensis	145
	Helianthus heterophyllus	321
	Ichthyothere terminalis	*
	Montanoa pteropoda	123
	Perymenium klattianum	103
	Ruilopezia lindenii	*
	Ruilopezia marcescens	*
	Ruilopezia ruizii	*
	Tetrachyron manicata	*
	Viguiera excelsea	242
	Viguiera latibracteata	
	Viguiera linearis	454
	Wedelia calycina	328
	Wedelia hookeriana	193
695.	Ambrosia hispida	322
	Coespeletia lutescens	144
	Coespeletia marcana	145
	Espeletiopsis glandulosa	145
	Espeletiopsis guacharaca	145
	Espeletiopsis purpurascens	145
	Espeletiopsis tachirensis	145

TABLE 3. (*contd.*)

Compound: No.	Species	Reference
	Helianthus heterophyllus	321
	Ichthyothere terminalis	*
	Ruilopezia jahnii	*
	Ruilopezia lindenii	*
	Ruilopezia marcescens	*
	Ruilopezia ruizii	*
	Viguiera excelsa	242
	Viguiera hypargyrea	5
696.	Coespeletia lutescens	144
	Espeletiopsis tachirensis	145
	Ichthyothere terminalis	*
	Mikania arrojadoi	*
	Smallanthus fruticosus	211
697.	Baccharis minutiflora	117
698.	Baccharis minutiflora	117
	Helichrysum dendroideum	374
699.	Baccharis quitensis	116
700.	Espeletia figueirasii	480
	Espeletia floccosa	480
	Espeletia moritziana	480
	Helianthus occidentalis var. dowellianus	105
701.	Austroeupatorium inulaefolium	406
	Coespeletia lutescens	144
	Coespeletia marcana	145
	Coespeletia moritziana	145
	Espeletia uribei	144
	Espeletiopsis guacharaca	145
	Espeletiopsis purpurascens	145
	Espeletiopsis tachirensis	145
	Helianthus annuus	387
	Helianthus debilis Nutt. subsp. cucumerifolius	332
	Helianthus grosseserratus	332
	Helianthus nuttallii subsp. nuttallii	367
	Helianthus occidentalis	332
	Helianthus petiolaris	327
	Helianthus radula	326
	Mikania luetzelburgii	48
	Othonna sedifolia	*
	Ruilopezia jahnii	144
	Ruilopezia lindenii	144
	Ruilopezia marcescens	*
	Ruilopezia paltonioides	*
	Ruilopezia ruizii	*
	Viguiera latibracteata	
	Viguiera porteri	331
	Wedelia helianthoides	*
702.	Coespeletia moritziana	145
	Espeletiopsis guacharaca	145
	Helianthus debilis subsp.	332

TABLE 3. (*contd.*)

Compound: No.	Species	Reference
	cucumerifolius	
	Helianthus occidentalis	332
	Helianthus radula	326
	Helianthus simulans	332
	Mikania arrojadoi	142
	Ruilopezia paltonioides	*
	Ruilopezia ruizii	*
703.	Helianthus debilis subsp. cucumerifolius	332
	Helianthus occidentalis	332
	Helianthus occidentalis var. dowellianus	105
	Viguiera latibracteata	
704.	Polymnia canadensis	187
705.	Polymnia canadensis	187
706.	Polymnia canadensis	187
707.	Polymnia canadensis	187
708.	Polymnia canadensis	187
709.	Smallanthus uvedalia	114
710.	Smallanthus uvedalia	117
711.	Smallanthus uvedalia	117
712.	Mikania sessilifolia	48
713.	Baccharis elliptica	*
	Baccharis minutiflora	117
	Disynaphia multicrenulata	137
	Goyazianthus tetrastichus	60
	Mikania arrojadoi	142
	Mikania sessilifolia	48
	Othonna sedifolia	*
	Wedelia helianthoides	*
714.	Baccharis elliptica	*
	Helianthus occidentalis var. dowellianus	105
	Mikania arrojadoi	142
	Mikania luetzelburgii	48
	Mikania sessilifolia	48
715.	Perymenium ecuadoricum	163
	Smallanthus fruticosus	211
716.	Espeletia uribei	144
717.	Aspilia parvifolia	213
	Aspilia pluriseta	
718.	Aspilia parvifolia	213

TABLE 3. (*contd.*)

Compound: No.	Species	Reference
719.	Ichthyothere terminalis	107
720.	Arctotis revoluta	119
	Aspilia parvifolia	213
	Austroeupatorium inulaefolium	177
	Baccharis minutiflora	117
	Cacalia bulbifera	257
	Coespeletia marcana	145
	Enhydra fluctuans	422
	Espeletia humbertii	479
	Espeletia littlei	
	Espeletia schultzii	132,219
	Espeletiopsis guacharaca	145
	Espeletiopsis purpurascens	145
	Espeletiopsis tachirensis	145
	Helianthus annuus	446
	Helianthus argophyllus	493
	Helianthus occidentalis	332
	Helianthus salicifolius	324
	Helichrysum pallidum	171
	Helichrysum refluxum	*
	Perymenium ecuadoricum	162
	Perymenium serratum	*
	Oyedaea buphthalmoides	
	Robinsonia gayana	*
	Ruilopezia jahnii	144
	Ruilopezia lindenii	144
	Ruilopezia margarita	481
	Steiractinia mollis	115
	Steiractinia sodiroi	124
	Verbesina oncophora	160
	Viguiera bishopii	92
	Wedelia grandiflora	120
	Zaluzania cinarescens	*
721.	Ichthyothere terminalis	107
722.	Helichrysum davyi	*
	Helichrysum pallidum	171
	Ichthyothere terminalis	107
723.	Coespeletia marcana	145
	Helianthus debilis subsp. cucumerifolius	332
	Perymenium discolor	
	Ruilopezia margarita	481
	Steiractinia mollis	*
	Viguiera bishopii	92
	Viguiera linearis	
	Viguiera stenoloba	
	Zaluzania angustifolia	
724.	Helianthus angustifolius	409
	Helianthus argophyllus	493
	Helianthus radula	326
	Helianthus rigidus	324
	Helianthus salicifolius	324
725.	Morithamnus crassus	102

TABLE 3. (*contd.*)

Compound: No.	Species	Reference
	Peteravenia malvaefolia	*
726.	Trichogoniopsis morii	192
727.	Hymenopappus newberryi	207
728.	Helianthus occidentalis	332
	Helianthus petiolaris	327
	Helianthus radula	326
	Helianthus simulans	332
	Helichrysum diosmifolium	366
	Sigesbeckia pubescens	285
	Steiractinia mollis	*
	Viguiera excelsa	242
	Viguiera latibracteata	
	Wedelia calycina	328
729.	Baccharis minutiflora	117
730.	Baccharis minutiflora	117
731.	Baccharis minutiflora	117
732.	Baccharis minutiflora	117
733.	Baccharis minutiflora	117
734.	Baccharis minutiflora	117
735.	Baccharis minutiflora	117
736.	Baccharis minutiflora	117
737.	Baccharis minutiflora	117
738.	Baccharis minutiflora	117
739.	Baccharis minutiflora	117
	Ichthyothere terminalis	*
740.	Baccharis quitensis	116
741.	Sigesbeckia pubescens	397
742.	Sigesbeckia pubescens	397
743.	Sigesbeckia pubescens	397
744.	Sigesbeckia pubescens	397
745.	Sigesbeckia pubescens	397
746.	Helianthus debilis subsp. cucumerifolius	332
	Viguiera latibracteata	
	Wedelia calycina	328
747.	Adenostemma caffrum	127

TABLE 3. (*contd.*)

Compound: No.	Species	Reference
748.	Adenostemma lavenia	230,336
749.	Helianthus radula	326
750.	Coespeletia marcana	145
	Coespeletia moritziana	145
	Espeletia hartwegiana	*
	Espeletiopsis guacharaca	145
	Espeletiopsis purpurascens	145
	Espeletiopsis tachirensis	145
	Libanothamnus granatesianus	174
	Libanothamnus tamanus	174
	Libanothamnus wurdackii	174
	Ruilopezia figueirasii	*
	Ruilopezia marcescens	*
	Ruilopezia paltonioides	*
	Ruilopezia ruizii	*
	Tamania chardonii	*
751.	Coespeletia marcana	145
	Coespeletia moritziana	145
	Espeletiopsis glandulosa	145
	Espeletiopsis guacharaca	145
	Espeletiopsis purpurascens	145
	Espeletiopsis tachirensis	145
	Ruilopezia lindenii	144
	Ruilopezia paltonioides	*
752.	Coespeletia marcana	145
	Coespeletia moritziana	145
	Espeletia hartwegiana	*
	Espeletiopsis guacharaca	145
	Espeletiopsis purpurascens	145
	Libanothamnus granatesianus	174
	Libanothamnus tamanus	174
	Libanothamnus wurdackii	174
	Ruilopezia figueirasii	*
	Ruilopezia marcescens	*
	Ruilopezia paltonioides	*
	Ruilopezia ruizii	*
	Tamania chardonii	*
753.	Coespeletia lutescens	144
	Coespeletia marcana	145
	Coespeletia moritziana	145
	Espeletiopsis purpurascens	145
	Espeletiopsis tachirensis	145
	Ruilopezia jahnii	144
	Ruilopezia lindenii	144
	Tamania chardonii	*
754.	Athrixia phylicoides	164
755.	Athrixia arachnoidea	150
	Athrixia elata	164
	Athrixia phylicoides	164
756.	Athrixia phylicoides	164

TABLE 3. (*contd.*)

Compound: No.	Species	Reference
757.	Athrixia phylicoides	164
758.	Atractylis gummifera	239
759.	Ichthyothere terminalis	107
760.	Ageratum fastigiatum	126
	Ichthyothere terminalis	107
	Wedelia trilobata	213
761.	Ichthyothere terminalis	107
762.	Ichthyothere terminalis	107
763.	Gnaphalium undulatum	213
764.	Helichrysum dendroideum	373
	Viguiera insignis	245
765.	Baccharis tola	450
	Dimorphotheca pseudoaurantiaca	199
	Helichrysum dendroideum	373
	Nidorella anomala	153
	Viguiera grammatoglossa	199
	Wedelia hookeriana	193
	Viguiera pazensis	*
766.	Baccharis tola	450
767.	Helichrysum dendroideum	374
	Nidorella anomala	153
	Viguiera insignis	245
768.	Viguiera insignis	245
769.	Nidorella anomala	153
770.	Helichrysum dendroideum	374
771.	Baccharis tola	451
	Helichrysum dendroideum	374
772.	Baccharis tola	451
773.	Montanoa tomentosa subsp. xanthiifolia	351
	Viguiera grammatoglossa	199,425
774.	Dimorphotheca aurantiaca	199
	Dimorphotheca pseudoaurantiaca	199
	Helianthopsis bishopii	198
	Helianthopsis microphylla	198
	Helianthopsis utcubambensis	198
	Helianthus annuus	264
	Helichrysum fulvum	208
	Helichrysum refluxum	86
	Montanoa tomentosa subsp. xanthiifolia	351

TABLE 3. (*contd.*)

Compound:		
No.	Species	Reference
	Nidorella anomala	153
	Perymenium klattianum	103
	Senecio asperulus	*
	Stevia aristata	*
	Stevia myriadenia	191
	Viguiera bishopii	92
	Viguiera cordifolia	199
	Viguiera grammatoglossa	199
	Viguiera insignis	245
	Viguiera pazensis	*
	Wedelia hookeriana	193
775.	Nidorella anomala	153
	Peteravenia malvaefolia	*
	Peteravenia schultzii	*
776.	Dimorphotheca aurantiaca	199
	Dimorphotheca pseudoaurantiaca	199
777.	Viguiera bishopii	92
778.	Dimorphotheca aurantiaca	199
779.	Baccharis tola	451
780.	Baccharis tola	450
781.	Nidorella anomala	153
782.	Baccharis tola	451
783.	Helichrysum chionosphaerum	45
784.	Helianthus occidentalis	331
785.	Garuleum sonchifolium	78
	Helianthus decapetalus	31,98
786.	Helianthus decapetalus	98
787.	Helianthus decapetalus	98
788.	Helianthus decapetalus	98
789.	Garuleum sonchifolium	78
790.	Garuleum sonchifolium	78
791.	Helianthopsis bishopii	198
	Helianthopsis microphylla	198
	Helianthopsis utcubambensis	198
	Helianthus annuus	31,392,491
	Helianthus debilis subsp. cucumerifolius	31
	Helianthus giganteus	31
	Helianthus hirsutus	31
	Helianthus occidentalis var. dowellianus	105

TABLE 3. (*contd.*)

Compound: No.	Species	Reference
	Helianthus rigidus	31
	Helianthus tomentosus	31
	Helichrysum fulvum	208
	Helichrysum refluxum	86
	Viguiera bishopii	92
	Viguiera dentata	275
	Viguiera lanceolata	92
	Viguiera pazensis	92
	Viguiera procumbens	92
	Viguiera trichopylla	*
792.	Helianthus annuus	387
	Helianthus argophyllus	493
	Helianthus californicus	282
	Helianthus ciliaris	37
	Helianthus grosseserratus	332
	Helianthus niveus subsp. canescens	410
	Helianthus nuttallii subsp. nuttallii	367
	Helianthus occidentalis	331
	Helianthus petiolaris	327
	Helianthus rigidus	332
	Helianthus salicifolius	324
793.	Helianthus radula	326
794.	Viguiera bishopii	92
	Viguiera trichophylla	*
795.	Helianthus radula	326
796	Helichrysum chionosphaerum	45
	Mikania arrojadoi	*
797.	Helichrysum chionosphaerum	45
798.	Helichrysum fulvum	208
799.	Helichrysum fulvum	208
800.	Helianthus debilis subsp. debilis	411
	Montanoa tomentosa subsp. xanthiifolia	
	Tetrachyron orizabaensis var. websteri	411
	Viguiera latibracteata	
801.	Inula royleana	255
802.	Inula royleana	255
803.	Inula royleana	255
804.	Eupatorium inulaefolium	23
805.	Eupatorium album	323
806.	Eupatorium album	323

TABLE 3. (*contd.*)

Compound: No.	Species	Reference
807.	Piqueria trinervia	354
808.	Stevia origanoides	221
809.	Stevia andina Stevia origanoides	* 221
810.	Stevia origanoides	221
811.	Stevia origanoides	221
812.	Eupatorium petaloideum	336
813.	Mikania banisteriae	228
814.	Eupatorium petaloideum	336
816.	Viguiera insignis	244
817.	Viguiera insignis	244
818.	Wedelia calycina	328
819.	Wedelia calycina	328
820.	Helianthus strumosus Mikania triangularis	327 360
821.	Helianthus strumosus	327
822.	Helianthus petiolaris	327
823.	Helianthus hirsutus	327
824.	Helianthus hirsutus	327
825.	Helianthus annuus	264
826.	Helianthus annuus Viguiera pazensis	264 205
827.	Helianthus annuus Viguiera pazensis	264 205
828.	Austroeupatorium inulaefolium	406
829.	Austroeupatorium inulaefolium	406
830.	Critonia daleoides Critonia hebebotrya Gutierrezia sarothrae	217 * 197
831.	Gutierrezia sarothrae	197
832.	Gutierrezia sarothrae	197
833.	Gutierrezia sarothrae	197
834.	Gutierrezia sarothrae	197

TABLE 3. (*contd.*)

Compound: No.	Species	Reference
835.	Gutierrezia sarothrae	197
836.	Gutierrezia sarothrae	197
837.	Gutierrezia sarothrae	197
838.	Gutierrezia sarothrae	197
839.	Gutierrezia sarothrae	197
840.	Gutierrezia sarothrae	197
841.	Gutierrezia sarothrae	197
842.	Gutierrezia grandis Gutierrezia sarothrae	271 197
843.	Gutierrezia sarothrae	197
844.	Gutierrezia sarothrae	197
845.	Gutierrezia sarothrae	197
846.	Baccharis incanum Baccharis kingii Conyza scabrida	 196 83
847.	Viguiera maculata	241
848.	Ophryosporus chilca	152
849.	Ophryosporus chilca	152
850.	Helianthus annuus	446
851.	Helianthus annuus	446
852.	Helianthus annuus	446
853.	Helianthus annuus	446
854.	Helianthus annuus	446
855.	Helianthus annuus	446
856.	Helianthus annuus	446
857.	Fleischmannia viscidipes	81
858.	Fleischmannia viscidipes	81
859.	Fleischmannia viscidipes	81
860.	Fleischmannia viscidipes	81
861.	Fleischmannia viscidipes Oxylobus canescens	81 *
862.	Fleischmannia viscidipes	81

TABLE 3. (*contd.*)

Compound: No.	Species	Reference
863.	Fleischmannia viscidipes	81
864.	Acritopappus longifolius	52
865.	Acritopappus longifolius	52
866.	Wedelia regis	75
867.	Wedelia regis	75
868.	Wedelia regis	75
869.	Aspilia floribunda	75
870.	Aristeguieta pseudoarborea	194
871.	Aristeguietia pseudoarborea	194
872.	Aristeguietia pseudoarborea	194
873.	Aristeguietia pseudoarborea	194
874.	Aristeguietia pseudoarborea	194
875.	Villanova titicaensis	195
876.	Villanova titicaensis	195
877.	Villanova titicaensis	195
878.	Villanova titicaensis	195
879.	Villanova titicaensis	195
880.	Villanova titicaensis	195
881.	Trichogonia salvaiaefolia Trichogonia villosa	183 183
882.	Trichogonia salvaiaefolia	183
883.	Trichogonia villosa	183
884.	Trichogonia salvaiaefolia Trichogonia villosa	183 183
885.	Trichogonia salvaiaefolia	183
886.	Trichogonia salvaiaefolia	183
887.	Trichogonia salviaefolia Trichogonia villosa	* 183
888.	Trichogonia villosa	183
889.	Trichogonia villosa	183
890.	Trichogonia salvaiaefolia Trichogonia villosa	183 183

TABLE 3. (*contd.*)

Compound: No.	Species	Reference
891.	Trichogonia villosa	183
892.	Aspilia mossambicensis Viguiera pazensis	* 205
893.	Viguiera pazensis	205
894.	Viguiera pazensis	205
895.	Stevia ovata Stevia paniculata	355,464 501
896.	Stevia ovata Stevia paniculata	355 501
897.	Stevia ovata Stevia paniculata	355 501
898.	Stevia ovata Stevia paniculata	355 501
899.	Stevia ovata Stevia paniculata	355 501
900.	Conyza scabrida	83
901.	Conyza scabrida	83
902.	Conyza scabrida	83
903.	Conyza scabrida	83
904.	Baccharis salicifolia Conyza scabrida	492 83
905.	Conyza scabrida	83
906.	Conyza scabrida	83
907.	Conyza scabrida	83
908.	Conyza scabrida	83
909.	Conyza scabrida	83
910.	Conyza scabrida	83
911.	Conyza scabrida	83
912.	Conyza scabrida	83
913.	Conyza scabrida	83
914.	Conyza scabrida	83
915.	Conyza scabrida	83
916.	Conyza scabrida	83

TABLE 3. (*contd.*)

Compound: No.	Species	Reference
917.	Conyza scabrida	83
918.	Conyza scabrida	83
919.	Conyza scabrida	83
920.	Conyza scabrida Pulicaria angustifolia	83 463
921.	Sigesbeckia pubescens	285
922.	Sigesbeckia pubescens	285
923.	Sigesbeckia pubescens	285
924.	Sigesbeckia pubescens	285
925.	Sigesbeckia pubescens	285
926.	Lasianthaea podocephala Melampodium perfoliatum Mikania banisteriae	124 159 228
927.	Zexmenia phyllocephala	124
928.	Zexmenia phyllocephala	124
929.	Zexmenia phyllocephala	124
930.	Zexmenia phyllocephala	124
931.	Zexmenia phyllocephala	124
932.	Zexmenia phyllocephala	124
933.	Zexmenia phyllocephala	124
934.	Zexmenia phyllocephala	124
935.	Zexmenia phyllocephala	124
936.	Zexmenia phyllocephala	124
937.	Zexmenia phyllocephala	124
938.	Eupatorium villosum	379
939.	Melampodium divaricatum Melampodium paludosum Melampodium perfoliatum Smallanthus glabratus	* * 159 108
940.	Eupatorium jhanii	285
941.	Eupatorium jhanii	285
942.	Montanoa tomentosa subsp. xanthiifolia Perymenium klattianum	227 103

TABLE 3. (*contd.*)

Compound: No.	Species	Reference
943.	Montanoa tomentosa subsp. xanthiifolia	227
944.	Sigesbeckia pubescens	225
945.	Dornoicum macrophyllum Doronicum microphyllum	3 285
946.	Espeletia wedellii	394
947.	Solidago altissima	500
948.	Balsamorhiza sagittata	130
949.	Balsamorhiza sagittata	130
950.	Heteropappus altaicus	181
951.	Heteropappus altaicus	181
952.	Heteropappus altaicus	181
953.	Heteropappus altaicus	181
954.	Heteropappus altaicus	181
955.	Grindelia squarrosa	473
956.	Grindelia squarrosa	473
957.	Helianthopsis bishopii	198
958.	Macowania glandulosa Oxylobus arbutifolius	* 87
959.	Oxylobus adscendens	87
960.	Solidago altissima	143
961.	Solidago altissima	143
962.	Baccharis tucumanensis	442
963.	Baccharis grandicapitulata	57
964.	Baccharis scoparia	57
965.	Baccharis eggersii	57
966.	Baccharis eggersii	57
967.	Baccharis hutchisonii	57
968.	Baccharis nitida	57
969.	Baccharis scoparia	57
970.	Baccharis scoparia	57

TABLE 3. (*contd.*)

Compound: No.	Species	Reference
971.	Aster alpinus	97
972.	Aster alpinus	97
973.	Tridax peruviensis	97
974.	Gutierrezia gilliesii Gutierrezia grandis Gutierrezia sobrigii	348 271 347
975.	Gutierrezia grandis	271
976.	Helichrysum refluxum	86
977.	Rudbeckia fulgida var. fulgida	329
978.	Smallanthus glabratus	108
979.	Smallanthus glabratus	108
980.	Smallanthus glabratus	108
981.	Smallanthus glabratus	108
982.	Smallanthus glabratus	108
983.	Smallanthus glabratus	108
984.	Ageratina tristis	58
985.	Pulicaria angustifolia	463
986.	Pulicaria angustifolia	463
987.	Pulicaria angustifolia	463
988.	Solidago altissima Solidago virgaurea	401 378
989.	Viguiera potosina	276
990.	Viguiera potosina	276
991.	Brickellia vernicosa Eupatorium maretiana	2 2
992.	Brickellia vernicosa	2
993.	Brickellia vernicosa	2
994.	Brickellia vernicosa	2
995.	Brickellia vernicosa	2
996.	Denekia capensis Helichrysum confertum Viguiera dentata	180 180 275
997.	Disynaphia multicrenulata	137

TABLE 3. (*contd.*)

Compound: No.	Species	Reference
	Viguiera dentata	275
998.	Stevia eupatoria	416
	Viguiera excelsa	242
999.	Hebeclinium macrophyllum	77
1000.	Solidago missouriensis	369
1001.	Solidago missouriensis	369
1002.	Mikania congesta	330
1003.	Mikania congesta	330
1004.	Mikania congesta	330
1005.	Mikania congesta	330
1006.	Melampodium leucanthum	436
1007.	Viguiera stenoloba	243
1008.	Eupatorium inulaefolium	23
1010.	Symphyopappus reticulatus	190
1011.	Symphyopappus reticulatus	190
1012.	Piptothrix jaliscensis	269
1013.	Piptothrix jaliscensis	269
1014.	Piptothrix sinaloae	269
1015.	Austrobrickellia patens	349
1016.	Austrobrickellia patens	349
1017.	Austrobrickellia patens	349
1018.	Austrobrickellia patens	349
1019.	Austrobrickellia patens	349
1020.	Stevia lemmonia	170
1021.	Stevia berlandiera	170
1022.	Stevia berlandiera	170
1023.	Stevia andina	*
	Stevia berlandiera	170
1024.	Stevia salicifolia	170
1025.	Stevia salicifolia	170
1026.	Stevia salicifolia	170

TABLE 3. (*contd.*)

Compound: No.	Species	Reference
1027.	Chromolaena laevigata	390
1028.	Chromolaena laevigata	390
1029.	Chromolaena laevigata	390
1030.	Chromolaena laevigata	390
1031.	Chromolaena laevigata	390
1032.	Chromolaena laevigata	390
1033.	Baccharis salicifolia Baccharis sarothroides Baccharis vaccinoides	22 22 22
1034.	Baccharis sarothroides	22
1035.	Grindelia boliviana	348
1036.	Grindelia boliviana	348
1037.	Gutierrezia gilliesii Gutierrezia solbrigii	348 347
1038.	Gutierrezia gilliesii Gutierrezia solbrigii	348 347
1039.	Gutierrezia gilliesii	348
1040.	Gutierrezia spathulata	348
1041.	Gutierrezia spathulata	348
1042.	Gutierrezia spathulata	348
1043.	Gutierrezia spathulata	348
1044.	Gutierrezia spathulata	348
1045.	Haplopappus glutinosus	348
1046.	Haplopappus pectinatus	348
1047.	Haplopappus pectinatus	348
1048.	Haplopappus pectinatus	348
1049.	Haplopappus pectinatus	348
1050.	Haplopappus pectinatus	348
1051.	Haplopappus pectinatus	348
1052.	Haplopappus pectinatus	348
1053.	Haplopappus pectinatus	348
1054.	Solidago petradoria	348

TABLE 3. (*contd.*)

Compound: No.	Species	Reference
1055.	Haplopappus paucidentatus	348
1056.	Haplopappus paucidentatus	348
1057.	Haplopappus paucidentatus	348
1058.	Haplopappus paucidentatus	348
1059.	Haplopappus paucidentatus	348
1060.	Grindelia boliviana	348
1061.	Baccharis scoparia	57
1062.	Solidago gigantea var. serotina	317
1063.	Solidago gigantea var. serotina	317
1064.	Solidago gigantea var. serotina	317
1065.	Solidago gigantea var. serotina	317
1066.	Solidago gigantea var. serotina	317
1067.	Solidago gigantea var. serotina	317
1068.	Solidago gigantea var. serotina	317
1069.	Liatris spicata	29
1070.	Baccharis salicifolia	492
1071.	Solidago juncea	318
1072.	Solidago juncea	318
1073.	Solidago juncea	318
1074.	Solidago juncea	318
1075.	Solidago serotina	385
1076.	Solidago serotina	385
1077.	Solidago gigantea var. serotina	385
1078.	Solidago gigantea var. serotina	385
1079.	Solidago gigantea var. serotina	385
1080.	Solidago gigantea var. serotina	385
1081.	Solidago gigantea var. serotina	385
1082.	Solidago gigantea var. serotina	385
1083.	Gutierrezia texana	274
1083.	Gutierrezia texana	274

TABLE 3. (*contd.*)

Compound: No.	Species	Reference
1084.	Gutierrezia texana	274
1085.	Gutierrezia texana	274
1086.	Gutierrezia texana	274
1087.	Gutierrezia texana	274
1088.	Gutierrezia texana	274
1089.	Gutierrezia texana	274
1090.	Gutierrezia texana	274
1091.	Gutierrezia texana	274
1092.	Gutierrezia texana	274
1093.	Baccharis salicifolia	492
1094.	Montanoa tomentosa	432
1095.	Montanoa tomentosa	432
1096.	Montanoa tomentosa	432
1097.	Montanoa tomentosa subsp. tomentosa	431
1098.	Montanoa tomentosa subsp. tomentosa	431
1099.	Montanoa tomentosa subsp. tomentosa	417
1100.	Acritopappus confertus	184
1101.	Helichrysum acutatum	40
1102.	Denekia capensis Helichrysum confertum	180 180
1103.	Helichrysum confertum	180
1104.	Helichrysum confertum	180
1105.	Helichrysum albirosulatum	180
1106.	Helichrysum albirosulatum	180
1107.	Helichrysum albirosulatum	180
1108.	Helianthus decapetalus Helianthus decapetalus var. multiflorus Helianthus hirsutus Helianthus rigidus	31 31 31 31
1109.	Cronquistianthus bishopii	61

Table 3. (*contd.*)

Compound: No.	Species	Reference
1110.	Cronquistianthus bishopii	61
1111.	Solidago drummondii	65
1112.	Mikania periplocifolia	296
1113.	Mikania periplocifolia	296
1114.	Mikania periplocifolia	296
1115.	Mikania periplocifolia	296
1116.	Bejaranoa balansae	453
1117.	Bejaranoa balansae	453
1118.	Chiliotrichium rosarinifolium	346
1119.	Solidago drummondii	65
1120.	Aspilia parvifolia	182
1121.	Vernonia glabra var. glabra	55
1122.	Heterochrysum oreophilum	351
1123.	Heterochrysum oreophilum	351
1124.	Acanthospermum australe	138
1125.	Solidago flexicaulis Solidago racemosa Solidago drummondii	65 65 65
1126.	Solidago drummondii	65
1127.	Solidago drummondii	65
1128.	Solidago drummondii	65
1129	Solidago drummondii	65
1130.	Gutierrezia solbrigii	347
1131.	Gutierrezia solbrigii	347
1132.	Gutierrezia solbrigii	347
1133.	Nordophyllum lanatum	346
1134.	Nordophyllum lanatum	346
1135.	Gutierrezia solbrigii	347
1136.	Gutierrezia solbrigii	347
1137.	Solidago drummondii	65

TABLE 3. (*contd.*)

Compound: No.	Species	Reference
1138.	Helichrysum nudifolium var. nudifolium	351
1139.	Oxylobus glanduliferus	6
1140.	Grindelia discoidea	472
1141.	Grindelia discoidea	472
1142.	Grindelia discoidea	472
1143.	Eupatorium salvia	339
1144.	Eupatorium salvia	339
1145.	Grindelia discoidea	471
1146.	Chrysothamnus nauseosus Grindelia discoidea	471 471
1147.	Helichrysum confertum	180
1148.	Gutierrezia sphaerocephala	272
1149.	Gutierrezia sphaerocephala	272
1150.	Gutierrezia sphaerocephala	272
1151.	Helichrysum nudifolium var. nudifolium	351
1152.	Helichrysum nudifolium var. nudifolium	351
1153.	Isocoma coronopifolia	253
1154.	Isocoma coronopifolia	253
1155.	Lindheimera texana	337
1156.	Lindheimera texana	337
1157.	Stevia rebaudiana	418
1158.	Stevia rebaudiana	418
1159.	Stevia rebaudiana	418
1160.	Stevia rebaudiana	418
1161.	Palafoxia texana	290
1162.	Palafoxia texana	290
1163.	Eupatorium turbinatum	350
1164.	Grindelia aegialitis	414
1165.	Baccharis pingraea	503

TABLE 3. *(contd.)*

Compound: No.	Species	Reference
	Baccharis salicifolia	503
1166.	Baccharis pingraea	503
	Baccharis salicifolia	503
1167.	Baccharis pingraea	503
	Baccharis salicifolia	503
1168.	Baccharis pingraea	503
1169.	Baccharis pingraea	503
1170.	Baccharis salicifolia	503
1171.	Baccharis pingraea	503
1172.	Baccharis salicifolia	503
1173.	Baccharis trinervis	27
1174.	Gutierrezia solbrigii	347
1175.	Haplopappus pectinatus	348
1176.	Haplopappus pectinatus	348
1177.	Haplopappus pectinatus	348
1178.	Haplopappus pectinatus	348
1179.	Haplopappus pectinatus	348
1180.	Acritopappus morii	179
1181.	Gutierrezia solbrigii	347
1182.	Gutierrezia solbrigii	347
1183.	Gutierrezia solbrigii	347
1184.	Gutierrezia solbrigii	347
1185.	Mikania alvimii	47
1186.	Palafoxia rosea	26
1187.	Grindelia boliviana	348
1188.	Baccharis salicifolia	503
1189.	Gutierrezia solbrigii	347
1190.	Gutierrezia solbrigii	347
1191.	Gutierrezia solbrigii	347
1192.	Baccharis salicifolia	503
1193.	Baccharis salicifolia	503

TABLE 3. (*contd.*)

Compound: No.	Species	Reference
1194..	Gutierrezia spathulata	348
1195.	Gutierrezia spathulata	348
1196.	Baccharis salicifolia	503
1197.	Baccharis salicifolia	503
1198.	Palafoxia rosea	26
1199.	Palafoxia rosea	26
1200.	Palafoxia rosea	26
1201.	Ayapana amygdalina	113
1202.	Ayapana amygdalina	113
1203.	Baccharis pingraea	503
1204.	Helichrysum argolepis	*
1205.	Austroeupatorium inulaefolium	139
1206.	Austroeupatorium inulaefolium	139
1207.	Austroeupatorium inulaefolium	139
1208.	Haplopappus paucidentatus	348
1209.	Baccharis polifera	503
1210.	Baccharis polifera	503
1211.	Eupatorium turbinatum	350
1212.	Chiliotrichium rosmarinifolium	346
1213.	Chiliotrichium rosmarinifolium	346
1214.	Chiliotrichium rosmarinifolium	346
1215.	Chiliotrichium rosmarinifolium	346
1216.	Chiliotrichium rosmarinifolium Nardophyllum lanatum	346 346
1217.	Nardophyllum lanatum	346
1218.	Nardophyllum lanatum	346
1219.	Nardophyllum lanatum	346
1221.	Chiliotrichium rosmarinifolium	346
1222.	Baccharis incarum	449
1223.	Pulicaria salviifolia	403

TABLE 3. (*contd.*)

Compound: No.	Species	Reference
1224.	Baccharis salicifolia	503
1225.	Baccharis salicifolia	503
1226.	Baccharis salicifolia	503
1227.	Baccharis incarum	284
1228.	Goyazianthus tetrastichus	60
1229.	Baccharis rhomboidalis	452
1230.	Baccharis rhomboidalis	452
1231.	Baccharis rhetinodes	503
1232.	Baccharis incarum	449
1233.	Baccharis incarum	449
1234.	Pulicaria salviifolia	404
1235.	Pulicaria salviifolia	404
1236.	Baccharis incarum	284
1237.	Baccharis articulata	466
1238.	Baccharis articulata	466
1239.	Baccharis grandicapitulata	57
1240.	Baccharis incarum	284
1241.	Baccharis rhomboidalis	452
1242.	Baccharis rhomboidalis	452
1243.	Baccharis rhomboidalis	452
1244.	Baccharis magellanica	503
1245.	Baccharis magellanica	503
1246.	Chiliotrichium rosmarinifolium	346
1247.	Gochnatia glutinosa	278
1248.	Gochnatia glutinosa	278
1249.	Senecio hypochoerideus	71
1250.	Mikania triangularis	360
1251.	Helichrysum setosum	351
1252.	Helichrysum setosum	351

TABLE 3. (*contd.*)

Compound: No.	Species	Reference
1253.	Helichrysum setosum	351
1254.	Helichrysum setosum	351
1255.	Ruilopezia margarita [=Espeletia margarita]	481
1256.	Ruilopezia margarita	481
1257.	Baccharis incarum	284
1258.	Baccharis incarum	284
1259.	Helianthus grosseserratus Helianthus laciniatus Helianthus rigidus	* * *
1260.	Eupatorium petaloideum	336

I. Linear or Unicarbocyclic

1. Geranylnerol

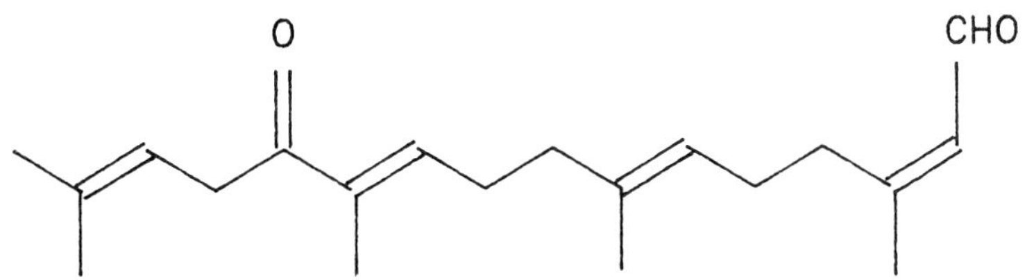

1     $R_1 = R_2 = R_3 = R_4 = H$

2     $R_1 = R_3 = R_4 = H$, $R_2 = OH$

3     $R_1 = R_2 = R_3 = H$, $R_4 = OH$

4     $R_1 = R_4 = OH$, $R_2 = R_3 = H$

984   $R_1 = R_2 = H$, $R_3 = OAc$, $R_4 = OH$

1109 $R_1 = R_2 = OH$, $R_3 = R_4 = H$

1110 $R_1 = R_2 = OH$, $R_3 = H$, $R_4 = OH$

5     $R_1 = CHO$, $R_2 = H$, $R_3 = R_4 = Me$

6     $R_1 = CH_2OH$, $R_2 = OH$, $R_3 = CO_2H$, $R_4 = Me$

1002 $R_1 = CH_2OH$, $R_2 = H$, $R_3 = Me$, $R_4 = CO_2H$

1003 $R_1 = CH_2OAc$, $R_2 = H$, $R_3 = Me$, $R_4 = CO_2H$

1111

FIGURE 1. Structures of diterpenes reported from the Compositae.

7

8 R_1 = CHO, R_2 = H, R_3 = OH, R_4 = OH

957 R_1 = CH_2OH, R_2 = OH, R_3 = OH, R_4 = H

1112 R_1 = CH_2OH, R_2 = R_3 = OAc, R_4 = OH

1113 R_1 = CH_2OH, R_2 = R_3 = R_4 = OAc

1114 R_1 = CH_2OH, R_2 = OAc, R_3 = H, R_4 = OH

1115 R_1 = CH_2OH, R_2 = H, R_3 = OAc, R_4 = OH

15 R_1 = H, R_2 = CH_2OAc, R_3 = CH_2OH

948 R_1 = OH, R_2 = CH_2OH, R_3 = Me

1116 R = H

1117 R = OH

FIGURE 1. (contd.)

978

949

979

1004 R_1 = R_2 = OH

1005 R_1 = R_2 = OAc

1095

FIGURE 1. (contd.)

1096

1020

14

2. Geranylgeraniol

11   R = OH

12   R = OAc

FIGURE 1. (*contd.*)

13 R_1 = OH, R_2 = Me, R_3 = H

954 R_1 = H, R_2 = CO_2H, R_3 = H (10,11E)

16

17 R_1 = CHO, R_2 = H, R_3 = Me, R_4 = H

18 R_1 = CHO, R_2 = OH, R_3 = Me, R_4 = H

19 R_1 = CH_2OAc, R_2 = H, R_3 = CO_2H, R_4 = H

1118 R_1 = CH_2OH, R_2 = H, R_3 = CH_2OH, R_4 = OAc

20

FIGURE 1. (*contd.*)

1119

21

22

1120

3. Geranyllinalool

23 $R_1 = R_2 = R_3 = H$

FIGURE 1. (*contd.*)

24 $R_1 = R_2 = H$, $R_3 = OH$

25 $R_1 = R_2 = H$, $R_3 = OAc$

26 $R_1 = H$, $R_2 = OH$, $R_3 = H$

27 $R_1 = H$, $R_2 = OAc$, $R_3 = H$

28 $R_1 = OH$, $R_2 = R_3 = H$

29 $R_1 = OH$, $R_2 = OAc$, $R_3 = H$

30 $R_1 = R_2 = OAc$, $R_3 = H$

31 $R_1 = OH$, $R_2 = H$, $R_3 = OAc$

1121 $R_1 = R_2 = H$, $R_3 = OAc$

32

1122 (10,11-dihydro-homolog of **32**)

33

34

FIGURE 1. (*contd.*)

35

1123

36

37 $R_1 = R_2 = H$

38 $R_1 = OH$, $R_2 = H$

39 $R_1 = H$, $R_2 = OAc$

4. Phytol

59

FIGURE 1. (*contd.*)

80. 14-Methylgeranylnerol

9   R = OH

10   R = OAc

980

981

982

983

FIGURE 1. (*contd.*)

1094

5. Acanthoaustralane

40	R_1 = OH, R_2 = R_3 = H
41	R_1 = OAc, R_2 = R_3 = H
42	R_1 = R_2 = OH, R_3 = H
43	R_1 = OH, R_2 = OAc, R_3 = H
1124	R_1 = OAc, R_2 = H, R_3 = Ac

6. Acanthoaustralane-6,11-Epoxide

44	R = OAc
45	R = OH

FIGURE 1. (*contd.*)

7. Isoacanthoaustralane

46 R_1 = R_2 = OH

47 R_1 = OH, R_2 = OAc

48 R_1 = OAc, R_2 = H

8. Geranylgeraniol-18,8-Lactone

49   R = H

50   R = OAc

51   R = OOH

52   R = OH

FIGURE 1. (*contd.*)

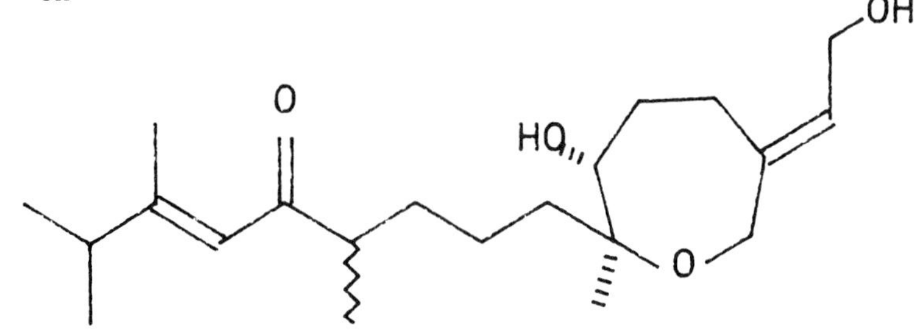

53   R = OOH

54   R = OH

9. Oxepane

55

56   R = OAc

1098 R = OH

57

FIGURE 1. (*contd.*)

1097

1099

86. Melcanthane

1006

10. Mikanofurane

58

FIGURE 1. (*contd.*)

11. Centipedane

60 $R_1 = CO_2H$, $R_2 = Me$, $R_3 = H$

61 $R_1 = R_2 = CH_2OH$, $R_3 = H$

1126 $R_1 = R_2 = Me$, $R_3 = OH$

1125 R = H

1127 R = OH

1128

12. Geranylgeraniol-1,20-Lactone

62 $R = CO_2H$

63 $R = CO_2Me$

FIGURE 1. (*contd.*)

1129

1130 R = H

1131 R = H (10,11E)

1132 R = Ac

64

1133 R = H

1134 R = Ac

FIGURE 1. (*contd.*)

1135 R = H

1136 R = Ac

1137

13. Geranylterpinene

65   R = CH$_2$OH

66   R = CHO

67   R = CO$_2$H

1138

FIGURE 1. (*contd.*)

68

69

70

71

72

FIGURE 1. *(contd.)*

1101

14. Isocembrene

73

15. Geranylnerol-11,14-α-Epoxide

74   R = OH

75   R = OAc

FIGURE 1. (*contd.*)

II. Normal-Bicarbocyclics and Derivatives

16. Normal-Labdane

79 R_1 = H, R_2 = CH_2OH

80 R_1 = OH, R_2 = CH_2OH

958 R_1 = H, R_2 = CO_2H (=121?)

86

87

81

FIGURE 1. (*contd.*)

1139

124 R = H

125 R = Mebut-2-OH

126 R = Ang

129

76 R_1 = R_2 = H, R_3 = CH_2OH

77 R_1 = OH, R_2 = Oβ-Gal, R_3 = CH_2OH

111 R_1 = R_2 = H, R_3 = CO_2H

FIGURE 1. (*contd.*)

1140 R_1 = OH, R_2 = α-OH

1141 R_1 = OAc, R_2 = α-OH

1142 R_1 = OH, R_2 = β-OH

1143 R = H

1144 R = Ac

1103

78   R_1 = H, R_2 = H, R_3 = CH_2OH

FIGURE 1. (*contd.*)

108 R_1 = H, R_2 = OAng, R_3 = CO_2H

109 R_1 = Ang, R_2 = OH, R_3 = CO_2H

82 R_1 = R_2 = H

83 R_1 = H, R_2 = Ac (Acetylation product)

84 R_1 = R_2 = Ac (Acetylation product)

85

88 R_1 = R_2 = R_3 = H

89 R_1 = OH, R_2 = R_3 = H

90 R_1 = R_2 = H, R_3 = Ac

91 R_1 = OH, R_2 = H, R_3 = Ac

92 R_1 = H, R_2 = Ac, R_3 = H

93 R_1 = OH, R_2 = Pro, R_3 = Ac

FIGURE 1. (*contd.*)

995

959

100 R_1 = OAng, R_2 = H, R_3 = R_4 = Me

116 R_1 = H, R_2 = OH, R_3 = R_4 = Me

101 R_1 = R_2 = H, R_3 = CH$_2$OAng, R_4 = Me

102 R_1 = R_2 = H, R_3 = CH$_2$OTig, R_4 = Me

103 R_1 = α-OH, R_2 = OAng, R_3 = R_4 = Me

104 R_1 = α-OH, R_2 = O-trans-Cinn, R_3 = R_4 = Me

105 R_1 = α-OH, R_2 = O-cis-Cinn, R_3 = R_4 = Me

106 R_1 = α-OH, R_2 = OAng, R_3 = R_4 = Me

FIGURE 1. (*contd.*)

107 R_1 = α-OAng, R_2 = OH, R_3 = R_4 = Me

110 R_1 = R_2 = H, R_3 = Me, R_4 = CO_2H

97

977

112 R_1 = OTig, R_2 = H, R_3 = Me

113 R_1 = H, R_2 = β-OAc, R_3 = Me

114 R_1 = H, R_2 = α-OH, R_3 = Me

115 R_1 = H, R_2 = β-OH, R_3 = Me

130 R_1 = R_2 = H, R_3 = CO_2H

FIGURE 1. (*contd.*)

117

98 R_1 = OAng or OMebut-2-OH, R_2 = OAng or OMebut-2-OH,
R_3 = O-trans-Cinn

99 R_1 = OAng or OMebut-2-OH, R_2 = OAng or OMebut-2-OH,
R_3 = O-cis-Cinn

1024 R_1 = R_2 = R_3 = H

118 R_1 = R_2 = H, R_3 = CO_2H, R_4 = Me

119 R_1 = H, R_2 = α-OH, R_3 = CO_2H, R_4 = Me (=**991**?)

120 R_1 = H, R_2 = β-OH, R_3 = CO_2H, R_4 = Me

FIGURE 1. (*contd.*)

122 $R_1 = R_2 = H$, $R_3 = CO_2H$, $R_4 = CH_2OAng$

123 $R_1 = R_2 = H$, $R_3 = CO_2H$, $R_4 = CH_2OTig$

127 $R_1 = OH$, $R_2 = \alpha\text{-}O\text{-}trans\text{-}Cinn$, $R_3 = CO_2H$, $R_4 = Me$

128 $R_1 = OH$, $R_2 = \alpha\text{-}O\text{-}cis\text{-}Cinn$, $R_3 = CO_2H$, $R_4 = Me$

131 $R_1 = R_2 = H$, $R_3 = R_4 = CO_2H$

870 $R_1 = OSen$, $R_2 = \beta\text{-}OAc$, $R_3 = CO_2Me$, $R_4 = Me$

871 $R_1 = OH$, $R_2 = H$, $R_3 = CH_2OH$, $R_4 = Me$

872 $R_1 = OH$, $R_2 = H$, $R_3 = CH_2OCoum$, $R_4 = Me$

873 $R_1 = OH$, $R_2 = H$, $R_3 = CH_2O3\text{-}Me\text{-}Coum$, $R_4 = Me$

874 $R_1 = OH$, $R_2 = H$, $R_3 = CH_2O\text{-}p\text{-}OH\text{-}Cinn$, $R_4 = Me$

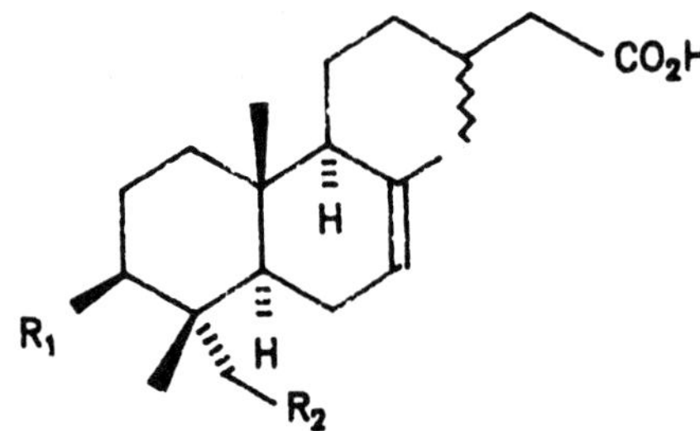

1145 $R_1 = H$, $R_2 = OH$

1146 $R_1 = OH$, $R_2 = H$ (=**120**)

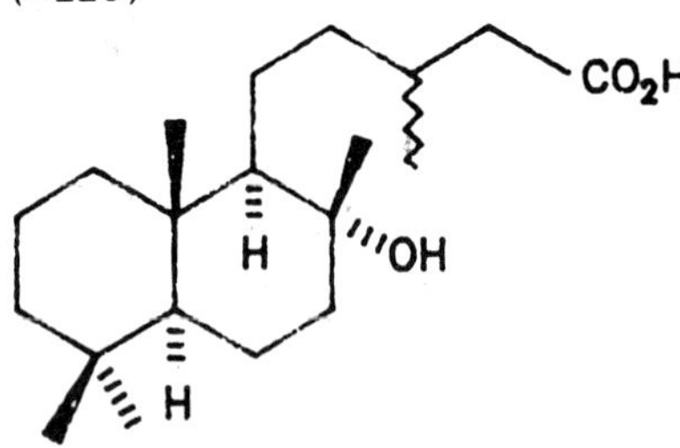

121 (=**958**?)

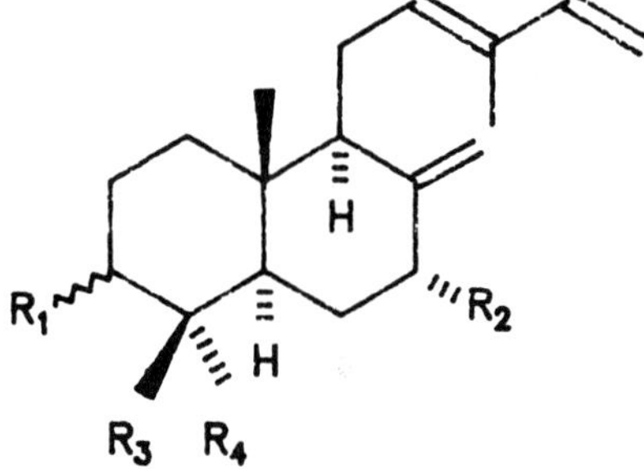

132 $R_1 = R_2 = H$, $R_3 = R_4 = Me$

FIGURE 1. (contd.)

133 R_1 = α-OH, R_2 = H, R_3 = R_4 = Me

134 R_1 = β-OH, R_2 = H, R_3 = R_4 = Me

135 R_1 = β-OAc, R_2 = H, R_3 = R_4 = Me

137 R_1 = R_2 = H, R_3 = CO_2H, R_4 = Me

138 R_1 = H, R_2 = α-OAc, R_3 = CO_2H, R_4 = Me

1104 R_1 = R_2 = H, R_3 = Me, R_4 = CO_2Me

1147

991 R = CO_2H (=**119**?)

992 R = CO_2Me

993 R = CH_2OH

994 R = CH_2O2-Mebut

136

FIGURE 1. (*contd.*)

139 $R_1 = R_2 = H$

140 $R_1 = H$, $R_2 = OH$

1021 $R_1 = H$, $R_2 = OAc$

1023 $R_1 = R_2 = OH$

141 $R_1 = R_2 = H$, $R_3 = OH$, $R_4 = Me$

142 $R_1 = H$, $R_2 = \alpha\text{-}OH$, $R_3 = OH$, $R_4 = Me$

143 $R_1 = H$, $R_2 = \alpha\text{-}OAng$, $R_3 = OH$, $R_4 = Me$

144 $R_1 = H$, $R_2 = \beta\text{-}OH$, $R_3 = H$, $R_4 = Me$

145 $R_1 = R_2 = H$, $R_3 = OAc$, $R_4 = Me$

146 $R_1 = H$, $R_2 = \beta\text{-}OH$, $R_3 = OH$, $R_4 = Me$

147 $R_1 = H$, $R_2 = \beta\text{-}OH$, $R_3 = H$, $R_4 = CH_2OH$

148 $R_1 = H$, $R_2 = \beta\text{-}OH$, $R_3 = OAc$, $R_4 = Me$

149 $R_1 = OAc$, $R_2 = R_3 = H$, $R_4 = CH_2OArac$

150 $R_1 = OAc$, $R_2 = R_3 = H$, $R_4 = CH_2OCoum$

151 $R_1 = OAc$, $R_2 = R_3 = H$, $R_4 = CH_2O\text{-}cis\text{-}Coum$

1022 $R_1 = H$, $R_2 = \alpha\text{-}OH$, $R_3 = OAc$, $R_4 = Me$

FIGURE 1. (*contd.*)

152

94 $R_1 = R_2 = H$

95 $R_1 = H$, $R_2 = Ac$ (Acetylation product)

96 $R_1 = R_2 = Ac$ (Acetylation product)

153 $R = \alpha\text{-OAng}$

1102 $R = H$

FIGURE 1. (*contd.*)

2. Compound Names, Structures and Sources

996

154

1105 R = H

1106 R = OH

95 R = Tig

96 R = H

FIGURE 1. (*contd.*)

1148 R_1 = CHO, R_2 = Ac

1149 R_1 = CO_2Me, R_2 = Ac

1150 R_1 = CO_2Me, R_2 = H

1107

155 R_1 = R_2 = R_3 = R_4 = H

156 R_1 = OH, R_2 = OAc, R_3 = β-O-OH, R_4 = H

157 R_1 = OH, R_2 = OAc, R_3 = α-O-OH, R_4 = H

1151 R_1 = R_2 = R_3 = H, R_4 = OAc

FIGURE 1. (*contd.*)

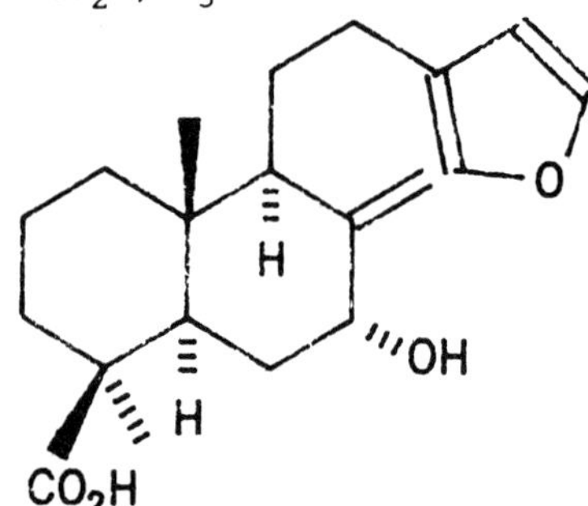

1152

158

159 R_1 = H, R_2 = CH_2OH, R_3 = CO_2H

160 R_1 = O2-Mebut, R_2 = CO_2H, R_3 = Me

161 R_1 = Oi-But, R_2 = CO_2H, R_3 = Me

162 R_1 = Oi-Val, R_2 = CO_2H, R_3 = Me

163

FIGURE 1. (contd.)

17. Grindelane

165 R = α-OH

1036 R = β-OH

166

167 $R_1 = R_2 = R_3 = H$, $R_4 = Me$

168 $R_1 = OH$, $R_2 = R_3 = H$, $R_4 = Me$

169 $R_1 = H$, $R_2 = \beta$-OH, $R_3 = H$, $R_4 = Me$

170 $R_1 = R_2 = H$, $R_3 = \alpha$-OH, $R_4 = Me$

171 $R_1 = R_2 = H$, $R_3 = \alpha$-OCHO, $R_4 = Me$

172 $R_1 = R_2 = H$, $R_3 = \beta$-OH, $R_4 = Me$

173 $R_1 = R_2 = R_3 = H$, $R_4 = CH_2OH$

FIGURE 1. (*contd.*)

174 $R_1 = R_2 = R_3 = H$, $R_4 = CH_2OAc$

175 $R_1 = R_2 = R_3 = H$, $R_4 = CH_2OPro$

176 $R_1 = R_2 = R_3 = H$, $R_4 = CH_2Oi\text{-}But$

177 $R_1 = R_2 = R_3 = H$, $R_4 = CH_2Oi\text{-}Val$

178 $R_1 = R_2 = R_3 = H$, $R_4 = CH_2O2\text{-}Mebut$

179 $R_1 = R_2 = R_3 = H$, $R_4 = CH_2OMe$

1035 $R_1 = H$, $R_2 = \alpha\text{-}OH$, $R_3 = H$, $R_4 = Me$

180 $R_1 = Me$, $R_2 = CH_2OH$, $R_3 = H$

181 $R_1 = Me$, $R_2 = CH_2OAc$, $R_3 = H$

182 $R_1 = Me$, $R_2 = CH_2Oi\text{-}Val$, $R_3 = H$

183 $R_1 = Me$, $R_2 = CH_2Oi\text{-}But$, $R_3 = H$

184 $R_1 = Me$, $R_2 = CH_2O2\text{-}Mebut$, $R_3 = H$

185 $R_1 = Me$, $R_2 = CH_2OSucc$, $R_3 = H$

186 $R_1 = Me$, $R_2 = CHO$, $R_3 = H$

187 $R_1 = Me$, $R_2 = CO_2H$, $R_3 = H$

188 $R_1 = CH_2OH$, $R_2 = Me$, $R_3 = H$

1153 $R_1 = Me$, $R_2 = CO_2H$, $R_3 = Me$

189 $R = Me$

190 $R = CH_2OAc$

FIGURE 1. (*contd.*)

191

192

193

194

FIGURE 1. *(contd.)*

955

1060

956

1187

FIGURE 1. (*contd.*)

18. Normal-Labdane-8,13-Epoxide

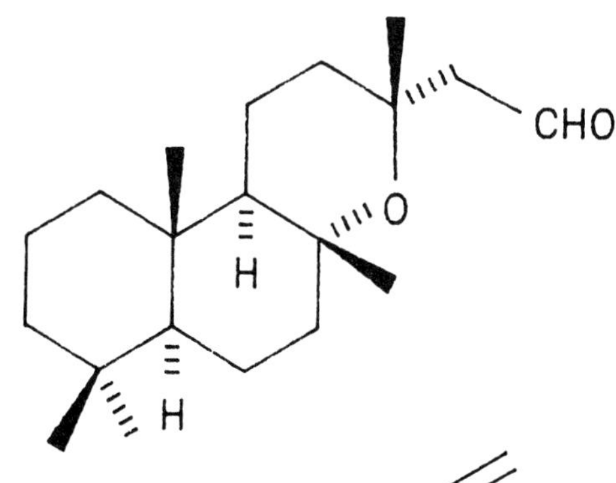

272 $R_1 = R_2 = H$, $R_3 = R_4 = Me$

195 $R_1 = R_2 = H$, $R_3 = Me$, $R_4 = CH_2OH$

196 $R_1 = R_2 = H$, $R_3 = Me$, $R_4 = CH_2OAc$

197 $R_1 = OH$, $R_2 = H$, $R_3 = Me$, $R_4 = CH_2OH$

198 $R_1 = OH$, $R_2 = H$, $R_3 = Me$, $R_4 = CH_2OAc$

199 $R_1 = OAc$, $R_2 = H$, $R_3 = Me$, $R_4 = CH_2OAc$

200 $R_1 = H$, $R_2 = OAc$, $R_3 = R_4 = Me$

201 $R_1 = H$, $R_2 = OH$, $R_3 = CH_2OBenz$, $R_4 = Me$

964

273 R = Me

202 R = CH_2OH

FIGURE 1. (*contd.*)

19. Normal-Labdane-8,12-Epoxide

203   R_1 = R_2 = CH_2OH

204   R_1 = CH_2OH, R_2 = CH_2OAc

205   R_1 = R_2 = CH_2OAc

206   R_1 = CH_2OH, R_2 = CHO

207   R_1 = R_2 = CHO

210   R_1 = CH_2OH, R_2 = Me

211   R_1 = CO_2H, R_2 = Me

212

213

FIGURE 1. (*contd.*)

208

209

20. Normal-Labdane-15,16-Lactone

164 R_1 = Me, R_2 = CO_2H

965 R_1 = CH_2OH, R_2 = Me

966 R_1 = CHO, R_2 = Me

FIGURE 1. (*contd.*)

21. Normal-Labdane-20,12-Lactone

214

22. Normal-Labdane Spiroketal

215 R_1 = Ang, R_2 = H

216 R_1 = Ang, R_2 = α-OH

217 R_1 = Ang, R_2 = β-OH

218

FIGURE 1. (*contd.*)

88. Normal-Labdane-15,12-Lactone

938

23. Normal-Labdane, 7,8-Seco-

219

Normal-Labdane, 3,4-Seco-

220

25. Grindelane, 8,9-Seco-

223

FIGURE 1. (*contd.*)

89. Grindelane, 19-Nor-

221 R = H

222 R = CHO

1154

26. Grindelane, 8,17-Bis-Nor-8,9-Seco-

224

FIGURE 1. (*contd.*)

27. Normal-Labdane-20,12-Lactone, 5,10-Seco-

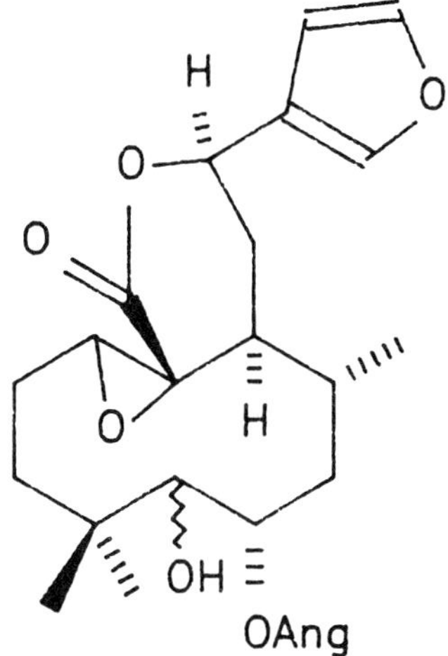

230 (Reaction product)

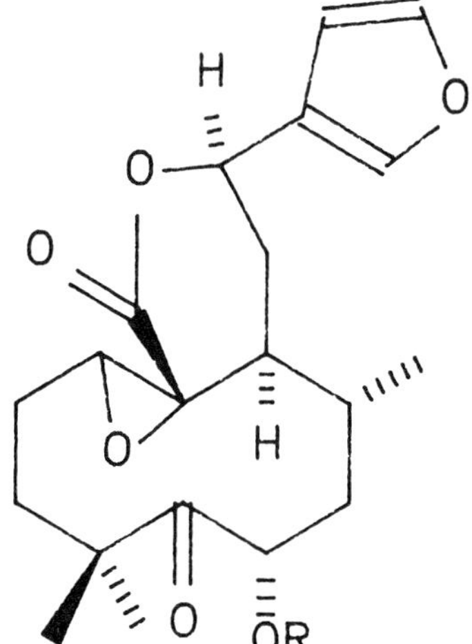

225 R = H

226 R = Ang

227 R = i-But

228 R = 2-Mebut

229 R = i-Val

FIGURE 1. (*contd.*)

100. Normal-Labdane, 14,15-Bisnor-

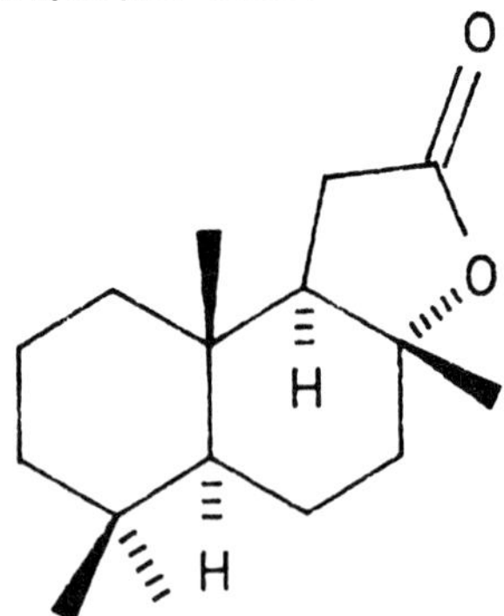

1155 R_1 = R_2 = H

1157 R_1 = R_2 = OH

1158 R_1 = OAc, R_2 = OH

1159 R_1 = OH, R_2 = OAc

1160 R_1 = H, R_2 = OH

101. Normal-Labdane, 13,14,15,16-Tetranor-

1156

87. Hexanorlabdane

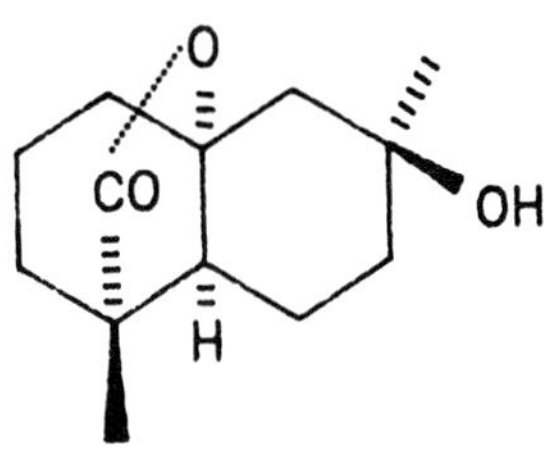

941

FIGURE 1. (*contd.*)

28. Relhaniane

231

232

233

234

FIGURE 1. (*contd.*)

235

29. <u>trans</u>- Normal-Clerodane

    (No proven structures of this skeletal type)

30. Clerodane, 5,10-Seco-

255   R = CO_2H (Revised to 9α-Me; = **573**)

256   R = CO_2Me (Revised to 9α-Me; = **573** methyl derivative)

31. <u>cis</u>-Normal-Clerodane

238 R_1 = Me, R_2 = CH_2OH (= **241**)

FIGURE 1. (contd.)

239 R_1 = Me, R_2 = CO_2H

240 R_1 = Me, R_2 = CHO (= **1062**)

241 R_1 = Me, R_2 = CH_2OH (= **1064**)

242 R_1 = R_2 = CH_2OH

243 R_1 = R_2 = CHO (= **1063**)

244 R_1 = CH_2OAng, R_2 = CO_2H

1065 R_1 = CH_2OH, R_2 = Me

253 R_1 = R_2 = H, R_3 = CH_2OH

258 R_1 = H, R_2 = OAng, R_3 = Me

259 R_1 = H, R_2 = OTig, R_3 = Me

263 R_1 = OH, R_2 = OAng, R_3 = Me

264 R_1 = OH, R_2 = OTig, R_3 = Me

545 R_1 = H, R_2 = OH, R_3 = Me

546 R_1 = H, R_2 = OAc, R_3 = Me

551 R_1 = OH, R_2 = H

552 R_1 = H, R_2 = OAng

FIGURE 1. (*contd.*)

2. Compound Names, Structures and Sources

254

257

260 R = β-OAng

553 R = β-OAc

988 R = β-OTig

556 R = H

261

FIGURE 1. (*contd.*)

262 R = H

265 R = OH

266 R = Ang

267 R = Tig

268 R = Ang

269 R = Tig

246

FIGURE 1. (*contd.*)

247

270

557

1025

FIGURE 1. (*contd.*)

1066

1067

1068 (= 247?)

1083 R_1 = OH, R_2 = CH_2OH, R_3 = Me

1084 R_1 = H, R_2 = R_3 = CH_2OH

1091 R_1 = H, R_2 = Me, R_3 = CO_2Arab

FIGURE 1. (*contd.*)

1085

1086

1087

1088

FIGURE 1. (*contd.*)

1089

1090

1092

90. Tricycloclerodane

947

Figure 1. *(contd.)*

72. Chrysolane

271

III. Normal Tricarbocyclics and Derivatives

32. Sandaracopimaranes (Normal-Pimaranes)

274 R_1 = R_2 = R_3 = H

927 R_1 = H, R_2 = OH, R_3 = H

928 R_1 = R_2 = OH, R_3 = H

929 R_1 = OAc, R_2 = OH, R_3 = H

930 R_1 = OH, R_2 = OAc, R_3 = H

931 R_1 = OH, R_2 = H, R_3 = OH

932 R_1 = OAc, R_2 = H, R_3 = OAc

933 R_1 = OH, R_2 = OAc, R_3 = OH

934 R_1 = OAc, R_2 = OH, R_3 = OAc

935 R_1 = R_2 = R_3 = OAc

FIGURE 1. (*contd.*)

936

937

275 R_1 = H, R_2 = Me

301 R_1 = H, R_2 = CO_2H

303 R_1 = OAc, R_2 = CO_2H

276 R_1 = R_2 = R_3 = H, R_4 = Me

FIGURE 1. (*contd.*)

277 R_1 = OH, R_2 = R_3 = H, R_4 = Me

278 R_1 = OAc, R_2 = R_3 = H, R_4 = Me

279 R_1 = H, R_2 = OH, R_3 = H, R_4 = Me

280 R_1 = H, R_2 = OSen, R_3 = H, R_4 = Me

281 R_1 = H, R_2 = OTig, R_3 = H, R_4 = Me

282 R_1 = H, R_2 = H, R_3 = OAc, R_4 = Me

284 R_1 = R_2 = H, R_3 = OAc, R_4 = Me

285 R_1 = R_2 = R_3 = H, R_4 = CH_2OH

288 R_1 = H, R_2 = R_3 = OH, R_4 = Me

289 R_1 = H, R_2 = OAc, R_3 = OH, R_4 = Me

290 R_1 = H, R_2 = OSen, R_3 = OH, R_4 = Me

291 R_1 = H, R_2 = OTig, R_3 = OH, R_4 = Me

292 R_1 = H, R_2 = OH, R_3 = OAc, R_4 = Me

293 R_1 = H, R_2 = OH, R_3 = OCoum, R_4 = Me

294 R_1 = H, R_2 = R_3 = OAc, R_4 = Me

298 R_1 = R_2 = OH, R_3 = H, R_4 = Me

299 R_1 = OAc, R_2 = OH, R_3 = H, R_4 = Me

283 R = H

296 R = OH

297 R = OAc

FIGURE 1. (*contd.*)

286

287

295

300 R = H

302 R = OAc

FIGURE 1. (*contd.*)

807

921

304 R_1 = OH, R_2 = H

305 R_1 = H, R_2 = OH

306 R_1 = OH, R_2 = H, R_3 = CH_2OH (Revised to rosane)

307 R_1 = R_2 = OH, R_3 = Me (Revised to rosane)

FIGURE 1. (*contd.*)

891

308

922 R_1 = β-OGlu, R_2 = Me (Probably <u>ent</u>-pimarane; = **593**?)

923 R_1 = α-OH, R_2 = CH$_2$OH (Probably <u>ent</u>-pimarane; = **944**?)

FIGURE 1. (*contd.*)

34. Cassane

310   R = H

311   R = OH

312

35. Rosane

313 R_1 = H, R_2 = Me

314 R_1 = H, R_2 = CH_2OH

1161 R_1 = Ac, R_2 = CH_2OAc

1162 R_1 = Ac, R_2 = CH_2OH

FIGURE 1. (*contd.*)

315

881 R_1 = H, R_2 = Me

882 R_1 = H, R_2 = CH_2OH

883 R_1 = H, R_2 = CH_2OAc

884 R_1 = R_2 = Me

885 R_1 = H, R_2 = CH_2OTig

886 R_1 = OH, R_2 = CH_2OH

887 R_1 = R_2 = H

888 R_1 = Ac, R_2 = H

889 R_1 = H, R_2 = Ac

FIGURE 1. (*contd.*)

890

36. Sandaracopimarane, 19-Nor-13-Epi

309

37. Normal-Abietane

316 R_1 = OAc, R_2 = H (**316-318** probable artifacts; dehydration

products ?)

317 R_1 = H, R_2 = OH

FIGURE 1. (contd.)

316 R$_1$ = OAc, R$_2$ = H (Possible dehydration precursor)

318

319

320

FIGURE 1. (*contd.*)

321

322

323

FIGURE 1. (*contd.*)

324

1072 R_1 = OH, R_2 = Sen

1073 R_1 = OSen, R_2 = H

1074 R_1 = R_2 = H

1163

FIGURE 1. *(contd.)*

94. Stevisalane

1026

IV. Normal-Tetracarbocyclics

73. Normal-Stachane

325 (Probable <u>ent</u>-stachane; = **774**?)

V. <u>ent</u>-Bicarbocyclics and Derivatives

38. <u>ent</u>-Labdane

326 R_1 = R_2 = R_3 = H, R_4 = CH$_2$OH, R_5 = Me

327 R_1 = R_2 = R_3 = H, R_4 = CH$_2$Obern, R_5 = Me

FIGURE 1. (*contd.*)

848 R_1 = OH, R_2 = R_3 = H, R_4 = CH_2OH, R_5 = Me

328 R_1 = R_2 = H, R_3 = OAc, R_4 = CH_2OH, R_5 = Me

844 R_1 = H, R_2 = OAng, R_3 = H, R_4 = R_5 = CH_2OH

329 R_1 = R_2 = R_3 = H, R_4 = CH_2OH, R_5 = CO_2H

330 R_1 = R_2 = R_3 = H, R_4 = CH_2OAc, R_5 = CO_2H

331 R_1 = OH, R_2 = R_3 = H, R_4 = CHO, R_5 = Me

335 R_1 = R_2 = R_3 = H, R_4 = CO_2H, R_5 = Me (= **1040**)

1014 R_1 = H, R_2 = OAc, R_3 = H, R_4 = CH_2OH, R_5 = CH_2OTig

1041 R_1 = R_2 = R_3 = H, R_4 = CO_2H, R_5 = CH_2OH

1164 R_1 = R_2 = R_3 = H, R_4 = R_5 = CO_2H

1165 R_1 = R_2 = R_3 = H

1166 R_1 = H, R_2 = Ac, R_3 = H

1167 R_1 = Ac, R_2 = R_3 = H

1168 R_1 = H, R_2 = Ac, R_3 = OH

1169

FIGURE 1. (*contd.*)

2. Compound Names, Structures and Sources

1170

1171

1172

849   R = CH₂OH

333   R = CHO

FIGURE 1. (*contd.*)

332 R_1 = OH, R_2 = R_3 = H, R_4 = CHO

336 R_1 = R_2 = R_3 = H, R_4 = CO_2H

337 R_1 = H, R_2 = OH, R_3 = H, R_4 = CO_2H

338 R_1 = H, R_2 = R_3 = OH, R_4 = CO_2H

1173

334

FIGURE 1. (*contd.*)

339 R_1 = OH, R_2 = H, R_3 = CHO

340 R_1 = OH, R_2 = CHO, R_3 = H

343 R_1 = OH, R_2 = CHO, R_3 = OH

344 R_1 = H, R_2 = H, R_3 = CO_2H

341 R_1 = H, R_2 = CHO

342 R_1 = CHO, R_2 = H

345

FIGURE 1. (*contd.*)

346

347

348 R = CH$_2$OH

349 R = CHO

1174

FIGURE 1. (*contd.*)

350

351 R_1 = H, R_2 = OH, R_3 = CH_2OH

355 R_1 = OAng, R_2 = H, R_3 = CO_2H (13,14 Z or E)

352

353

FIGURE 1. (*contd.*)

354

356

i-ValO

358

R_1

R_2

357 R_1 = OH, R_2 = CH_2OH

857 R_1 = OH, R_2 = CO_2H

FIGURE 1. (*contd.*)

858 R_1 = OAng, R_2 = CO_2H

859 R_1 = OTig, R_2 = CO_2H

860 R_1 = O2-Mebut, R_2 = CO_2H

359 R = CH_2OH

861 R = CO_2H

862 R = OAng

863 R = O2-Mebut

360 R_1 = H, R_2 = CH_2OMe, R_3 = Me

361 R_1 = β-OAng, R_2 = R_3 = Me

FIGURE 1. (*contd.*)

1046 R_1 = H, R_2 = CH_2OH, R_3 = Me

1047 R_1 = H, R_2 = CH_2OAc, R_3 = Me

1048 R_1 = H, R_2 = CHO, R_3 = Me

1049 R_1 = α-OH, R_2 = CH_2OH, R_3 = Me

1050 R_1 = α-OH, R_2 = CH_2OAc, R_3 = Me

1051 R_1 = α-OH, R_2 = CHO, R_3 = Me

1052 R_1 = H, R_2 = CO_2H, R_3 = Me

1053 R_1 = H, R_2 = CH_2OAc, R_3 = CH_2OH

1179 R_1 = OH, R_2 = CH_2OMe, R_3 = Me

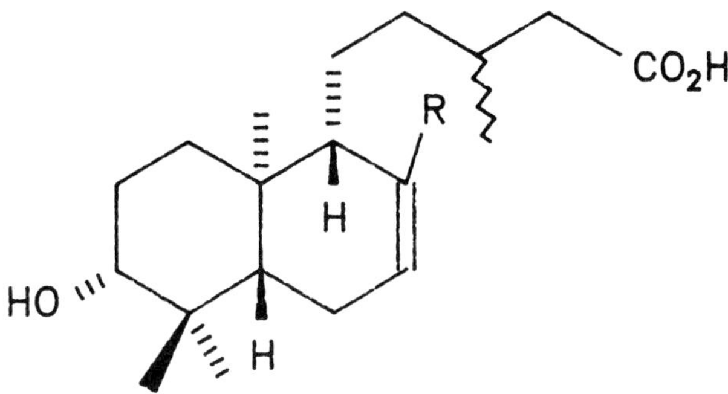

1175 R = CH_2OH

1176 R = CH_2OAc

1177 R = CHO

1178 R = CH_2OMe

362 R_1 = R_2 = R_3 = CH_2OH

363 R_1 = R_2 = CH_2OAc, R_3 = Me

FIGURE 1. (*contd.*)

364 $R_1 = R_2 = R_3 = CH_2OAc$

366 $R_1 = CO_2H$, $R_2 = CH_2OH$, $R_3 = Me$

367 $R_1 = CO_2H$, $R_2 = R_3 = CH_2OH$

369 $R_1 = CO_2H$, $R_2 = CH_2OH$, $R_3 = CHO$

370 $R_1 = CO_2H$, $R_2 = CH_2OAc$, $R_3 = CH_2OH$

1180 $R_1 = CO_2H$, $R_2 = CH_2OH$, $R_3 = CH_2OMe$

365

368

FIGURE 1. (contd.)

371 R_1 = H, R_2 = CH$_2$OAc

372 R_1 = α-OH, R_2 = CH$_2$OAc

373 R_1 = β-OH, R_2 = CH$_2$OAc

374 R = CH$_2$OAc

375 R = CHO

376 R = CH$_2$OH

377 R = CH$_2$OAc

378 R = CH$_2$OMe

FIGURE 1. (*contd.*)

379   R_1 = OH, R_2 = H, R_3 = Me

380   R_1 = H, R_2 = Ac, R_3 = CHO

381   R_1 = H, R_2 = Ac, R_3 = CO_2H

382

383

FIGURE 1. (contd.)

384

385

1181 R_1 = R_2 = H, R_3 = OH

1182 R_1 = H, R_2 = OAng, R_3 = H

1183 R_1 = Glu, R_2 = OAng, R_3 = H

1184 R_1 = H, R_2 = OAng, R_3 = OH

FIGURE 1. (*contd.*)

386 $R_1 = R_2 = H$

387 $R_1 = OH$, $R_2 = H$

388 $R_1 = H$, $R_2 = \alpha\text{-}OH$

389 $R_1 = H$, $R_2 = \beta\text{-}OH$

390 $R_1 = R_2 = OH$

391 $R_1 = OH$, $R_2 = \alpha\text{-}OH$

392 $R_1 = OH$, $R_2 = \beta\text{-}OH$

393 $R = H$

394 $R = OH$

395 $R = \alpha\text{-}OH$

396 $R = \beta\text{-}OH$

FIGURE 1. (*contd.*)

397 R_1 = OH, R_2 = Me

843 R_1 = H, R_2 = CO_2H

398 R_1 = R_2 = H

399 R_1 = OH, R_2 = β-OH

400 R_1 = OH, R_2 = α-OH

401 R_1 = R_2 = H, R_3 = OH, R_4 = R_5 = Me

402 R_1 = R_2 = H, R_3 = OAc, R_4 = R_5 = Me

403 R_1 = R_2 = R_3 = H, R_4 = Me, R_5 = CH_2OH

FIGURE 1. (*contd.*)

404 R_1 = H, R_2 = OH, R_3 = H, R_4 = Me, R_5 = CH_2OH

839 R_1 = R_2 = R_3 = H, R_4 = Me, R_5 = CO_2H

840 R_1 = OH, R_2 = R_3 = H, R_4 = Me, R_5 = CO_2H

841 R_1 = R_2 = R_3 = H, R_4 = R_5 = CH_2OH

842 R_1 = OH, R_2 = R_3 = H, R_4 = CH_2OH, R_5 = Me

974 R_1 = OAng, R_2 = R_3 = H, R_4 = CH_2OH, R_5 = Me

975 R_1 = OH, R_2 = R_3 = H, R_4 = CH_2OAng, R_5 = Me

830 R_1 = R_2 = R_3 = H, R_4 = CO_2H, R_5 = Me

831 R_1 = OH, R_2 = R_3 = H, R_4 = CO_2H, R_5 = Me

832 R_1 = OAng, R_2 = R_3 = H, R_4 = CO_2H, R_5 = Me

833 R_1 = O2-Mebut, R_2 = R_3 = H, R_4 = CO_2H, R_5 = Me

834 R_1 = Oi-But, R_2 = R_3 = H, R_4 = CO_2H, R_5 = Me

835 R_1 = R_2 = R_3 = H, R_4 = CO_2H, R_5 = CH_2OH

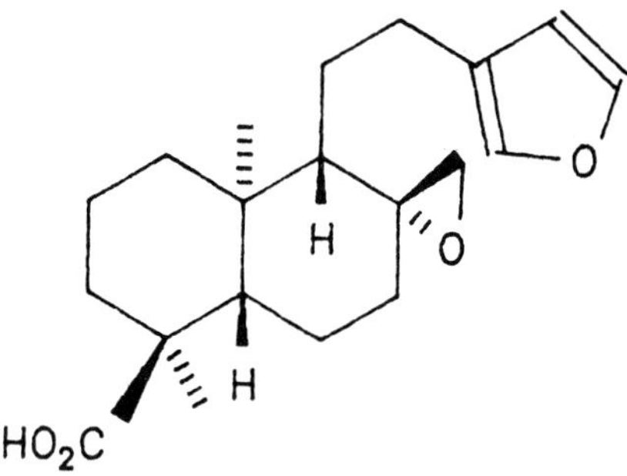

836

837

FIGURE 1. (*contd.*)

838

405

406 R = H

413 R = OH

407 R = H

412 R = OH

FIGURE 1. (*contd.*)

408 R_1 = OAc, R_2 = CH_2OH

409 R_1 = OAc, R_2 = CH_2OAc

410 R_1 = OH, R_2 = CHO (Saponification product)

411 R_1 = OAc, R_2 = CHO

414 R_1 = CH_2OH, R_2 = Me

415 R_1 = Me, R_2 = CO_2H

416

FIGURE 1. (*contd.*)

417 R = α-OH

418 R = β-OH

421

422 R_1 = H, R_2 = Me

423 R_1 = H, R_2 = CHO

424 R_1 = H, R_2 = CO_2H

426 R_1 = OH, R_2 = CO_2H

FIGURE 1. (*contd.*)

425

427

428 $R_1 = R_2 = R_3 = OH$

429 $R_1 = OAc$, $R_2 = R_3 = OH$

808 $R_1 = OAng$, $R_2 = OAc$, $R_3 = \beta$-OH

FIGURE 1. (*contd.*)

430 $R_1 = R_2 = H$, $R_3 = R_4 = Me$

431 $R_1 = R_2 = H$, $R_3 = Me$, $R_4 = CH_2OH$

432 $R_1 = H$, $R_2 = OH$, $R_3 = Me$, $R_4 = CH_2OH$

433 $R_1 = R_2 = H$, $R_3 = CH_2OH$, $R_4 = Me$

434 $R_1 = R_2 = H$, $R_3 = R_4 = CH_2OH$

435 $R_1 = OAc$, $R_2 = H$, $R_3 = CO_2H$, $R_4 = Me$

1185 $R_1 = R_2 = H$, $R_3 = R_4 = CH_2OAc$

436 $R_1 = OH$, $R_2 = R_3 = H$, $R_4 = Me$

809 $R_1 = H$, $R_2 = R_3 = OH$, $R_4 = Me$

810 $R_1 = H$, $R_2 = OAc$, $R_3 = OH$, $R_4 = Me$

811 $R_1 = H$, $R_2 = OAng$, $R_3 = OAc$, $R_4 = Me$

1015 $R_1 = R_2 = R_3 = H$, $R_4 = CH_2OH$

1186

FIGURE 1. (*contd.*)

437

438

439

440

FIGURE 1. *(contd.)*

442   R = H

443   R = β-OH

444   R = α-OH

445   R = β-OSucc

446

447   R = H

448   R = OH (Revised to normal-labdane; = **1106**?)

FIGURE 1. (*contd.*)

2. Compound Names, Structures and Sources

449 (Revised to normal-labdane; = 1107?)

450

460

461

FIGURE 1. (*contd.*)

1012 R = OAc

1013 R = OH

1016

1017 R = CH$_2$OH

1018 R = CHO

1019 R = CO$_2$H

FIGURE 1. (*contd.*)

1037 R_1 = H, R_2 = OAng

1038 R_1 = OH, R_2 = OAng

1188

1039

1189 R_1 = CO_2H, R_2 = H, R_3 = OH, R_4 = H

FIGURE 1. (*contd.*)

1190 R_1 = CO_2H, R_2 = H, R_3 = OAng, R_4 = H

1191 R_1 = CH_2OH, R_2 = OH, R_3 = OAng, R_4 = OH

1192 R = α-OAng

1193 R = β-OH

1194 R = H

1195 R = OH

1196 R = CO_2H

1197 R = CH_2OH

FIGURE 1. (*contd.*)

1042 R$_1$ = OH, R$_2$ = CHO

1043 R$_1$ = O-i-Val, R$_2$ = CO$_2$H

1044

1045

39. <u>ent</u>-Labdane-19,6-Lactone

462

FIGURE 1. (*contd.*)

40. <u>ent</u>-Labdane-8,13-Epoxide-15,16-Hemiacetal

463 R = H

464 R = Ac

97. <u>ent</u>-Labdane Dimer

419 R = α-H

420 R = β-H

FIGURE 1. (*contd.*)

98. <u>ent</u>-Manoyl Oxide, 13-Epi-

451   R$_1$ = H, R$_2$ = Me

452   R$_1$ = OH, R$_2$ = Me

455   R$_1$ = H, R$_2$ = CH$_2$OH

1198 R$_1$ = OAc, R$_2$ = Me

1199 R$_1$ = OH, R$_2$ = CH$_2$OCinn

1200 R$_1$ = OH, R$_2$ = CH$_2$OH

453

454   R = Me

456   R = CH$_2$OH

FIGURE 1. (*contd.*)

457

99. <u>ent</u>-Manoyl Oxide

458   R = H

459   R = OH

41. <u>ent</u>-Labdane, 17-Nor-

465

FIGURE 1. (*contd.*)

466 R_1 = H, R_2 = CH_2OAc

467 R_1 = OH, R_2 = Me

1201 R_1 = OH, R_2 = Me (6,7-dihydro)

468

1202 (6,7-dihydro)

42. _ent_-Labdane, 15-Nor-

469

FIGURE 1. (*contd.*)

43. <u>ent</u>-**Labdane, 14,15-Bis-Nor-**

470

1203

44. <u>ent</u>-**Labdane, 13,14,15,16-Tetra-Nor-**

471 R_1 = CHO, R_2 = Me

472 R_1 = CO_2H, R_2 = CH_2OH

FIGURE 1. *(contd.)*

45. _ent_-Labdane, 7,8-Seco-

473

74. _ent_-Labdane, 18-Nor-

828 (Structure revised; = 1206)

1205 R = Ac

1206 R = H

829 R = H

1207 R = Ac

FIGURE 1. (contd.)

95. <u>ent</u>-Friedolabdane

1059

1208 (Illustrated in report as 13,14-dihydro-homolog of **1059**,

but text indicates that **1208** is identical to **1059**.)

1209 X = O

1210 X = α-OH,H

1211

FIGURE 1. (*contd.*)

1212

1213

1214

102. _ent_-Friedolabdane, 3,4-seco-

1215 $R_1 = R_2 = H$

FIGURE 1. (*contd.*)

1216 R_1 = OH, R_2 = H

1217 R_1 = R_2 = OH

1218 R_1 = OMe, R_2 = OH

1219

1220

103. <u>ent</u>-Friedolabdane, rearranged-

1221

FIGURE 1. (*contd.*)

46. <u>trans</u>-<u>ent</u>-Clerodane

1222

248 R_1 = CO_2H, R_2 = Me (= **512** = **250**?)

251 R_1 = CO_2Me, R_2 = Me

911 R_1 = CO_2H, R_2 = CH_2OAc

1070 R_1 = CO_2H, R_2 = CH_2OH

475 R_1 = R_2 = H, R_3 = CH_2OH, R_4 = Me

476 R_1 = R_2 = H, R_3 = R_4 = CH_2OH

1010 R_1 = R_2 = H, R_3 = CH_2OArac, R_4 = CH_2OH

1011 R_1 = R_2 = H, R_3 = $CH_2OBehen$, R_4 = CH_2OH

FIGURE 1. (*contd.*)

495 $R_1 = R_2 = H$, $R_3 = CO_2Me$, $R_4 = Me$

496 $R_1 = \beta\text{-OAc}$, $R_2 = H$, $R_3 = CO_2Me$, $R_4 = Me$

497 $R_1 = \beta\text{-OAng}$, $R_2 = H$, $R_3 = CO_2Me$, $R_4 = Me$

498 $R_1 = \alpha\text{-OAng}$, $R_2 = H$, $R_3 = CO_2H$, $R_4 = Me$

499 $R_1 = \alpha\text{-OTig}$, $R_2 = H$, $R_3 = CO_2H$, $R_4 = Me$

500 $R_1 = H$, $R_2 = OAc$, $R_3 = CO_2H$, $R_4 = Me$

535 $R_1 = R_2 = H$, $R_3 = CHO$, $R_4 = CH_2OH$

537 $R_1 = R_2 = H$, $R_3 = CHO$, $R_4 = CO_2H$

1223

1224 $R_1 = R_2 = H$

1225 $R_1 = Ac$, $R_2 = OH$

1226

1227

236

477 R_1 = H, R_2 = Me

478 R_1 = H, R_2 = CH$_2$OH

1228 R_1 = Ac, R_2 = CH$_2$OH

479 R = H

480 R = OAng

FIGURE 1. (*contd.*)

481 R_1 = H, R_2 = R_3 = CH_2OH, R_4 = Me

482 R_1 = H, R_2 = CH_2OArac, R_3 = CH_2OH, R_4 = Me

483 R_1 = H, R_2 = $CH_2OBehen$, R_3 = CH_2OH, R_4 = Me

484 R_1 = H, R_2 =CH_2OH, R_3 = Me, R_4 = CH_2OSucc

486 R_1 = H, R_2 = CH_2OH, R_3 = CHO, R_4 = Me

488 R_1 = OH, R_2 = CH_2OH, R_3 = CH_2OAc, R_4 = CH_2OH

489 R_1 = OH, R_2 = CH_2OH, R_3 = $CH_2OPhe-Ac$, R_4 = CH_2OH

490 R_1 = OH, R_2 = CH_2OH, R_3 = $CH_2OPhe-Ac$, R_4 = CHO

491 R_1 = OH, R_2 = CH_2OH, R_3 = CH_2OAc, R_4 = CHO

492 R_1 = H, R_2 = CHO, R_3 = CH_2OH, R_4 = Me

503 R_1 = H, R_2 = CO_2H, R_3 = R_4 = Me

504 R_1 = H, R_2 = CO_2H, R_3 = Me, R_4 = CH_2OAc

1229 R = H

1230 R = OAc

FIGURE 1. (*contd.*)

485

487 R_1 = CH_2OH, R_2 = Ac

522 R_1 = CO_2H, R_2 = Phe-Ac

493 R_1 = H, R_2 = CO_2H, R_3 = Me

494 R_1 = OAc, R_2 = CO_2H, R_3 = Me

534 R_1 = H, R_2 = CHO, R_3 = CH_2OH

536 R_1 = H, R_2 = CHO, R_3 = CO_2H

FIGURE 1. (*contd.*)

FIGURE 1. *(contd.)*

506

507 R_1 = H, R_2 = R_3 = R_4 = Me

508 R_1 = H, R_2 = Me, R_3 = CH_2OMal, R_4 = Me

509 R_1 = OH, R_2 = Me, R_3 = CH_2OH, R_4 = Me

510 R_1 = H, R_2 = Me, R_3 = CH_2OMal, R_4 = CH_2OAc

846 R_1 = H, R_2 = Me, R_3 = R_4 = CH_2OH

511 R_1 = H, R_2 = Me, R_3 = CO_2H, R_4 = Me (= **245**? = **249**?)

967 R_1 = H, R_2 = CHO, R_3 = R_4 = Me

518 R_1 = H, R_2 = Me, R_3 = CO_2H, R_4 = CH_2OH (=**1033**)

900 R_1 = H, R_2 = Me, R_3 = CO_2H, R_4 = CH_2OAc

901 R_1 = H, R_2 = Me, R_3 = CO_2H, R_4 = CH_2OMe

902 R_1 = H, R_2 = Me, R_3 = CO_2H, R_4 = CH_2OAng

903 R_1 = H, R_2 = Me, R_3 = CO_2H, R_4 = $CH_2Oi\text{-}Val$

904 R_1 = H, R_2 = Me, R_3 = CO_2H, R_4 = CHO

1231 R_1 = H, R_2 = Me, R_3 = CH_2OH, R_4 = CH_2OMal

1232 R_1 = H, R_2 = Me, R_3 = CH_2OH, R_4 = Me

1233 R_1 = H, R_2 = Me, R_3 = R_4 = CH_2OAc

FIGURE 1. (*contd.*)

953 R$_1$ = H, R$_2$ = O2-Mebut, R$_3$ = CO$_2$H, R$_4$ = Me

513 R$_1$ = α-OH, R$_2$ = H, R$_3$ = CO$_2$H, R$_4$ = Me

519 R$_1$ = β-OH, R$_2$ = H, R$_3$ = CO$_2$H, R$_4$ = CH$_2$OH

1034 R$_1$ = β-OH, R$_2$ = H, R$_3$ = CO$_2$H, R$_4$ = CH$_2$OH (=519)

514 R = CH$_2$Oi-Val

515 R = CH$_2$O2-Mebut

516 R = CH$_2$OAng

524

FIGURE 1. (*contd.*)

520 R$_1$ = R$_2$ = R$_3$ = H

521 R$_1$ = OH, R$_2$ = R$_3$ = H

950 R$_1$ = R$_2$ = H, R$_3$ = OH

951 R$_1$ = H, R$_2$ = R$_3$ = OH

1234 R = H

1235 R = OH

523

FIGURE 1. (*contd.*)

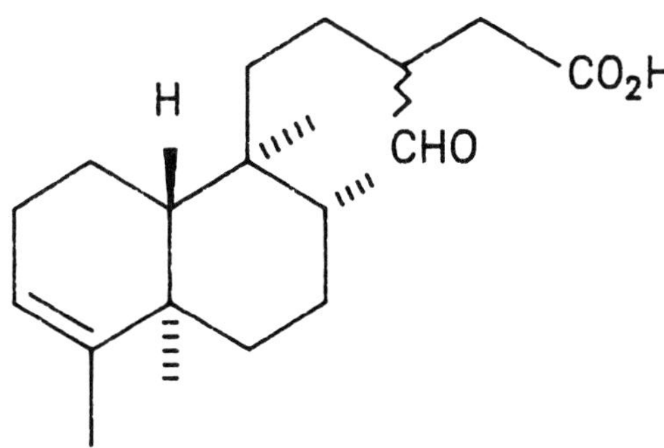

865

538 R_1 = H, R_2 = CH_2OH

539 R_1 = Ac, R_2 = CH_2OH

540 R_1 = H, R_2 = CHO

541 R_1 = Ac, R_2 = CHO

542

FIGURE 1. (*contd.*)

543 $R_1 = R_2 = H$, $R_3 = R_4 = Me$

544 $R_1 = OH$, $R_2 = H$, $R_3 = R_4 = Me$

864 $R_1 = H$, $R_2 = OH$, $R_3 = R_4 = Me$

550 $R_1 = R_2 = H$, $R_3 = CH_2OH$, $R_4 = Me$

558 $R_1 = R_2 = H$, $R_3 = CO_2H$, $R_4 = Me$

909 $R_1 = H$, $R_2 = \alpha\text{-}OH$, $R_3 = CO_2H$, $R_4 = CH_2OH$

910 $R_1 = H$, $R_2 = \beta\text{-}OH$, $R_3 = CO_2H$, $R_4 = CH_2OH$

1236 $R_1 = H$, $R_2 = OMe$, $R_3 = CH_2OAc$, $R_4 = Me$

559 $R_1 = OH$, $R_2 = H$, $R_3 = H$

560 $R_1 = H$, $R_2 = OH$, $R_3 = H$

1237 $R_1 = R_2 = H$, $R_3 = OH$

1238 $R_1 = R_2 = H$, $R_3 = OAc$

FIGURE 1. (*contd.*)

561

562 R = OH

563 R = H

564 R_1 = H, R_2 = Me

565 R_1 = OH, R_2 = Me

963 R_1 = OMe, R_2 = CH$_2$OAc

FIGURE 1. (*contd.*)

566 R_1 = H, R_2 = R_3 = Me, R_4 = CHO

905 R_1 = α-OH, R_2 = CO_2H, R_3 = CH_2OH, R_4 = Me

906 R_1 = β-OH, R_2 = CO_2H, R_3 = CH_2OH, R_4 = Me

1240 R_1 = H, R_2 = CH_2OAc, R_3 = R_4 = Me

907

908

567

FIGURE 1. (*contd.*)

568

569

1239 R_1 = OAc, R_2 = Me

1241 R_1 = OH, R_2 = Me

1242 R_1 = OH, R_2 = CH$_2$OH

1243 R_1 = OAc, R_2 = CH$_2$OAc

1027 R_1 = i-But, R_2 = Ac

FIGURE 1. (*contd.*)

1028 R_1 = Ang, R_2 = Ac

1029 R_1 = R_2 = Ac

1030

1031

1033

1071 R = CO_2H

1076 R = Me

FIGURE 1. (*contd.*)

1075

1077

1078

1079

FIGURE 1. (*contd.*)

2. Compound Names, Structures and Sources

1080

1081

1082

570

FIGURE 1. (*contd.*)

525

526

527 R$_1$ = R$_2$ = H

1069 R$_1$ = OH, R$_2$ = H

FIGURE 1. (*contd.*)

1257 R_1 = OAc, R_2 = H
1258 R_1 = H, R_2 = OH

528

969

970

FIGURE 1. (*contd.*)

971

972

973

FIGURE 1. (*contd.*)

529 $R_1 = R_2 = H$

530 $R_1 = H, R_2 = OAng$

968 $R_1 = OAng, R_2 = H$

532 $R_1 = OSen, R_2 = H$

533 $R_1 = O2\text{-}Mebut, R_2 = H$

531 $R_1 = Ang, R_2 = Sen$

trans-*ent*-Clerodane Dimer

1244 R = Mal

1245 R = Succ

FIGURE 1. (*contd.*)

47. Printziane

571

572

913 R = H

914 R = Me

915

FIGURE 1. (*contd.*)

916

76. Conyscabrane

917

919

918

FIGURE 1. (*contd.*)

920

48. **<u>ent</u>-Clerodane, 5,10-Seco-**

573 $R_1 = R_2 = H$

912 $R_1 = H, R_2 = OAc$

952 $R_1 = O2\text{-Mebut}, R_2 = H$

985

986

FIGURE 1. (*contd.*)

CO_2H

987

96. <u>trans</u>-<u>ent</u>-Clerodane, 17-Nor-

1061

49. <u>trans</u>-<u>ent</u>-Clerodane, 13,14,15,16-Tetra-Nor-

574 $R_1 = R_2 = H$, $R_3 = R_4 = CH_2OH$

575 $R_1 = R_2 = H$, $R_3 = CO_2H$, $R_4 = CH_2OH$

1032 $R_1 = R_2 = OH$, $R_3 = CO_2H$, $R_4 = Me$

FIGURE 1. (*contd.*)

1093

Undetermined

237

50. <u>cis</u>-<u>ent</u>-Clerodane

578 R_1 = H, R_2 = Me

579 R_1 = H, R_2 = CH$_2$OH

580 R_1 = H, R_2 = CH$_2$OAc

581 R_1 = OH, R_2 = CH$_2$OH

582 R_1 = OH, R_2 = CH$_2$OAc

FIGURE 1. (*contd.*)

576 R = CO_2H

577 R = CO_2Me

583

584 R = H

585 R = OH

1055 R = CH_2OAc

1056 R = CH_2OH

FIGURE 1. (*contd.*)

1057 R_1 = CH_2OH, R_2 = Me

1058 R_1 = Me, R_2 = CH_2OH

1246

82. Solidagonane

960

FIGURE 1. (*contd.*)

VI. _ent_-Tricarbocyclics and Derivatives

51. _ent_-Labdane, Cyclopropyl-

586

77. Gutierreziane

845

52. _ent_-Pimarane

587 R_1 = R_2 = H

588 R_1 = OH, R_2 = OAc

FIGURE 1. (_contd._)

589 R = OH

595 R = β-OH

590 R$_1$ = H, R$_2$ = Me, R$_3$ = CO$_2$H

1247 R$_1$ = OH, R$_2$ = Me, R$_3$ = CH$_2$OH

1248 R$_1$ = OH, R$_2$ = CH$_2$OH, R$_3$ = Me

820

FIGURE 1. (*contd.*)

821   R = CO_2H

1249 R = CH_2OH

823 R_1 = OAc, R_2 = CO_2H

1250 R_1 = H, R_2 = CH_2OH

599 R_1 = CH_2OH, R_2 = Me, R_3 = H

600 R_1 = CO_2H, R_2 = Me, R_3 = H

1251 R_1 = Me, R_2 = CH_2OH, R_3 = H

1252 R_1 = Me, R_2 = CH_2OMal, R_3 = H

1253 R_1 = CH_2OMal, R_2 = Me, R_3 = H

1254 R_1 = CH_2OH, R_2 = Me, R_3 = OH

FIGURE 1. (*contd.*)

591 R_1 = OH, R_2 = H, R_3 = R_4 = Me

592 R_1 = OGlu, R_2 = H, R_3 = R_4 = Me

593 R_1 = H, R_2 = OGlu, R_3 = R_4 = Me

594 R_1 = H, R_2 = OH, R_3 = R_4 = Me

944 R_1 = H, R_2 = OH, R_3 = CH_2OH, R_4 = Me

53. <u>ent</u>-Pimarane-8,15-Tetrahydrofuran

596 R_1 = H, R_2 = CH_2OH

597 R_1 = CH_2OH, R_2 = H

598

FIGURE 1. (*contd.*)

824

54. Cleistanthane

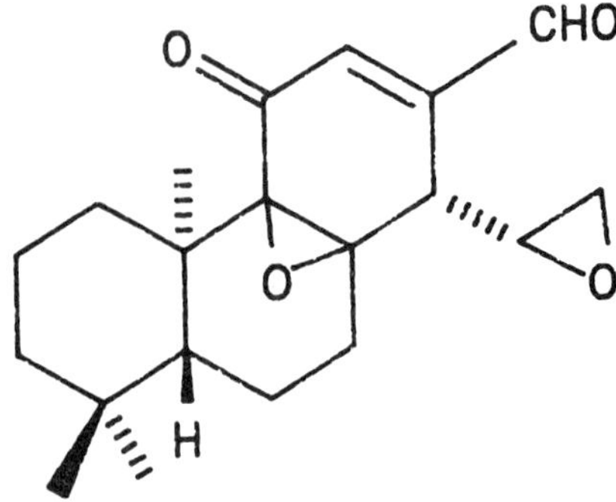

601 R_1 = H, R_2 = H, R_3 = Me

602 R_1 = H, R_2 = H, R_3 = CHO

603 R_1 = H, R_2 = H, R_3 = CH_2OAc

604 R_1 = OAng, R_2 = H, R_3 = Me

605 R_1 = H, R_2 = OH, R_3 = CH_2OAc

606 R_1 = OAng, R_2 = H, R_3 = CH_2OAc

607

FIGURE 1. (*contd.*)

608

85. Cleistanthane, Iso-

609

610

55. _ent_-Abietane

611

FIGURE 1. (*contd.*)

612 R_1 = H, R_2 = OH

613 R_1 = OAc, R_2 = H

614 R_1 = H, R_2 = R_3 = Me

615 R_1 = OAc, R_2 = R_3 = Me

616 R_1 = H, R_2 = Me, R_3 = CO_2H

1054 R_1 = H, R_2 = Me, R_3 = CH_2OSucc

56. Acritoconfertane

617

FIGURE 1. (*contd.*)

57. Acritoconfertane-Acetal

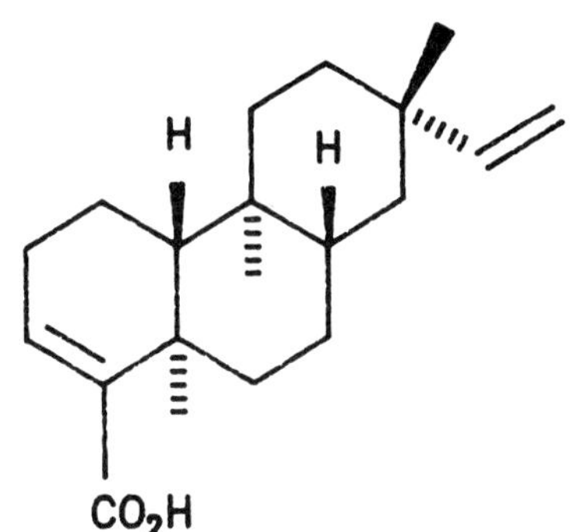

1100

83. Erythroxane

976

VII. _ent_-Tetracarbocyclics and Derivatives

58. _ent_-Kaurane

618    $R_1 = R_2 = H$, $R_3 = R_4 = Me$

619    $R_1 = R_2 = H$, $R_3 = Me$, $R_4 = CH_2OH$

620    $R_1 = R_2 = H$, $R_3 = Me$, $R_4 = CH_2OAc$

621    $R_1 = R_2 = H$, $R_3 = Me$, $R_4 = CH_2OMeform$

FIGURE 1. (*contd.*)

622 $R_1 = R_2 = H$, $R_3 = Me$, $R_4 = CH_2O$-p-OH-H-Cinn

623 $R_1 = OH$, $R_2 = H$, $R_3 = Me$, $R_4 = CH_2OH$

624 $R_1 = R_2 = H$, $R_3 = Me$, $R_4 = CHO$

625 $R_1 = R_2 = H$, $R_3 = Me$, $R_4 = CO_2H$

626 $R_1 = R_2 = H$, $R_3 = CH_2OAc$, $R_4 = Me$

635 $R_1 = R_2 = H$, $R_3 = CHO$, $R_4 = Me$

926 $R_1 = R_2 = H$, $R_3 = CO_2H$, $R_4 = Me$

939 $R_1 = R_2 = H$, $R_3 = CO_2H$, $R_4 = CH_2OAng$

627 $R_1 = R_2 = H$, $R_3 = Me$, $R_4 = CO_2Gly$

628 $R_1 = OH$, $R_2 = H$, $R_3 = Me$, $R_4 = CO_2H$

629 $R_1 = OAc$, $R_2 = H$, $R_3 = Me$, $R_4 = CO_2H$

630 $R_1 = OAng$, $R_2 = H$, $R_3 = Me$, $R_4 = CO_2H$

631 $R_1 = OTig$, $R_2 = H$, $R_3 = Me$, $R_4 = CO_2H$

632 $R_1 = OCinn$, $R_2 = H$, $R_3 = Me$, $R_4 = CO_2H$

633 $R_1 = OSen$, $R_2 = H$, $R_3 = Me$, $R_4 = CO_2H$

634 $R_1 = Oi$-Val, $R_2 = H$, $R_3 = Me$, $R_4 = CO_2H$

643 $R_1 = R_2 = H$, $R_3 = CH_2OAng$, $R_4 = CO_2H$

644 $R_1 = R_2 = H$, $R_3 = CH_2OSen$, $R_4 = CO_2H$

645 $R_1 = R_2 = H$, $R_3 = CH_2Oi$-Val, $R_4 = CO_2H$

1000 $R_1 = H$, $R_2 = OH$, $R_3 = Me$, $R_4 = CO_2H$

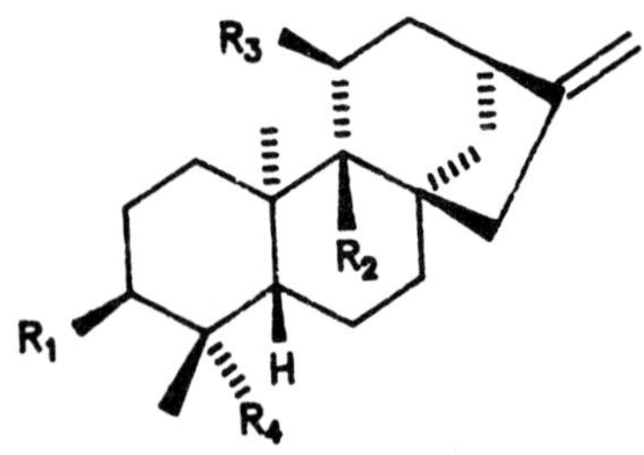

639 $R_1 = H$, $R_2 = OH$, $R_3 = H$, $R_4 = CO_2Me$

640 $R_1 = R_2 = H$, $R_3 = OAc$, $R_4 = CO_2H$

FIGURE 1. (*contd.*)

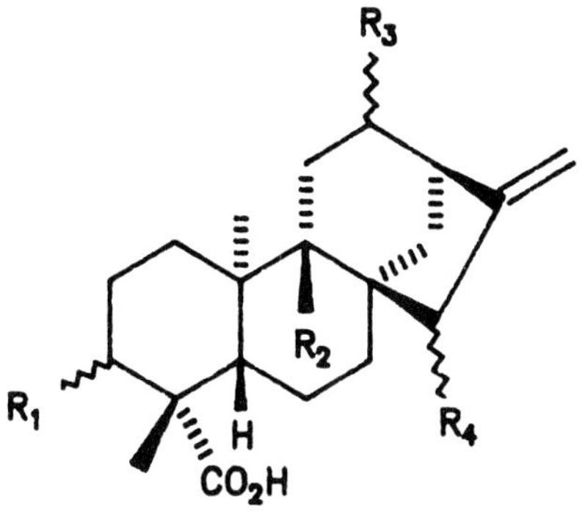

641    $R_1 = R_2 = H$, $R_3 = \alpha\text{-OH}$, $R_4 = H$

642    $R_1 = R_2 = H$, $R_3 = \alpha\text{-OAc}$, $R_4 = H$

796    $R_1 = R_2 = H$, $R_3 = \beta\text{-OAc}$, $R_4 = H$

646    $R_1 = R_2 = R_3 = H$, $R_4 = \alpha\text{-OH}$

647    $R_1 = R_2 = R_3 = H$, $R_4 = \beta\text{-OH}$

648    $R_1 = R_2 = R_3 = H$, $R_4 = \alpha\text{-OAc}$

649    $R_1 = R_2 = R_3 = H$, $R_4 = \alpha\text{-OAng}$

650    $R_1 = R_2 = R_3 = H$, $R_4 = \alpha\text{-OTig}$

651    $R_1 = R_2 = R_3 = H$, $R_4 = \alpha\text{-OSen}$

652    $R_1 = R_2 = R_3 = H$, $R_4 = \alpha\text{-Oi-Val}$

653    $R_1 = R_2 = R_3 = H$, $R_4 = \alpha\text{-O2-Mac}$

654    $R_1 = R_2 = R_3 = H$, $R_4 = \alpha\text{-Oi-But}$

655    $R_1 = R_2 = R_3 = H$, $R_4 = \alpha\text{-OCinn}$

656    $R_1 = R_2 = R_3 = H$, $R_4 = \alpha\text{-OEpoxyang}$

657    $R_1 = R_2 = R_3 = H$, $R_4 = \alpha\text{-OBenz}$

818    $R_1 = R_2 = R_3 = H$, $R_4 = \alpha\text{-O2-Mebut-2,3-OH}$

892    $R_1 = R_2 = R_3 = H$, $R_4 = \alpha\text{-O2Mebut}$

658    $R_1 = R_2 = R_3 = H$, $R_4 = \beta\text{-OAng}$

659    $R_1 = R_2 = R_3 = H$, $R_4 = \beta\text{-OTig}$

660    $R_1 = R_2 = R_3 = H$, $R_4 = \beta\text{-OSen}$

662    $R_1 = \alpha\text{-OAng}$, $R_2 = OH$, $R_3 = R_4 = H$

FIGURE 1. (*contd.*)

663 R_1 = α-OCinn, R_2 = OH, R_3 = R_4 = H

664 R_1 = H, R_2 = OH, R_3 = H, R_4 = α-OH

665 R_1 = H, R_2 = OH, R_3 = H, R_4 = α-OAng

666 R_1 = H, R_2 = OH, R_3 = H, R_4 = α-OTig

667 R_1 = H, R_2 = OH, R_3 = H, R_4 = α-OAc

668 R_1 = H, R_2 = OH, R_3 = H, R_4 = α-OSen

669 R_1 = H, R_2 = OH, R_3 = H, R_4 = α-Oi-Val

670 R_1 = H, R_2 = OH, R_3 = H, R_4 = α-OCinn

819 R_1 = H, R_2 = OH, R_3 = H, R_4 = α-O2-Mebut-3-OH

675 R_1 = β-OH, R_2 = OH, R_3 = H, R_4 = α-OAng

676 R_1 = β-OH, R_2 = OH, R_3 = H, R_4 = α-OSen

677 R_1 = β-OH, R_2 = OH, R_3 = H, R_4 = α-OTig

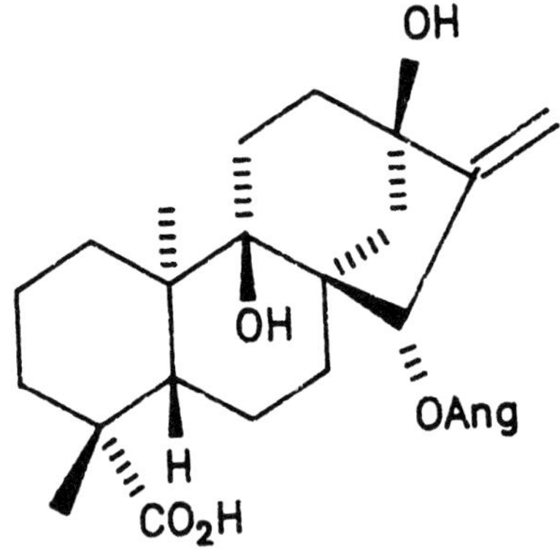

636

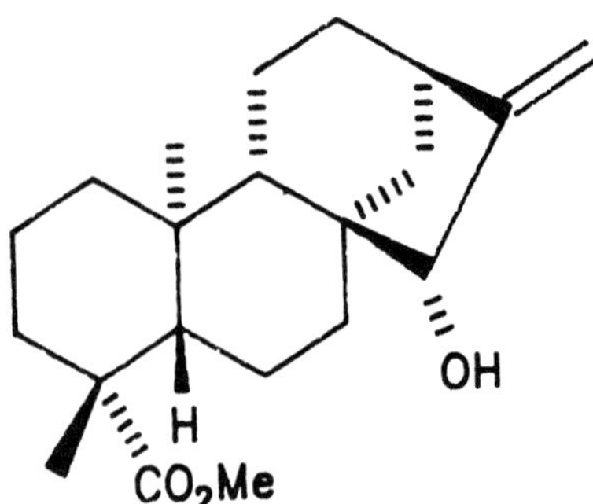

661

FIGURE 1. (*contd.*)

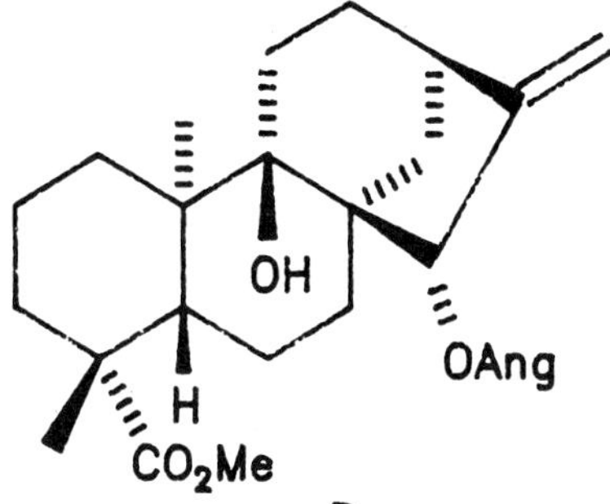

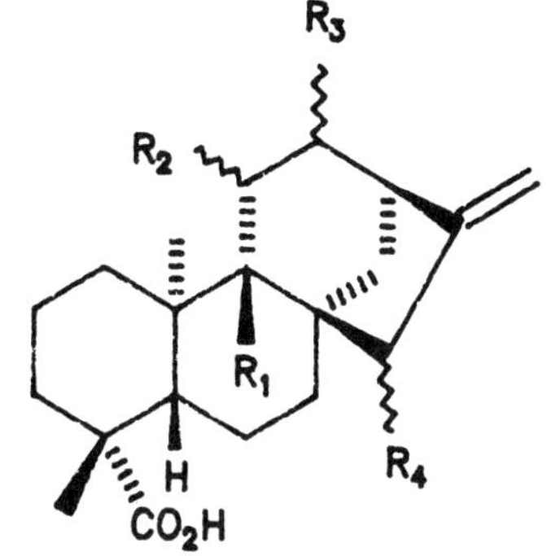

671 R₁ = H, R₂ = β-OH, R₃ = H, R₄ = β-OH

672 R₁ = H, R₂ = β-OH, R₃ = H, R₄ = β-OAc

637 R₁ = H, R₂ = β-OH, R₃ = H, R₄ = α-OH

869 R₁ = OH, R₂ = β-OH, R₃ = H, R₄ = α-OAng

1260 R₁ = H, R₂ = β-OH, R₃ = β-OH, R₄ = β-OH

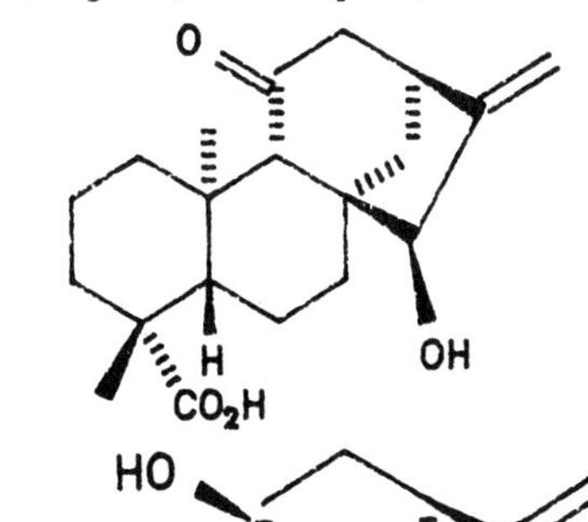

673 (Revised to **674**)

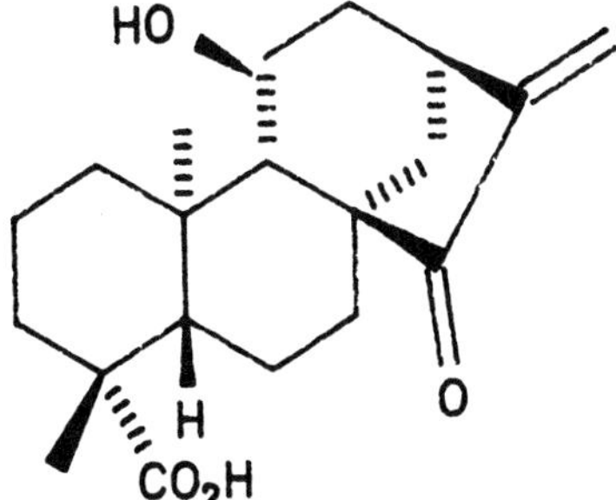

674

FIGURE 1. (*contd.*)

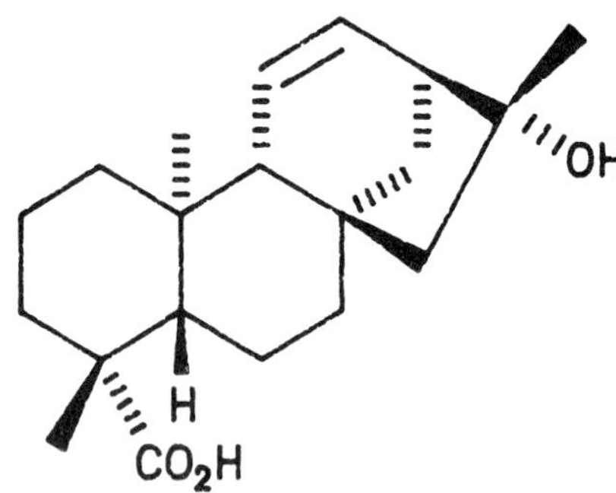

997 R_1 = H, R_2 = R_3 = Me (= 720)

721 R_1 = OAc, R_2 = R_3 = Me

722 R_1 = H, R_2 = Me, R_3 = CH_2OAc

1255 R_1 = H, R_2 = Me, R_3 = CHO

723 R_1 = H, R_2 = Me, R_3 = CO_2H

725 R_1 = H, R_2 = CH_2OH, R_3 = Me

726 R_1 = H, R_2 = R_3 = CH_2OH

728 R_1 = H, R_2 = CH_2OH, R_3 = CO_2H

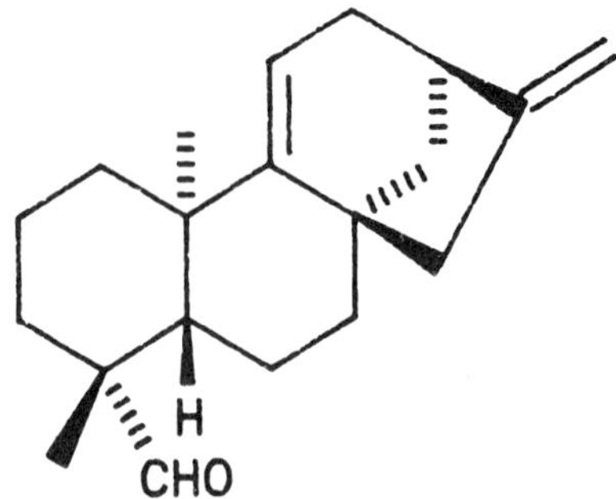

724

942

FIGURE 1. (*contd.*)

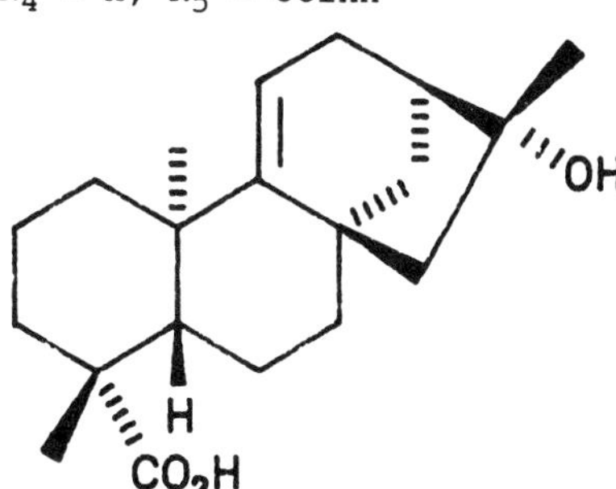

690 $R_1 = R_2 = R_3 = R_4 = R_5 = H$

691 $R_1 = OH, \ R_2 = R_3 = R_4 = R_5 = H$

692 $R_1 = H, \ R_2 = OH, \ R_3 = R_4 = R_5 = H$

693 $R_1 = R_2 = H, \ R_3 = OH, \ R_4 = R_5 = H$

694 $R_1 = R_2 = R_3 = H, \ R_4 = OH, \ R_5 = H$

998 $R_1 = R_2 = R_3 = H, \ R_4 = OEt, \ R_5 = H$

696 $R_1 = R_2 = R_3 = R_4 = H, \ R_5 = OH$

943 $R_1 = R_2 = R_3 = R_4 = H, \ R_5 = OCinn$

1256

695

FIGURE 1. (*contd.*)

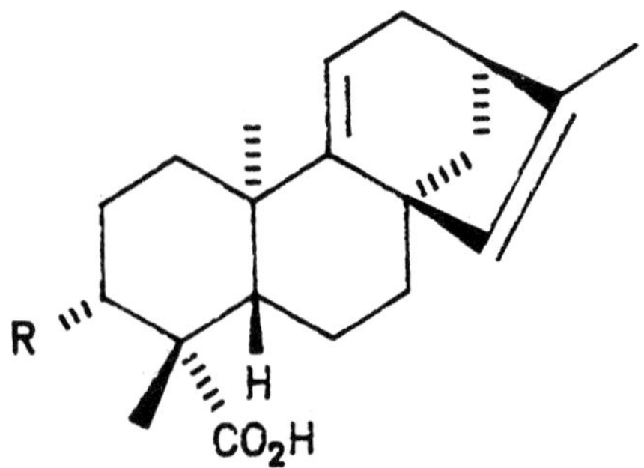

697 $R_1 = R_2 = H$, $R_3 = CH_2OH$, $R_4 = Me$

698 $R_1 = R_2 = H$, $R_3 = R_4 = CH_2OH$

699 $R_1 = R_2 = H$, $R_3 = CH_2OH$, $R_4 = CH_2Op-OH-H-Cinn$

946 $R_1 = R_2 = H$, $R_3 = Me$, $R_4 = CHO$

700 $R_1 = R_2 = H$, $R_3 = Me$, $R_4 = CO_2H$

701 $R_1 = R_2 = H$, $R_3 = CH_2OH$, $R_4 = CO_2H$

702 $R_1 = R_2 = H$, $R_3 = CHO$, $R_4 = CO_2H$

703 $R_1 = R_2 = H$, $R_3 = R_4 = CO_2H$

707 $R_1 = Oi-Val$, $R_2 = OH$, $R_3 = Me$, $R_4 = CO_2H$

708 $R_1 = Oi-But$, $R_2 = OH$, $R_3 = Me$, $R_4 = CO_2H$

704 $R = Oi-Val$

705 $R = Oi-But$

706 $R = OTig$

FIGURE 1. (*contd.*)

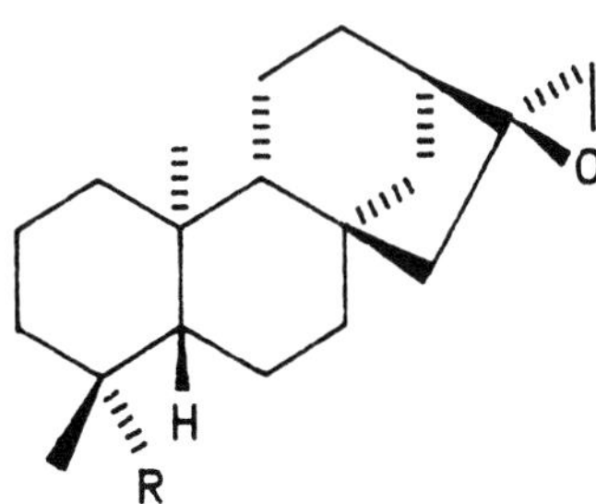

733 R_1 = H, R_2 = R_3 = Me

709 R_1 = OAng, R_2 = R_3 = Me

638 R_1 = H, R_2 = CH_2OH, R_3 = Me

734 R_1 = H, R_2 = Me, R_3 = CH_2OH

710 R_1 = H, R_2 = Me, R_3 = CH_2OAng

711 R_1 = OAng, R_2 = Me, R_3 = CHO

735 R_1 = H, R_2 = Me, R_3 = CHO

713 R_1 = H, R_2 = Me, R_3 = CO_2H

712 R = CHO

714 R = CO_2H

715 R = α-OAng

FIGURE 1. (*contd.*)

716   R = α-OAc

717   R = β-OAng

718   R = β-OTig

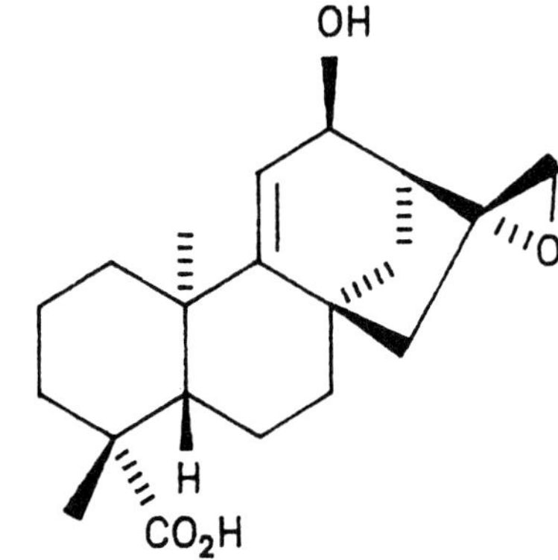

719

727

1001 R$_1$ = R$_2$ = Me (= **851**)

852   R$_1$ = Me, R$_2$ = CH$_2$OH

822   R$_1$ = CH$_2$OH, R$_2$ = CO$_2$H

FIGURE 1. (*contd.*)

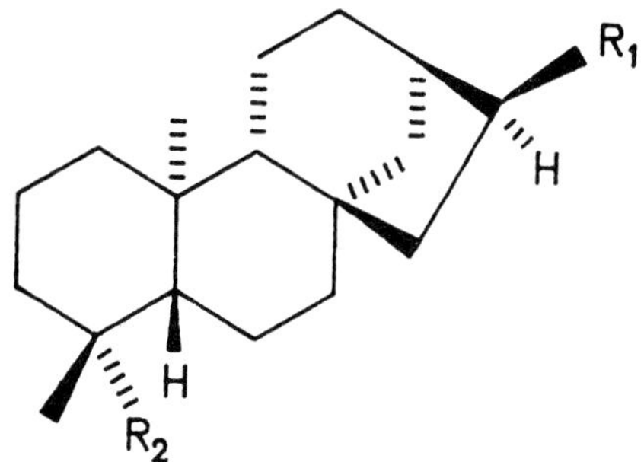

924 R_1 = CH$_2$OH, R_2 = CO$_2$H

925 R_1 = R_2 = CO$_2$H

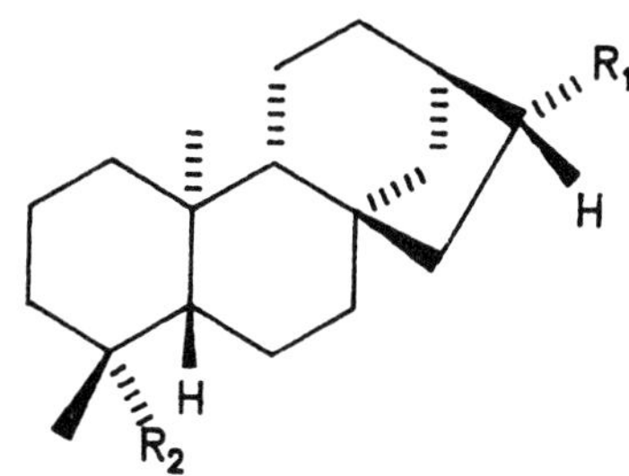

729 R_1 = CHO, R_2 = Me

730 R_1 = CHO, R_2 = CH$_2$OH

731 R_1 = CO$_2$H, R_2 = Me

732 R_1 = CO$_2$H, R_2 = CH$_2$OH

736 R_1 = COMe, R_2 = CH$_2$OH

737 R_1 = R_2 = CHO

738 R_1 = CH$_2$OH, R_2 = CO$_2$H

989 R_1 = CO$_2$H, R_2 = CHO

990 R_1 = R_2 = CO$_2$H

740 R_1 = CHO, R_2 = CH$_2$Op-OH-H-Cinn

741 R_1 = CH$_2$Oi-But, R_2 = CO$_2$H

742 R_1 = CH$_2$ODodec, R_2 = CO$_2$H

743 R_1 = CH$_2$OTetradec, R_2 = CO$_2$H

744 R_1 = CH$_2$OHexadec, R_2 = CO$_2$H

FIGURE 1. (*contd.*)

745 R_1 = CH$_2$OOctadec, R_2 = CO$_2$H

812 R_1 = CO$_2$Me, R_2 = CO$_2$H

739

746

747   R = β-OH (= 815)

748   R = β-OH

FIGURE 1. (*contd.*)

850

945

678 $R_1 = R_2 = CO_2H$, $R_3 = OCOCH_2CH(CH_3)_2$, $R_4 = R_5 = HO_3SO$, $R_6 = H$

FIGURE 1. (*contd.*)

679 R_1 = R_2 = CO_2H, R_3 = $NHCOCH_2CH(CH_3)_2$, R_4 = $CO(CH_2)_2C_6H_5$,
R_5 = OH, R_6 = OH

680 R_1 = R_2 = CO_2H, R_3 = $NHCOCH_2CH(CH_3)_2$, R_4 = $CO(CH_2)_2C_6H_5$,
R_5 = OL-Rha, R_6 = OH

681 R_1 = H, R_2 = CO_2H, R_3 = $OCOCH_2CH(CH_3)_2$, R_4 = R_5 = HO_3SO,
R_6 = H

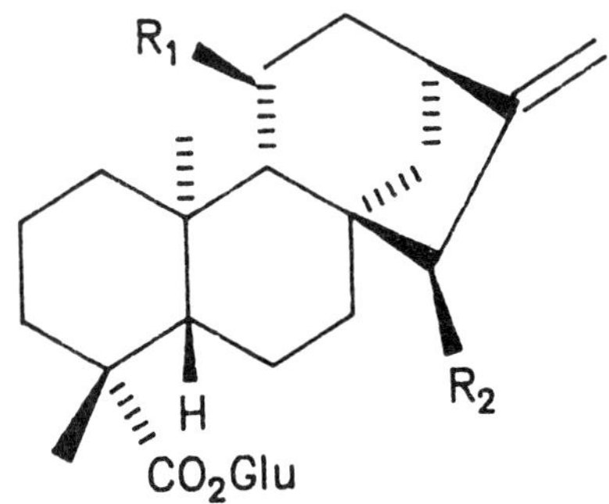

895 R_1 = H, R_2 = OH
896 R_1 = R_2 = OH

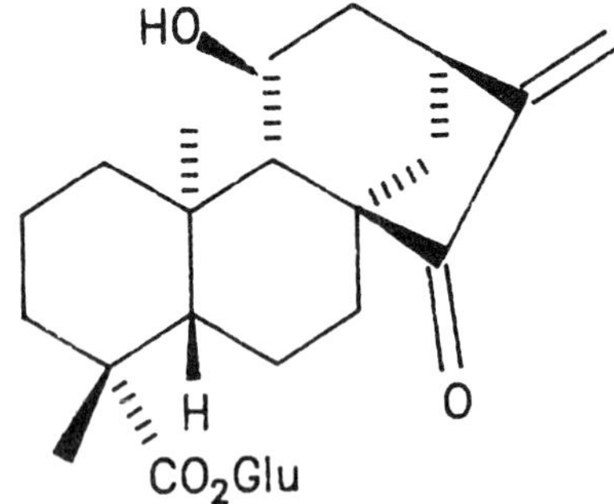

897

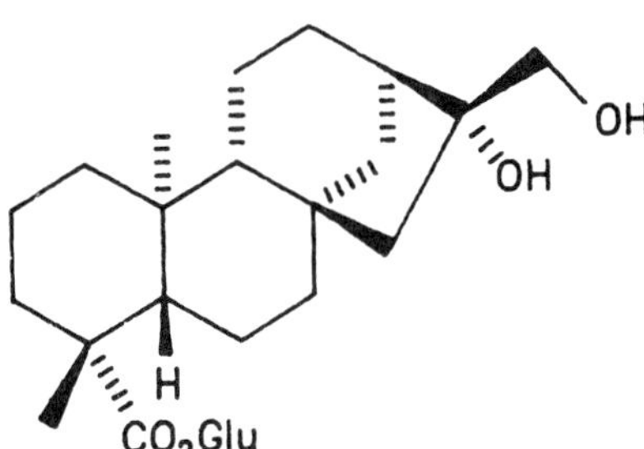

898

FIGURE 1. (contd.)

899

682 $R_1 = R_2 = H$, $R_3 = OH$

683 $R_1 = Glu$, $R_2 = H$, $R_3 = OH$

684 $R_1 = H$, $R_2 = Glu$, $R_3 = OH$

685 $R_1 = R_2 = Glu$, $R_3 = OH$

686 $R_1 = Glu$, $R_2 = R_3 = H$

687 $R_1 = R_2 = Glu$, $R_3 = H$

688 $R_1 = \beta Glu[2]-[1]Glu$, $R_2 = Glu$, $R_3 = H$

689 $R_1 = \beta Glu[2]-[1]Glu$, $R_2 = H$, $R_3 = OH$

FIGURE 1. (*contd.*)

59. _ent_-Kaurane, 17-Nor-

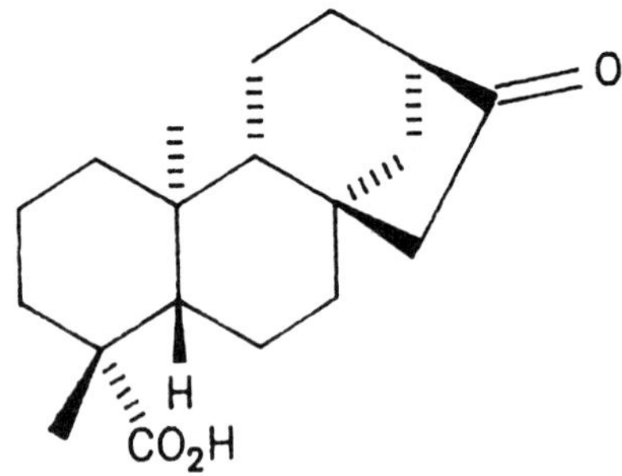

749

60. _ent_-Kaurane, 18-Nor-

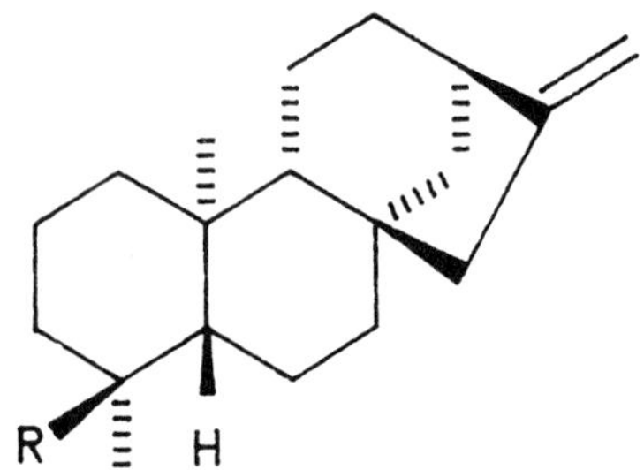

752   R = H

753   R = OH

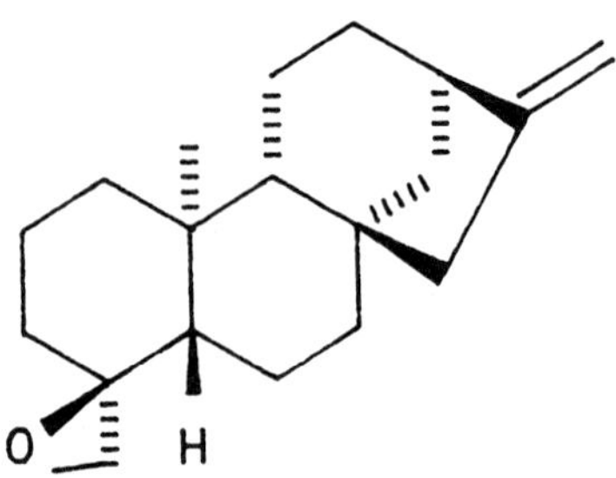

813

FIGURE 1. (*contd.*)

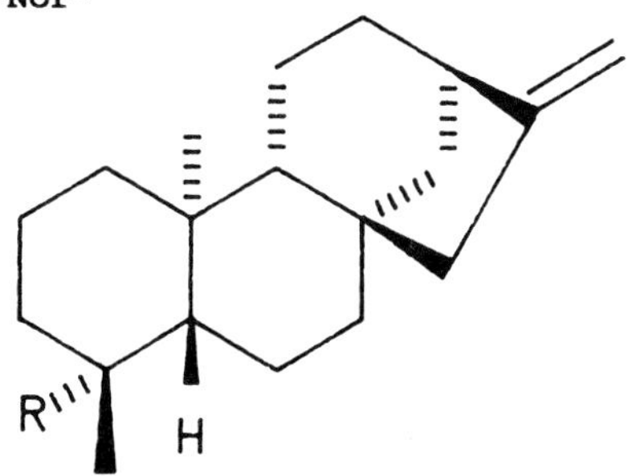

754 R = CH$_2$OH

755 R = CHO

756 R = CO$_2$Me

757

758

61. <u>ent</u>-Kaurane, 19-Nor-

750 R = H

751   R = OH

62. <u>ent</u>-Kaurane, 9,10-Seco-

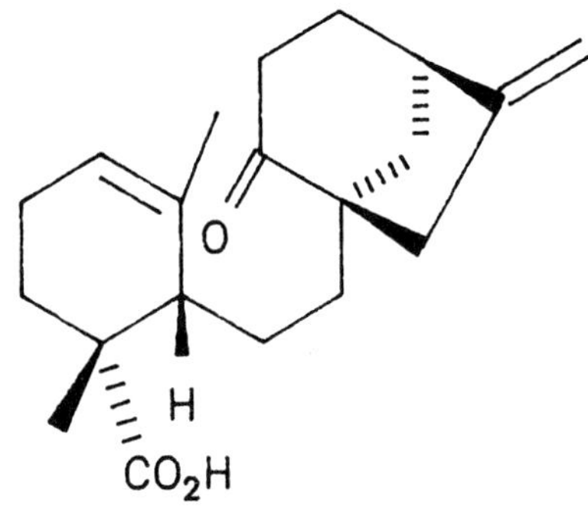

759

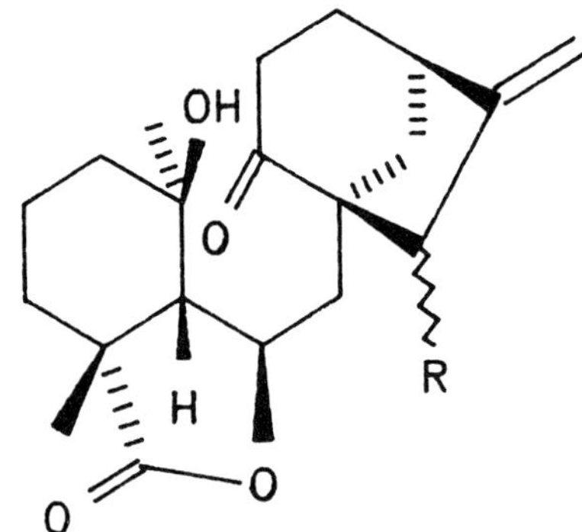

866   R = α-OH

867   R = β-OH

868   R = β-OAc

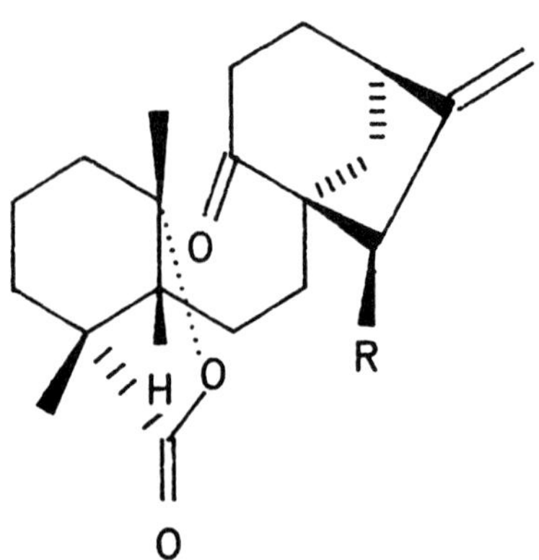

760   R = H

763   R = OH

FIGURE 1. (*contd.*)

761 R = α-OH

762 R = β-OH

64. _ent_-Stachane (_ent_-Beyerane)

764 R_1 = OH, R_2 = H, R_3 = R_4 = Me

765 R_1 = R_2 = H, R_3 = Me, R_4 = CH_2OH

766 R_1 = R_2 = H, R_3 = CH_2OH, R_4 = Me

767 R_1 = OH, R_2 = H, R_3 = Me, R_4 = CH_2OH

768 R_1 = R_2 = OH, R_3 = R_4 = Me

769 R_1 = R_2 = H, R_3 = Me, R_4 = CH_2OMal

773 R_1 = R_2 = H, R_3 = Me, R_4 = CHO

774 R_1 = R_2 = H, R_3 = Me, R_4 = CO_2H

775 R_1 = OH, R_2 = H, R_3 = Me, R_4 = CO_2H

776 R_1 = OTig, R_2 = H, R_3 = Me, R_4 = CO_2H

FIGURE 1. (*contd.*)

770 R_1 = OH, R_2 = CH_2OH, R_3 = R_4 = Me

771 R_1 = H, R_2 = CH_2OH, R_3 = Me, R_4 = CH_2OH

772 R_1 = H, R_2 = Me, R_3 = R_4 = CH_2OH

817

777 R = H

778 R = OTig

FIGURE 1. (*contd.*)

779

780

781

FIGURE 1. (*contd.*)

93. <u>ent</u>-Stachane, 19-Nor-

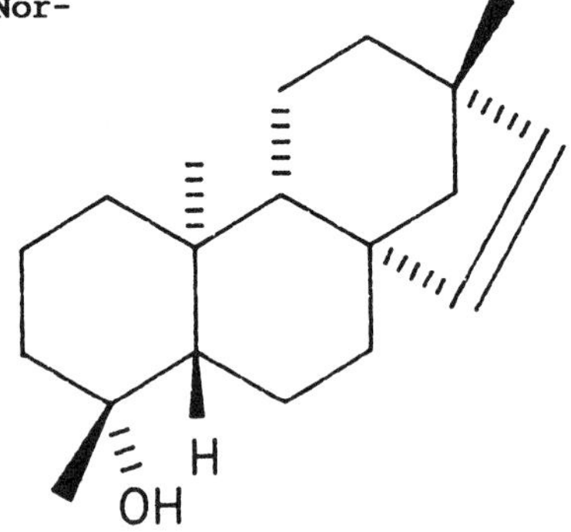

782

65. Atisirane (Atisane)

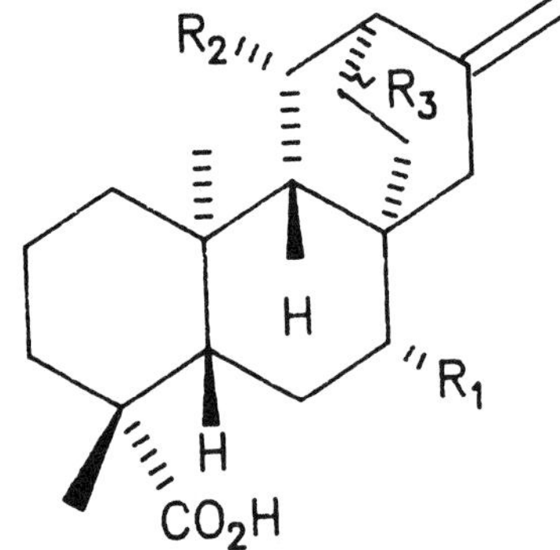

783 $R_1 = R_2 = R_3 = H$

784 $R_1 = OH$, $R_2 = R_3 = H$

785 $R_1 = H$, $R_2 = OAc$, $R_3 = H$

786 $R_1 = R_2 = H$, $R_3 = OAng$

787 $R_1 = R_2 = H$, $R_3 = Oi\text{-}Val$

788 $R_1 = R_2 = H$, $R_3 = Oi\text{-}But$

814 $R_1 = R_2 = H$, $R_3 = OH$

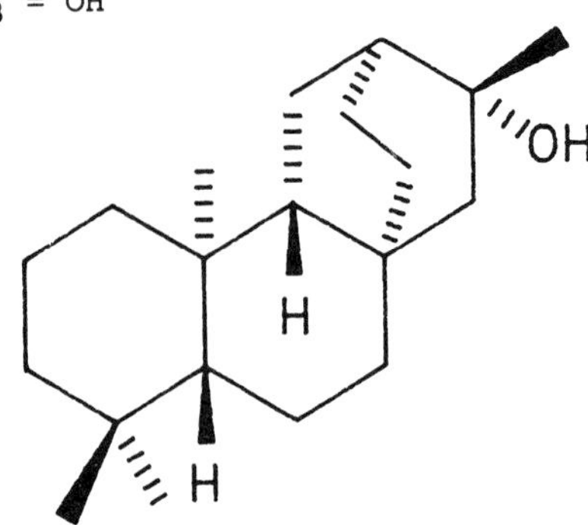

853

FIGURE 1. (*contd.*)

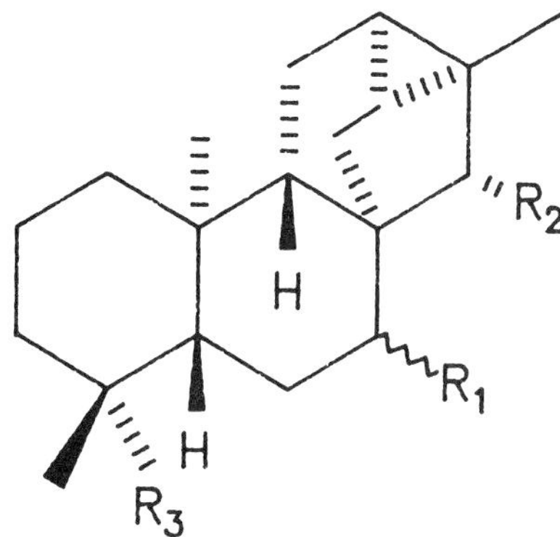

854

816

789 R_1 = OAc, R_2 = H

790 R_1 = H, R_2 = β-OAc

66. Trachylobane

FIGURE 1. (*contd.*)

855 $R_1 = R_2 = H$, $R_3 = CHO$

791 $R_1 = R_2 = H$, $R_3 = CO_2H$

825 $R_1 = H$, $R_2 = OAc$, $R_3 = CO_2H$

826 $R_1 = H$, $R_2 = OAng$, $R_3 = CO_2H$

827 $R_1 = H$, $R_2 = Oi\text{-}Val$, $R_3 = CO_2H$

893 $R_1 = H$, $R_2 = Oi\text{-}But$, $R_3 = CO_2H$

792 $R_1 = \alpha\text{-}OH$, $R_2 = H$, $R_3 = CO_2H$

1204 $R_1 = R_2 = H$, $R_3 = CO_2Me$

1259 $R_1 = \beta\text{-}OH$, $R_2 = H$, $R_3 = CO_2H$

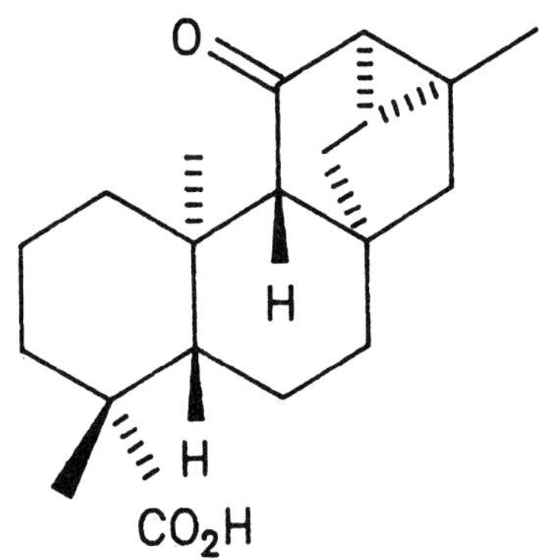

793

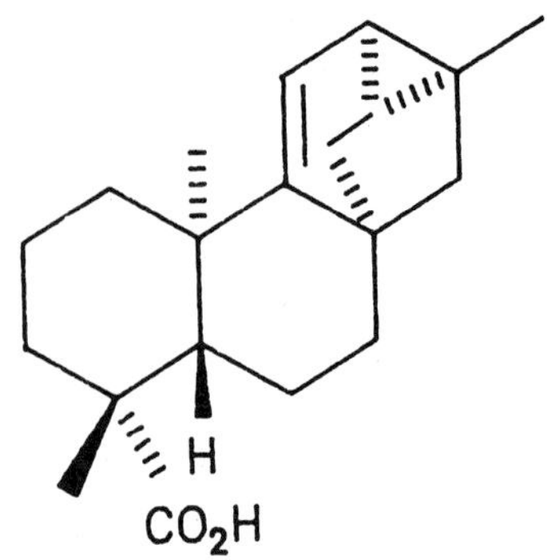

794

FIGURE 1. (*contd.*)

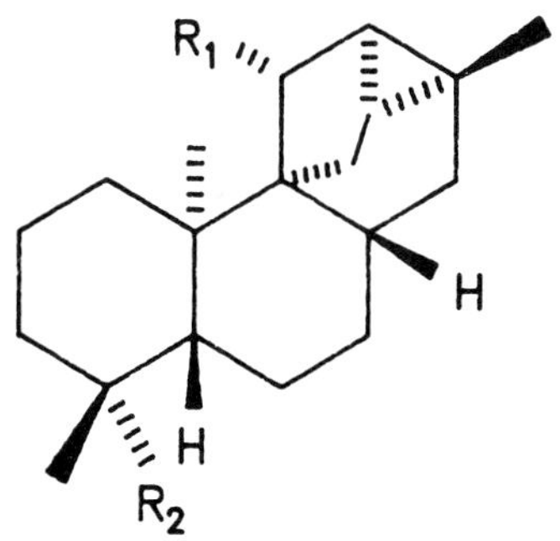

795

856

67. Helifulvane

FIGURE 1. (*contd.*)

796 R_1 = H, R_2 = CH_2OH

797 R_1 = H, R_2 = CO_2H

798 R_1 = OH, R_2 = CO_2H

799 R_1 = OAc, R_2 = CO_2H

68. Tetrachyrane

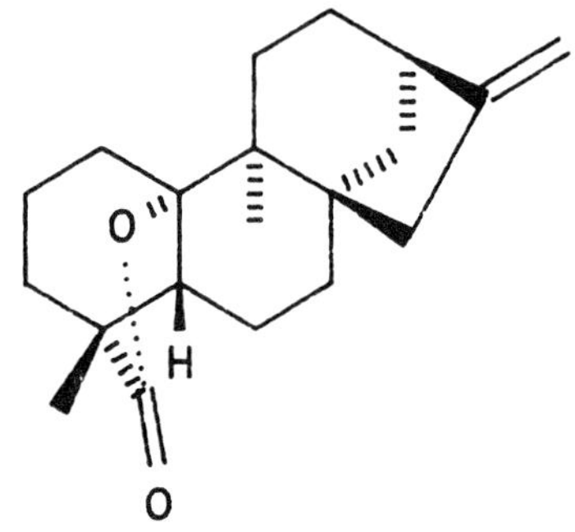

800

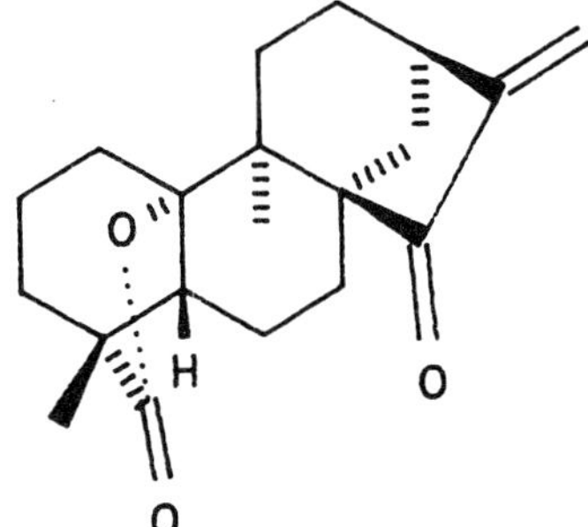

847

805 R_1 = H, R_2 = OH

806 R_1 = OH, R_2 = H

FIGURE 1. (*contd.*)

78. Villanovane

875 $R_1 = R_2 = H$, $R_3 = CH_2OH$, $R_4 = CH_2Oi\text{-}Val$

876 $R_1 = R_2 = H$, $R_3 = CH_2OAc$, $R_4 = CH_2Oi\text{-}Val$

877 $R_1 = R_2 = H$, $R_3 = CH_2OH$, $R_4 = CH_2OMeVal$

878 $R_1 = R_2 = H$, $R_3 = OAc$, $R_4 = CH_2OMeVal$

879 $R_1 = Oi\text{-}Val$, $R_2 = H$, $R_3 = CH_2OAc$, $R_4 = Me$

880 $R_1 = OMeVal$, $R_2 = H$, $R_3 = CH_2OAc$, $R_4 = Me$

894 $R_1 = H$, $R_2 = OH$, $R_3 = R_4 = Me$

69. Lycoctonine

801

FIGURE 1. (contd.)

802 R = H

803 R = Anth

VIII. Undetermined Stereochemistry

79. Inulaefane

804

84. Hebeclinane (9,10-Seco-Labdane)

940 (= 999)

FIGURE 1. (*contd.*)

CHAPTER 3

Biogenesis

"...apart from the carbon skeleton of abietic acid with an irregular sequence of 4 isoprene units, the carbon skeletons of all diterpenes are derived from a regular isoprene tetramer consisting of 4 isoprene units connected head-to-tail... This structural relationship may be termed the "phytol rule", in analogy to the farnesol rule of the sesquiterpenes."

Ruzicka (444, p. 358)

III.1. *ent*- and Normal-Labdanes

Cyclization of the linear diterpene precursor usually proceeds along two alternative parallel routes (Figure 2 [p. 392]), one leading to the bicyclic "normal" labdane (normal-labdane) absolute stereochemistry and the other resulting in its antipodal, enantiomeric (*ent*-labdane) absolute stereochemistry. As a consequence of these two parallel biosynthetic routes, diterpenes lack the homogeneous absolute stereochemistry of the triterpenes.

C-5, C-9, and C-10 substituents are oppositely oriented in the normal- and *ent*-labdane cyclization products. This distinction between normal- and *ent*-labdanes suggests that the linear precursor adopts one of two possible conformations at the onset of cyclization. The formation of the normal- and antipodal *ent*-AB ring junctions and the *trans*-anti- configuration of these junctions are governed by the relative orientation of the cyclizing double bonds within the chairlike conformation of the folded C20 linear precursor (Figure 2). These alternative cyclizations may be visualized as arising through two different coilings of the open-chain geranylgeranyl pyrophosphate (GGPP) precursor on an enzyme surface followed by (1) initiation of cyclization by proton addition to C-14 of the 14,15-double bond and (2) antiparallel cationic 1,2-additions resulting from the migration of electrons from the two internal 6,7- and 10,11-double bonds. Following cyclization to the *trans*-fused decalin ring system, the C-4 axial methyl group and H-5 are oriented in the same plane and arranged *trans* to one another, a relationship that also exists between the C-10 methyl and H-9. Likewise, the newly formed C-4-C-5 and C-9-C-10 bonds are in the same plane and are *trans* about the C-5-C-10 axis (Figure 2).

The hypothetical C-8 carbonium ion resulting from this initial cyclization is stabilized by several alternative events: (1) proton-loss yielding either the C-7-C-8 or C-8-C-17 double bond (e.g., Figure 2), (2) neutralization via hydration or lactone or ether formation and (3) rearrangement. Much of the diterpene structural variability (Figure 3 [pp. 393–415]) is a consequence of the variety of rearrangement products resulting from this last mode of stabilization.

In the following discussion, diterpene biogenesis is described in terms of a hypothetical reaction sequence leading from an appropriate labdane precursor to a product representative of one of the 103 skeletal types reported from the Compositae. These skeletal types are arranged in Figure 3 according to their putative biogenetic relationship.

Regardless of the extensive variability in leaf-resin chemistries found in the Compositae, all plants of this family presumably biosynthesize gibberellin diterpenes via a pathway outlined in Figure 4 (p. 416). As gibberellin biosynthesis (1) appears to be anatomically and functionally distinct from that of leaf-resin diterpenoids and (2) is regularly reviewed in the literature, it will not be discussed here.

III.1.1. Seco- and Nor-Labdane Skeletons

Ring cleavage and decarboxylation of labdanes have generated several normal- (Figure 3, Skeleton Nos. 23, 24, 25, 26, 27, 87 and 89) and *ent-* (Figure 3, Nos. 41, 42, 43, 44, 45 and 75) seco- and nor-labdane skeletons.

*ent-*Labdanes of *Ayapana amygdalina* and their C-17 nor-homologs (Figure 5 (p. 417); *113*) fit well into a biogenetic scheme originating from the proposed precursor **337**. 8,17-epoxidation and subsequent hydration to produce **343** facilitates loss of C-17 to give compounds with the C-17-nor labdane skeleton (i.e. **467** and **468**). An alternative route leads to compounds (**339–342, 344, 345**) with a C-17 aldehyde or carboxyl function.

Chrysothamnus paniculatus yielded a skeleton resembling the seco- (cleaved) product of a common grindelane precursor, **172** (Figure 6 (p. 418); *341*). This 8,9-seco-grindelane skeleton (Figure 3, No. 25) distinguishes **223**, a compound variously named chrysothame and strictanonic acid. Another skeleton of seco- origin reported from *Nidorella hottentotica* displays a ten-membered ring produced by the cleavage of the C-5-C-10 bond. The proposed relationship of representatives of this skeleton and structurally related, co-occurring normal-labdanes is depicted in Figure 7 (p. 419) (*155*).

III.1.2. Rearranged Labdanes

Rearranged labdane skeletons arise via a series of associated hydride and methyl shifts, usually involving concerted *trans*-migrations of the participating substituents.

III.1.2.1. Normal-Labdane Series

Among the normal-labdanes, the simplest rearrangement product, the relhaniane skeleton (Figure 3, No. 28), results from the concerted migration of C-9α-H to

C-8 and C-10β-methyl to C-9 (Figure 8 Scheme 2 [p. 420]). C-1 proton loss and C-1-C-10 double bond formation complete the biogenesis proposed for the *Relhania acerosa* compound, relhania acid, **231** (Figure 8, Scheme 1 (p. 420); *90*). Alternatively, migration of H-5 to C-10 and deprotonation of C-6 are the final steps of the proposed biogenesis for the *Koanophyllon conglobatum* compounds koanophyllic acids A-D (**232–235**; Scheme 2, Figure 8). As the absolute stereochemistry of the koanophyllic acid series is not established, their normal-labdane derivation is unproven and only follows the authors' tentative assignment (*42*).

The novel chrysolane skeleton (Figure 3, No. 72) probably results from the rearrangement of a precursor resembling 6β-hydroxy-grindelic acid (**172**). In this scheme, migration of the C-10 methyl sets in motion a complicated series of events terminating in the aromatic chrysolane skeleton reported from *Chrysothamnus paniculatus* (Figure 9 [p. 421]; *341*).

Examples of the clerodane skeleton, the most common rearrangement product, can be further subdivided into *cis-* and *trans*-fused decalin forms (Figure 10 [p. 422]). In the series of methyl and hydride shifts which make up the concerted "friedo" or "backbone" rearrangement process culminating in the clerodane skeleton, the migrating groups display a *trans*-diaxial relationship. For normal-labdanes, the expected product of a concerted series of *trans*-migrations (C-9α-H to C-8, C-10β-methyl to C-9, C-5α-H to C-10, and C-4β-methyl to C-5) would be the *trans*-clerodane (Figure 3, No. 29; Figure 10). There is as yet no proof that this normal-labdane-based pathway operates within the Compositae. However, the *cis*-clerodane skeleton (Figure 3, No. 31; Figure 10) which differs from the *trans*-version in the orientation of the C-5 methyl is well represented. Biogenetically, the existence of the *cis*-form is explained by a sequence of concerted *trans*-migrations (C-9α-H to C-8, C-10β-methyl to C-9 and C-5α-H to C-10) culminating in the formation of a carbonium ion at C-5. The AB *cis*-fused ring system would then result from the migration of the C-4α-methyl to C-5, followed by deprotonation at C-3.

III.1.2.2. *ent-Labdane Series*

Steps proposed for the biogenesis of rearranged *ent*-labdanes appear to correspond to the mechanisms indicated for the normal-labdanes, differing only in the opposite orientivity of the migrating groups.

The simplest rearranged *ent*-labdane is the *ent*-friedolabdane skeleton (Figure 3, No. 95). A proposed migration sequence for this structure begins with a hydride shift from C-9(β-H) to C-8, followed by a methyl shift from C-10(α-methyl) to C-9 and hydride shift from C-5(β-H) to C-10 and terminates in C-6 deprotonation and C-5-C-6 double bond formation (e.g., **1059**).

In contrast to the normal-clerodanes, both *cis-* and *trans*-fused *ent*-clerodanes are well documented (Figure 3, Nos. 50 and 46, respectively). Mechanistically, the proposed steps of *ent*-clerodane biogenesis differ from those of the normal-series only in the opposite orientation of the corresponding migrating hydride or methyl group (Figure 10).

III.1.3. Seco-, Nor- and Rearranged Clerodane Skeletons

The next level of skeletal complexity is expressed by seco-, rearranged and nor-derivatives of both normal- and *ent*-clerodanes. Such skeletons (e.g., Figure 3, No. 41) tend to have relatively narrow distributions, often being reported from a single source.

Rearrangement of an *ent*-clerodane precursor results in a series of skeletal types some of which are shared by two genera, the Astereae member, *Conyza* (*83*), and the Inuleae taxon, *Printzia* (*165*). These include the printzianes (Figure 3, No. 47; Figure 11 [p. 423], **571**, **572**, **915**), conyscabranes (Figure 3, No. 76; Figure 11, **917–920**) and 5,10-seco-clerodanes (Figure 3, No. 48; Figure 11, **573**, **912**). Another rearranged *ent*-clerodane skeleton, solidagonane, (Figure 3, No. 47; Figure 11 [p. 423], **571**, **572**, **915**), conyscabranes (Figure 3, solidagonal acid (**960**), which was probably derived from a kolavenic acid (**495**)-like precursor (Figure 12 [p. 424]).

III.2. Tricyclic Diterpenes

Instead of rearrangement to clerodanes and related bicyclic skeletons, the hypothetical product of labdane formation, labdadienyl pyrophosphate, can cyclize to produce either tricyclic pimara-8(9),15-diene or its 8(14)-isomer. The simplest mechanism for this process is as follows: attack at C-13 by electrons of the exocyclic double bond (S_N' reaction) is followed by migration of the double bond from C-13 and elimination of the pyrophosphate group from the allylic position. The resulting intermediate tricyclic cation at C-8 is stabilized by proton elimination at C-9 or C-14 (e.g., Figure 13 [p. 425] **274**). The resulting skeletons are referred to as sandaracopimaranes or normal-pimaranes (Figure 3, No. 32) if derived from the normal-labdane precursor or *ent*-pimaranes (Figure 3, No. 52) if formed from an *ent*-labdane.

Tricyclic compounds with the epimeric arrangement of methyl and vinyl substituents at C-13 often occur as pairs, e.g., rimuene and *ent*-rimuene (Figure 14A [p. 426]), suggesting that electron addition from the exocyclic methylene can occur at either face of the double bond depending on the stereoselectivity of the enzyme involved. (The two examples, rimuene and *ent*-rimuene, also illustrate that the prefix *ent*- does not always refer to compounds of *ent*-labdane origin.) Frequent reports of such C-13 epimeric pairs suggest that enzyme stereospecificity is low.

Instead of stabilization of the pimarane intermediate by proton elimination at C-9 or C-14, several further rearrangement processes can occur. For example, the cassane (Figure 3, No. 34) and abietane (Figure 3, No. 37) skeletons can result from alternative methyl migrations (Figure 13). The rosane skeleton (Figure 3, No. 35; Figure 14B [p. 426], **881–889**) can also be generated by methyl migration and stabilized by a variety of routes as demonstrated by the chemistry of *Trichogonia* (Figure 14B, *183*).

The abietane skeleton that Ruzicka excluded from the "phytol rule" (*444*) can be derived in a straightforward manner from either a normal- or *ent*-pimarane precursor (Figures 13 and 15 [p. 427], respectively). Representatives of both normal- and *ent*-abietanes (Figure 3, Nos. 37 and 55, respectively) have been reported from the Compositae.

In addition to the *ent*-abietane skeleton, other tricyclic *ent*-pimarane derived ring skeletons have been reported (Figure 15): Cleistanthanes and isocleistanthanes were detected in *Brickellia eupatoriedes* (*59*) and erythroxanes were identified in *Helichrysum refluxum* (*450*). The absolute configuration (5R, 10R) of the *Brickellia* cleistanthanes is identical to that reported for the cleistanthanes of the Euphorbiaceae, but antipodal to those (5S, 10S) of the Velloziaceae (*Vellozia*) (Figure 16 [p. 428]; *427*).

Acritopappus confertus (Eupatorieae) produces a pair of unusual tricarbocyclic compounds, **617** and **618**, that do no result from the cyclization processes described above. A biogenetic scheme for these constituents (Figure 17 [p. 429]) invokes an aldol condensation reaction to link C-11 and C-17 (*184*).

III.3. Tetracyclic Diterpenes

The hypothetical tricyclic pimarenyl cation can either undergo further cyclization to yield a tetracyclic cation which eliminates a proton to give an *ent*-stachane (*ent*-beyerane; Figure 18, Route I [p. 430]) or form alternative protonated-cyclopropane-containing intermediates (Figure 18, Routes II and III). Through either simple proton elimination or more complicated processes, a variety of tetra- and pentacyclic skeletons are produced via these cyclopropanoid intermediates. With one reported exception, normal-stachane (Figure 3, No. 73), all these polycyclics are members of the *ent*-series (Figure 18): *ent*-stachane, *ent*-atisirane, *ent*-kaurane, *ent*-trachylobane, *ent*-helifulvane, *ent*-tetrachyrane and *ent*-villanovane (Figure 3, Nos. 64, 65, 58, 66, 67, 68, and 78, respectively).

III.4. Biosynthetic Evidence

Assuming that similar enzymes and enzyme systems are involved in the biosynthesis of both gibberellins and resin diterpenes, the relative wealth of biosynthetic data concerning the *ent*-kaurane-derived gibberellins may provide some insight into resin diterpene biogenesis. Three aspects of gibberellin biosynthesis seem pertinent to understanding resin diterpene distribution in the Compositae. The first is the existence of two discrete and separable enzymatic processes, activities A and B. Activity A cyclizes geranylgeranyl pyrophosphate (GGPP) to give the labdane intermediate and B completes the cyclization process to yield *ent*-kaurene and/or other polycyclic structures. The second is the role of plastid compartmentation in the segregation of diterpene synthesis from other terpenoid biosynthetic processes. Further separation of biosynthetic processes is suggested by evidence that AB cyclization activity is a product of soluble enzymes (*313*) from which the diterpene is passed to membrane-bound mixed function oxygenases (*497*). The third is the role of GGPP as a key branch-point metabolite in polyisoprenoid metabolism. Each aspect will be discussed in detail below.

III.4.1. Diterpene Biosynthesis: AB Activity

The presence of two separable enzymatic activities associated with polycyclic diterpene (i.e., *ent*-kaurene) biosynthesis suggests that inactivation of the first (A inactivation) could result in the total absence of cyclic diterpenes and inactivation of the second (B inactivation) would lead to production of only labdanes and related bicyclic structures. Inactivation of A activity would prevent the cyclization of GGPP to the bicyclic intermediate. For the bicyclic product of A activity, absence of B activity would eliminate its role as precursor to *ent*-kaurane. If the enzymes of resin diterpene biosynthesis resemble those of gibberellin biosynthesis, the A and B activities serve as a model regulatory system accounting for many observed resin chemistry differences between plant groups. The largely tetracyclic diterpenoid chemistry of the tribe Heliantheae is radically different from the predominantly labdane and clerodane chemistries of the Astereae and Eupatorieae. Another large tribe, the Vernonieae yielded no significant reports of diterpenes despite intensive chemical analysis. Such widespread patters could be explained by a regulatory system based on either A or B inactivation.

In castor bean cell-free extracts, enzyme activity could be further differentiated into discrete (1) A activity, (2) AB activity which produced a range of polycyclic diterpenes including *ent*-sandaracopimara-8(14),16-diene, *ent*-beyerene, *ent*-trachylobane and *ent*-kaurene, and (3) discrete B activities that each correspond to distinct catalytic sites for the production of individual end products (e.g. sandaracopimara-8(14),16-diene) (*497*). The activation/inactivation of different B activities could account for major differences observed in the resin chemistries of large taxa. For example, a segment of the tribe Calenduleae is characterized by sandaracopimarane-type diterpenes.

III.4.2. Soluble Cyclization and Membrane-bound Oxgenation Enzymes

Kaurene synthetase (a term used collectively for A and B activities) appears to be a soluble stromal enzyme weakly associated with plastid membranes (*437*). Evidence from gibberellin biosynthesis suggests that subsequent to cyclization, the lipophilic *ent*-kaurene intermediate proceeds through a series of oxidation reactions catalyzed by a group of membrane-bound microsomal mixed function oxidases. Further oxidation of the more-polar product is evidently carried out by soluble oxidase enzymes to yield various gibberellins. The membrane-bound enzymes belong to the class of oxidases that require reduced pyridine nucleotide and O_2 as co-substrates and are cytochrome P-450 dependent. Members of this enzyme class may be more generally involved with diterpene oxygenation (*497*).

These results suggest a clear distinction between the biogenetic steps associated with the cyclization processes (AB activity) and subsequent oxidative modifications. Although the extent to which gibberellin biosynthesis can be generalized to other diterpenes is unknown, the organization of its biosynthesis provides a useful framework for interpreting resin diterpene structural variability.

III.4.3. GGPP and Diterpene Distribution

GGPP is a key branch point metabolite of polyisoprenoid metabolism. Apart from the cyclic diterpenes, it is the precursor of the tetraterpenes (carotenes, xanthophylls), and acyclic diterpenes (phytol). Cyclic diterpenes compete with compounds of established importance for either GGPP or its precursors (e.g., farnesyl pyrophosphate). The "spotty" nature of resin diterpene distribution may in part be explained by the obvious demand for carotenoids and phytyls and the relatively unestablished requirement for these nongibberellin diterpenes.

III.4.4. Compartmentation and Diterpene/Sesquiterpene Expression

GGPP biosynthesis and its conversion to acyclic, macrocyclic, bicyclic and polycyclic diterpenes may be localized within specific plastid compartments (*498*). This proposal fits into a broader scheme in which key isoprenoid metabolic branch points leading to mono-, sesqui-, and diterpenes are selectively activated only within certain membrane-bound sites. Compartmentation would then help to regulate the proximity of chain-length specific prenyl transferases to the appropriate C_{10}, C_{15}, and C_{20} cyclases, and modifying enzymes. Strict metabolic regulation would then be possible within a given compartment (*234*).

This model can be applied to the terpene-producing trichomes that typify Compositae species. Trichomes are clearly the primary sites of physical compartmentation of terpenoid biosynthesis. "Multiple compartmentalized biosynthetic sites" probably occur within the secretory cells of the trichomes (*234*). Thus, expression of a well-developed sesquiterpene and/or diterpene chemistry in a plant may only be the result of the activation of specific, preexisting "biosynthetic sites."

This view of compartmentation as an important regulatory factor in terpenoid biosynthesis (*498*) may help explain the extraordinary diversity of sesqui- and diterpene chemistries found even within groups of closely related species. A survey of glandular terpenoid chemistries reveals that some species are predominantly sesquiterpene producers, others are diterpene producers, and still others produce copious amounts of both. Such diversity even within clusters of related species suggests that the expression of the regulatory process controlling the activation/inactivation of compartmentalized sesqui- and diterpene biosynthesis may play a role in the speciation process.

FIGURE 2. Alternative cyclizations of geranylgeranyl pyrophosphate to produce normal- and *ent*-labdane products.

```
 I. Linear or Unicarbocyclic

 1.  Geranylnerol

 2.  Geranylgeraniol

 3.  Gerranyllinalool

 4.  Phytol-derived

80.  14-Methylgeranylnerol

 5.  Acanthoaustralane

 6.  Acanthoaustralane-6,11-epoxide

 7.  Isoacanthoaustralane

 8.  Geranylgeraniol-18,8-lactone

 9.  Oxepane

86.  Melcanthane

10.  Mikanofurane

11.  Centipedane

12.  Geranylgeraniol-1,20-lactone

13.  Geranylterpinene

14.  Isocembrene

15.  Geranylnerol-11,14-α-epoxide


II. Normal-Bicarbocyclics and Derivatives

 A. Bicyclic (regular)

16.  Normal-labdane

17.  Grindelane

18.  Normal-labdane, 8,13-epoxide-

19.  Normal-labdane, 8,12-epoxide-

20.  Normal-labdane-15,16-lactone
```

FIGURE 3. Linear diterpene skeletal types and cyclic diterpene skeletal types derived from normal and *ent*-labdane precursors.

21. Normal-labdane-20,12-lactone

22. Normal-labdane spiroketal

88. Normal-labdane-15,12-lactone

B. Bicyclic (seco- or nor-)

23. Normal-labdane, 7,8-seco-

24. Normal-labdane, 3,4-seco-

25. Grindelane, 8,9-seco-

89. Grindelane, 19-nor-

26. Grindelane, 8,17-bis-nor-8,9-seco-

27. Normal-labdane-20,12-lactone, 5,10-seco-

87. Hexanorlabdane

100. Normal labdane, 14,15-bisnor-

101. Normal labdane, 13,14,15,16-tetranor-

C. Bicyclic (irregular)

28. Relhaniane

29. <u>trans</u>-Normal-clerodane

30. Clerodane, 5,10-seco-

31. <u>cis</u>-Normal-clerodane

72. Chrysolane

90. Tricycloclerodane

III. Normal-Tricarbocyclics and Derivatives

32. Sandaracopimarane (Normal-pimarane)

33. Sandaracopimarane-20,8-hemiacetal

34. Cassane

35. Rosane

36. Sandaracopimarane, 19-nor-13-epi-

FIGURE 3. (contd.)

37. Normal-abietane

94. Stevisalane

IV. Normal Tetracarbocyclics

73. Normal-stachane

V. <u>ent</u>-Bicarbocyclics and Derivatives

A. Bicyclic (regular)

38. <u>ent</u>-Labdane

39. <u>ent</u>-Labdane-19,6-lactone

40. <u>ent</u>-Labdane-8,13-epoxide-15,16-hemiacetal

97. <u>ent</u>-Labdane dimer

98. <u>ent</u>-Manoyl oxide, 13-epi-

99. <u>ent</u>-Manoyl oxide

B. Bicyclic (seco- or nor-)

41. <u>ent</u>-Labdane, 17-nor-

42. <u>ent</u>-Labdane, 15-nor-

43. <u>ent</u>-Labdane, 14,15-bis-nor-

44. <u>ent</u>-Labdane, 13,14,15,16-tetra-nor-

45. <u>ent</u>-Labdane, 7,8-seco-

74. <u>ent</u>-Labdane, 18-nor-

C. Bicyclic (irregular)

95. Friedolabdane

46. <u>trans</u>-<u>ent</u>-Clerodane

47. Printziane

FIGURE 3. (*contd.*)

76. Conyscabrane

48. _ent_-Clerodane, 5,10-seco-

96. _trans_-_ent_-clerodane, 17-nor-

49. _trans_-_ent_-Clerodane, 13,14,15,16-tetra-nor-

50. _cis_-_ent_-Clerodane

82. Solidagonane

102. Friedolabdane, 3,4-seco-_ent_-

103. Friedolabdane, Rearranged 2,3-seco-_ent_-
VI. _ent_-Tricarbocyclics

51. _ent_-Labdane, cyclopropyl-

77. Gutierreziane

52. _ent_-Pimarane

53. _ent_-Pimarane-8,15-tetrahydrofuran

54. Cleistanthane (Cleistanthol)

85. Cleistanthane, iso-

55. _ent_-Abietane

56. Acritoconfertane

57. Acritoconfertane-acetal

83. Erythroxane

VII. _ent_-Tetracarbocyclics and Derivatives

58. _ent_-kaurane

59. _ent_-Kaurane, 17-nor-

60. _ent_-Kaurane, 18-nor-

61. _ent_-Kaurane, 19-nor-

62. _ent_-Kaurane, 9,10-seco-

63. _ent_-Kaurane-19,10-lactone, 9,10-seco-

FIGURE 3. (_contd._)

81. _ent_-Kaurane-19,6-lactone, 9,10-seco-

64. _ent_-Stachane (_ent_-Beyerane)

93. _ent_-Stachane, 19-nor-

65. Atisirane (Atisane)

66. Trachylobane

67. Helifulvane

68. Tetrachyrane

78. Villanovane

69. Lycoctonine-diterpene

VIII. Undetermined Stereochemistry

70. Pimarane, 18-nor-

79. Inulaefane

84. Hebeclinane (9,10-seco-labdane)

FIGURE 3. (_contd._)

9. OXEPANE

7. ISOACANTHOAUSTRALANE

6. ACANTHOAUSTRALANE—EPOXIDE

8. GERANYLNEROL—18,8—LACTONE

1. GERANYLNEROL

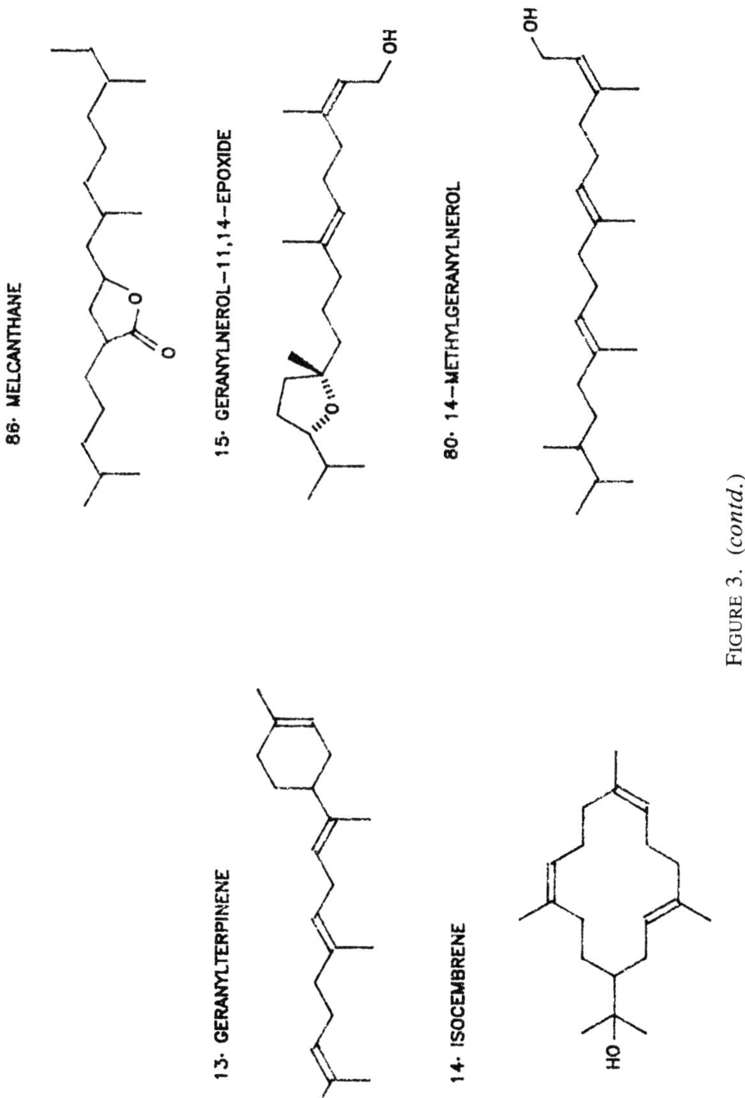

FIGURE 3. (*contd.*)

2· GERANYLGERANIOL

4· PHYTOL

5· ACANTHOAUSTRALANE

11· CENTIPEDANE

10· MIKANOFURANE

12· GERANYLGERANIOL—1,20—LACTONE

3· GERANYLLINALOOL

FIGURE 3. (contd.)

17· GRINDELANE

89· 19-NOR-GRINDELANE

25· 8,9-SECO-GRINDELANE

16· NORMAL-LABDANE

18· NORMAL-LABDANE-8,13-EPOXIDE

19· NORMAL-LABDANE-8,12-EPOXIDE

2· GERANYLGERANIOL

20· NORMAL-LABDANE-15,16-LACTONE

21· NORMAL-LABDANE-20,12-LACTONE

26· 8,17-BIS-NOR-8,9-SECO-
GRINDELANE

27· 5,10-SECO-NORMAL-
LABDANE-20,12-LACTONE

22· NORMAL-LABDANE SPIROKETAL

88· NORMAL-LABDANE-15,12-LACTONE

FIGURE 3. (*contd.*)

17· GRINDELANE

72· CHRYSOLANE

16· NORMAL—LABDANE

28· RELHANIANE

30· 5,10—SECO—CLERODANE

90· TRICYCLOCLERODANE

94· STEVISALANE

29· TRANS—NORMAL—CLERODANE

31· CIS—NORMAL—CLERODANE

FIGURE 3. *(contd.)*

 3. Biogenesis

36· 19-NOR-13-EPI-SANDARACOPIMARANE

34· CASSANE

2· GERANYLGERANIOL

OH

16· NORMAL-LABDANE

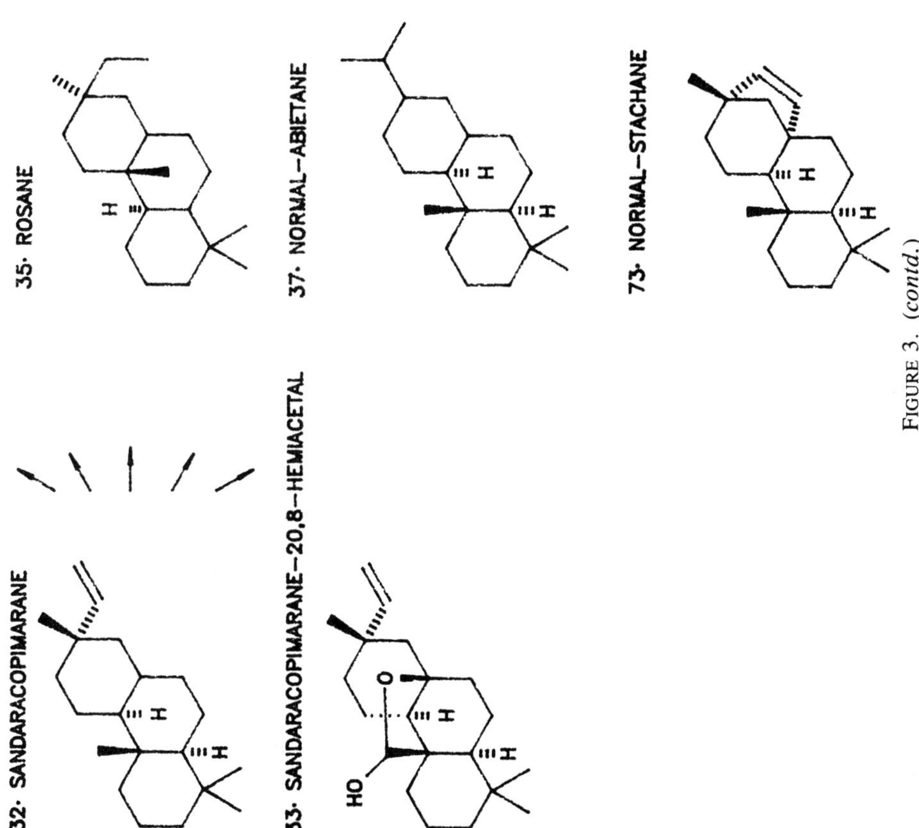

FIGURE 3. (*contd.*)

3. Biogenesis

42· 15-NOR-ENT-LABDANE

41· 17-NOR-ENT-LABDANE

74· 18-NOR-ENT-LABDANE

2· GERANYLGERANIOL

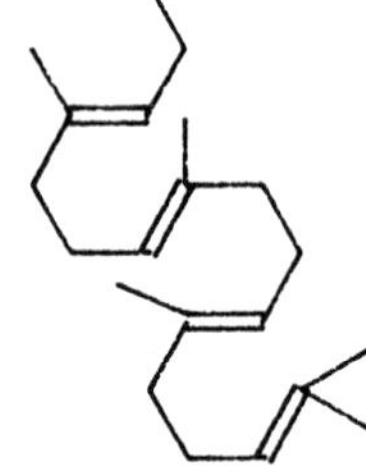

38· ENT-LABDANE

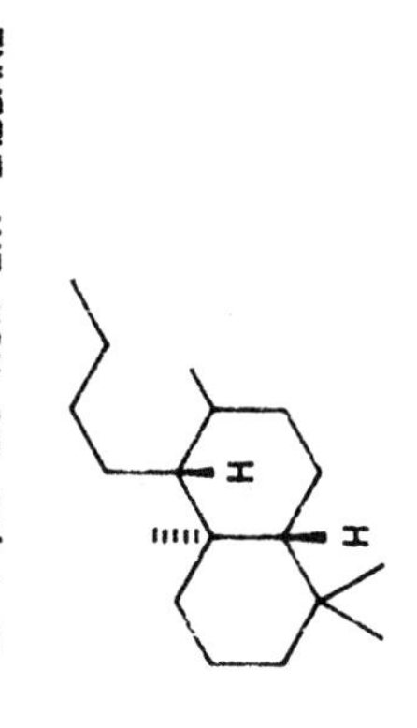

FIGURE 3. (contd.)

47· PRINTZIANE
76· CONYSCABRANE
48· 5,10–SECO–ENT–CLERODANE
38· ENT–LABDANE
95· ENT–FRIEDOLABDANE
46· TRANS–ENT–CLERODANE

82· SOLIDAGONANE

96· 17-NOR-TRANS-ENT-CLERODANE

49· 13,14,15,16-TETRA-NOR-TRANS-ENT-CLERODANE

50· CIS-ENT-CLERODANE

FIGURE 3. (*contd.*)

83· ERYTHROXANE

38· ENT—LABDANE

52· ENT—PIMARANE

53· ENT—PIMARANE—8,15—TETRAHYDROFURAN

51· CYCLOPROPYL—ENT—LABDANE

56· ACRITOCONFERTANE

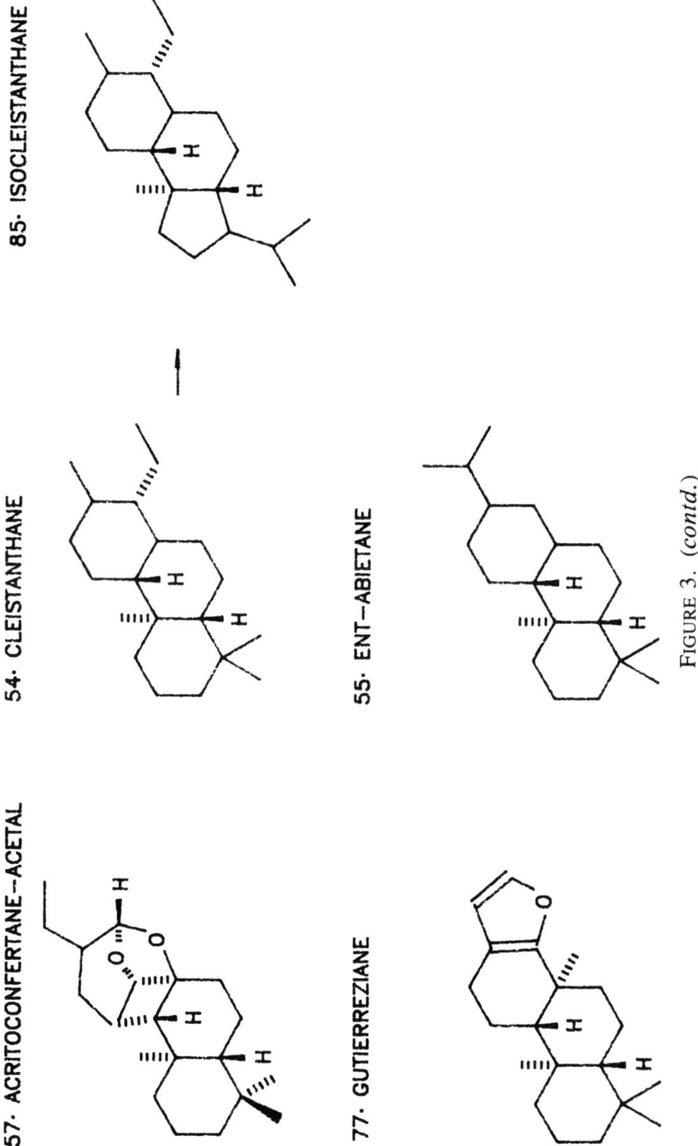

FIGURE 3. (contd.)

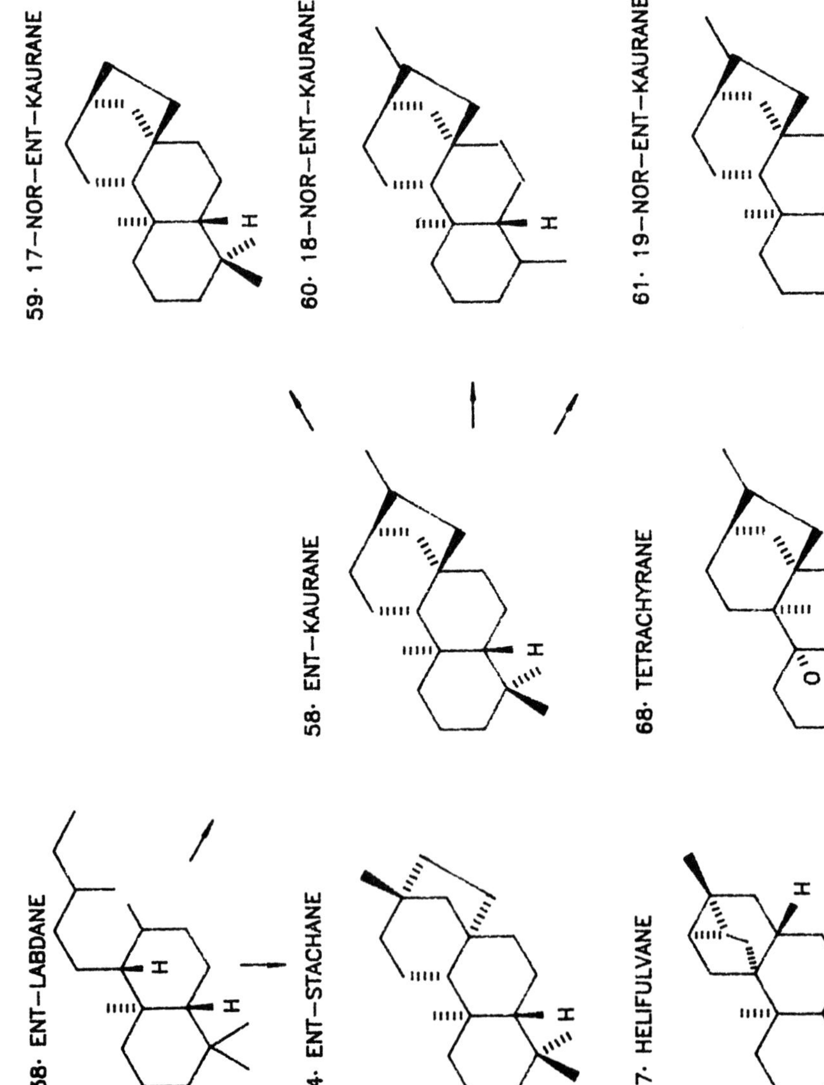
59· 17—NOR—ENT—KAURANE
60· 18—NOR—ENT—KAURANE
61· 19—NOR—ENT—KAURANE
58· ENT—KAURANE
68· TETRACHYRANE
38· ENT—LABDANE
64· ENT—STACHANE
67· HELIFULVANE

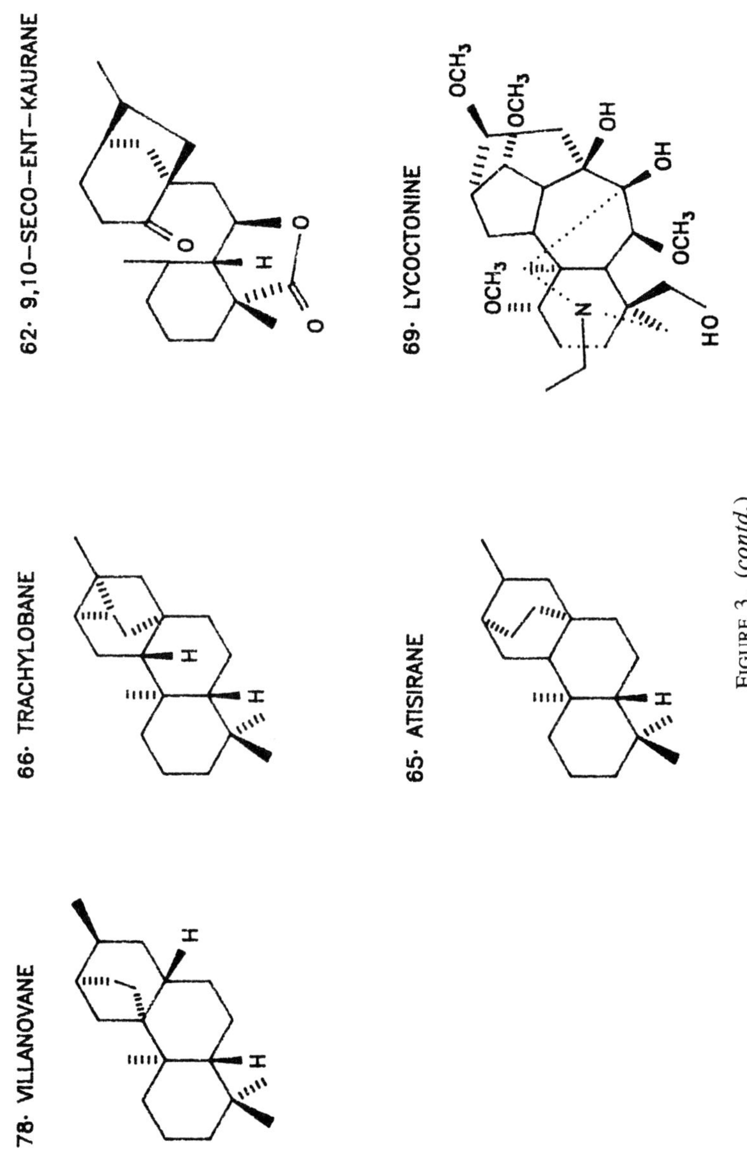

FIGURE 3. (contd.)

3. Biogenesis

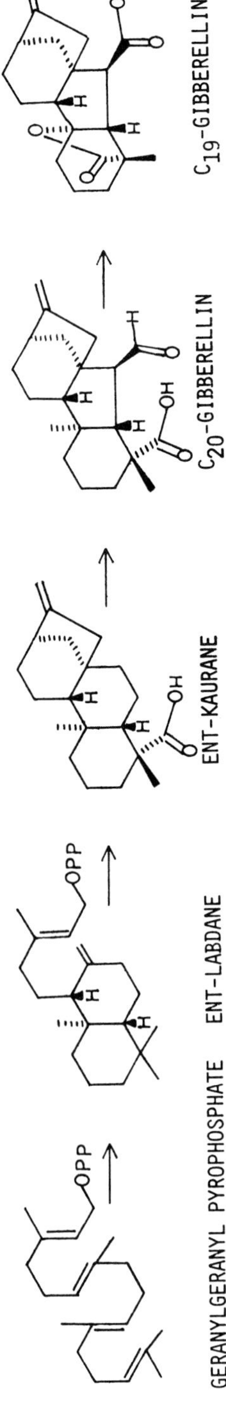

FIGURE 4. Generalized route of ''universal'' diterpene biogenesis culminating in C_{19}-gibberellin.

FIGURE 5. Proposed biogenesis of *ent*-labdanes and nor-homologs reported from *Ayapana* (Eupatorieae, *113*).

FIGURE 6. Proposed route of biogenesis of the 8,9-seco-grindelane chrysothame (223) from grindelic acid (167) in *Chrysothamnus paniculatus* (Astereae) (*341*).

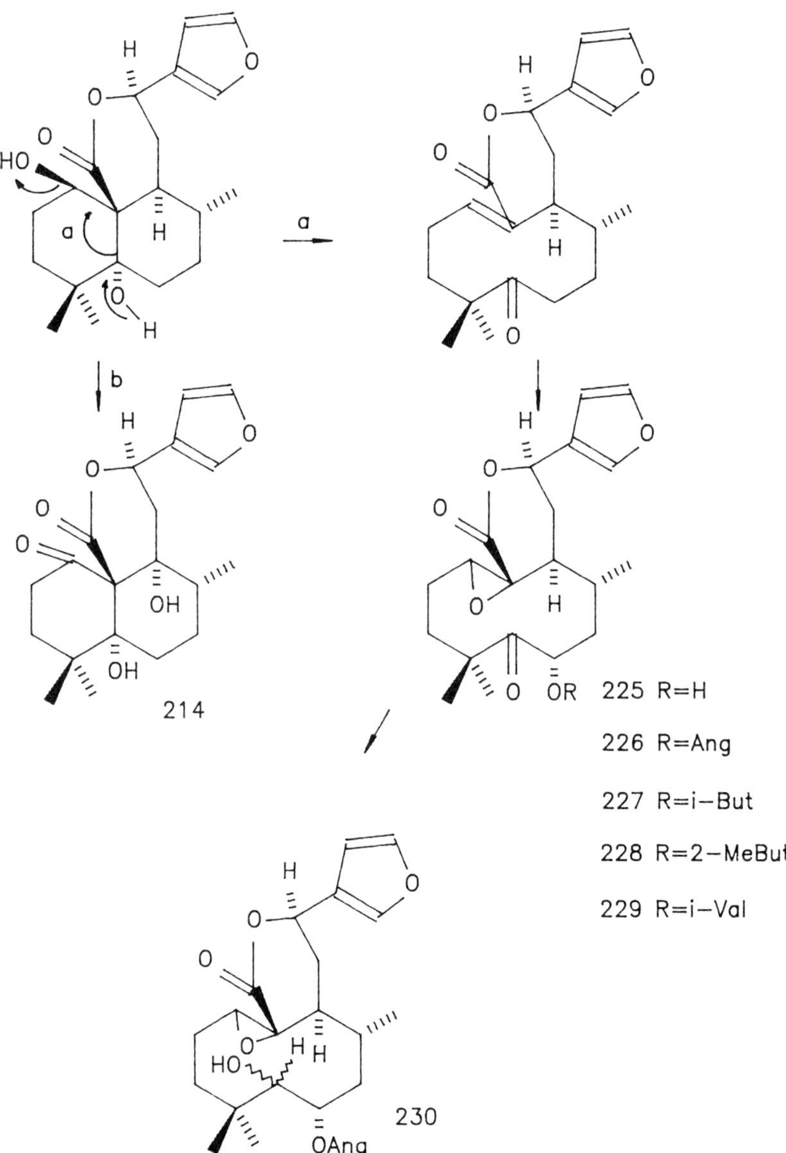

FIGURE 7. Proposed route of biogenesis of the 5,10-seco-normal-labdanes, 225–230, from a common precursor (similar to **214**) occurring in *Nidorella hottentotica* (*155*).

1· Relhania acerosa

2· Koanophyllon conglobatum

FIGURE 8. Diterpene skeleton products of the rearrangement of normal-labdane precursors: alternative routes for the biogenesis of diterpenes with the relhaniane skeleton, scheme 1 for compounds reported from *Relhania acerosa* (*90*) and scheme 2 for compounds from *Koanophyllon conglobatum* (*42*).

FIGURE 9. Proposed biogenesis of the chrysolane compound, 271, from the 6β-hydroxy-grindelic acid precursor, 172 in *Chrysothamnus paniculatus* (*341*).

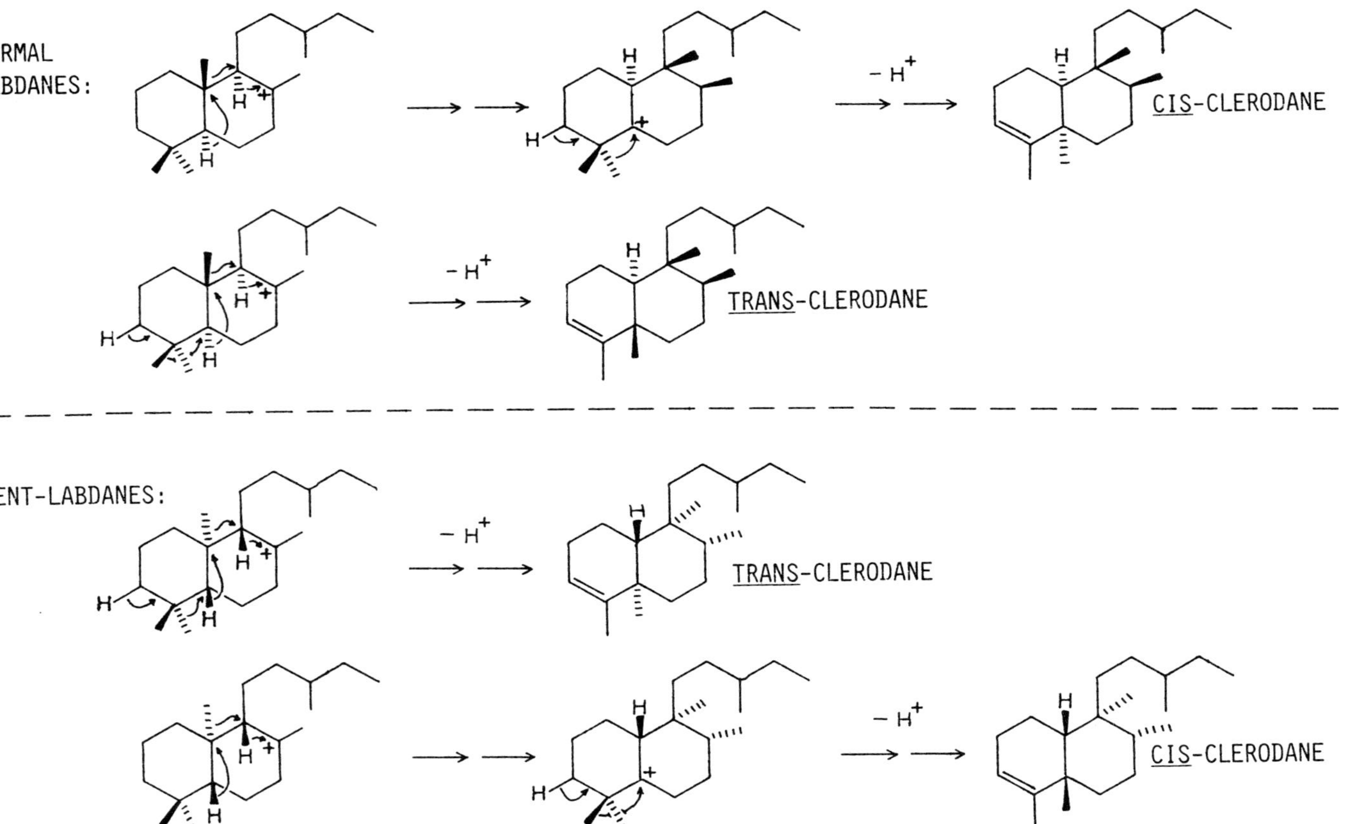

FIGURE 10. Proposed biogenetic routes for the four clerodane (kolavane) skeletal types.

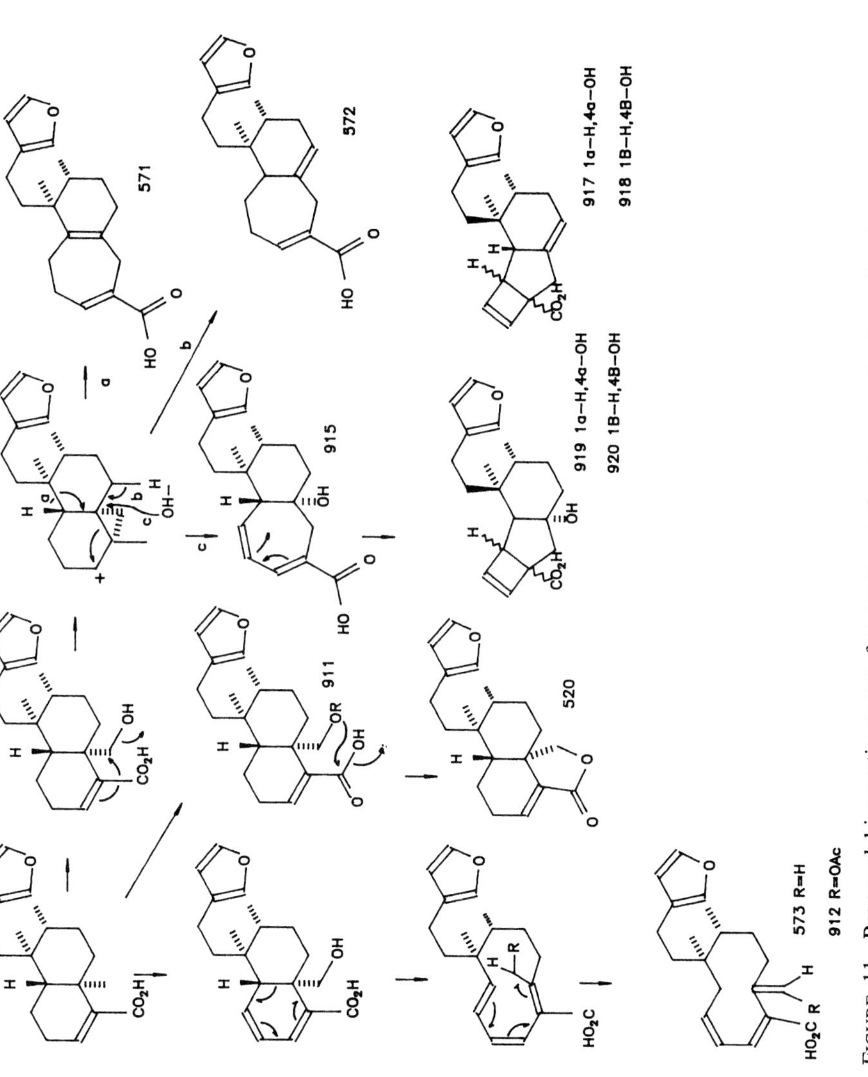

FIGURE 11. Proposed biogenetic routes for *trans-ent*-clerodanes and other related rearranged skeletons reported from *Conyza* (Astereae, *83*) and *Printzia* (Inuleae, *165*).

FIGURE 12. A proposed biogenetic route for solidagonane (**960**), a rearranged clerodane reported from *Solidago altissima* (*143*).

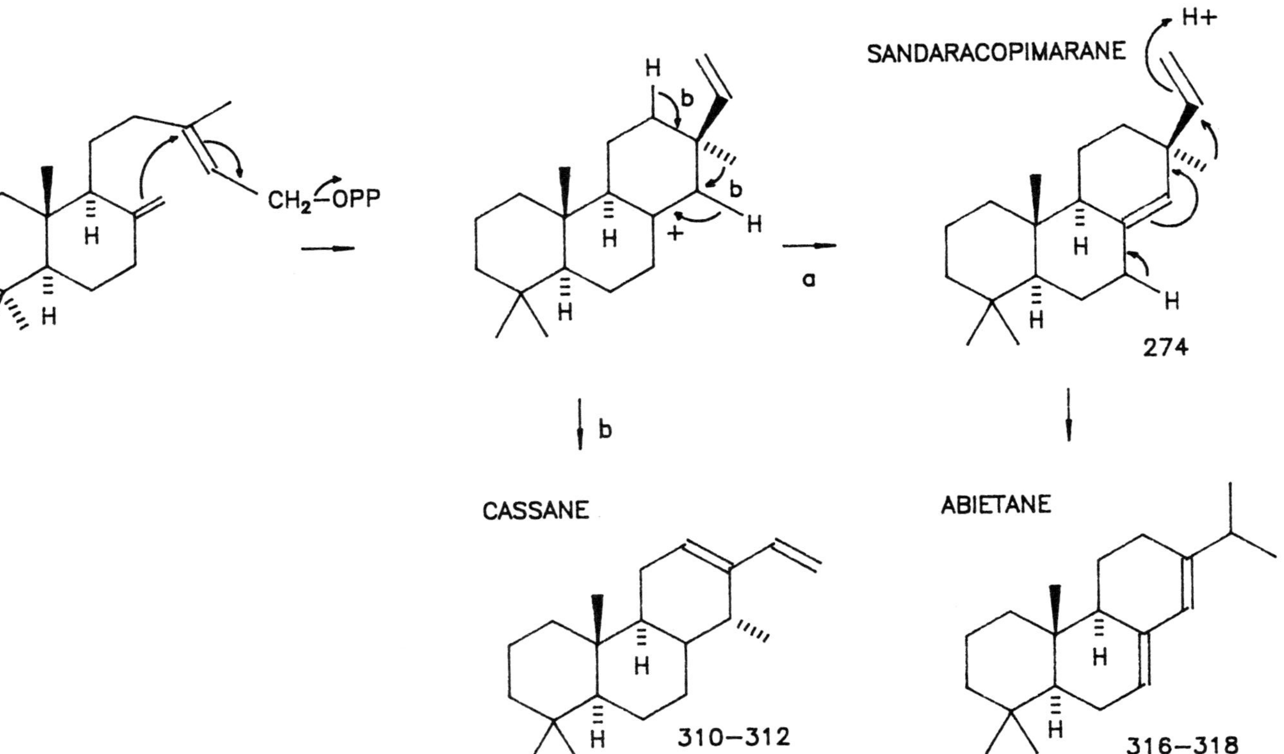

FIGURE 13. Proposed biogenetic routes for sandaracopimaranes (normal-pimaranes) and related skeletons.

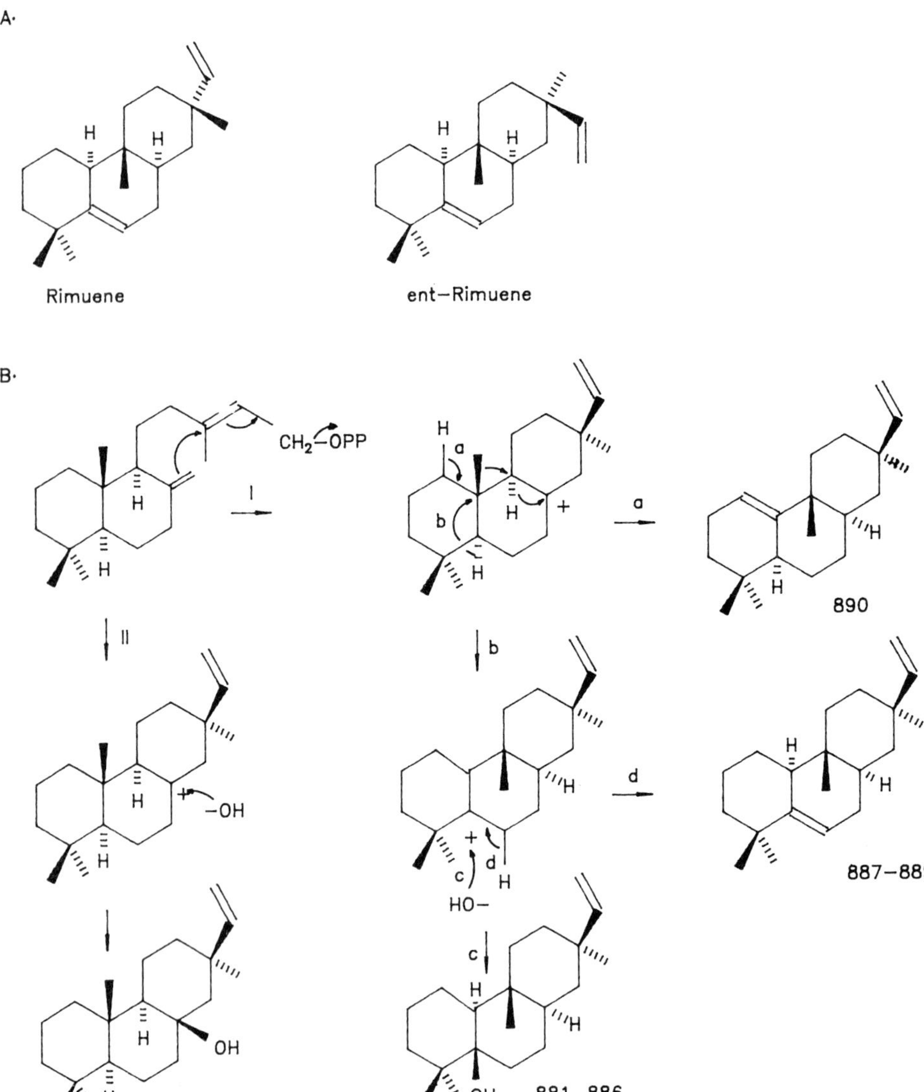

FIGURE 14. Proposed route of rosane biogenesis in *Trichogonia* (Eupatorieae) (*183*): A. an example of C-13 epimeric pair of rosane compounds (note that the prefix "*ent-*" refers not to an *ent*-labdane origin, bot to the C-13 orientivity relationship of this compound pair; B. biogenetic route for sandaracopimarane and rosane constituents of *Trichogonia*.

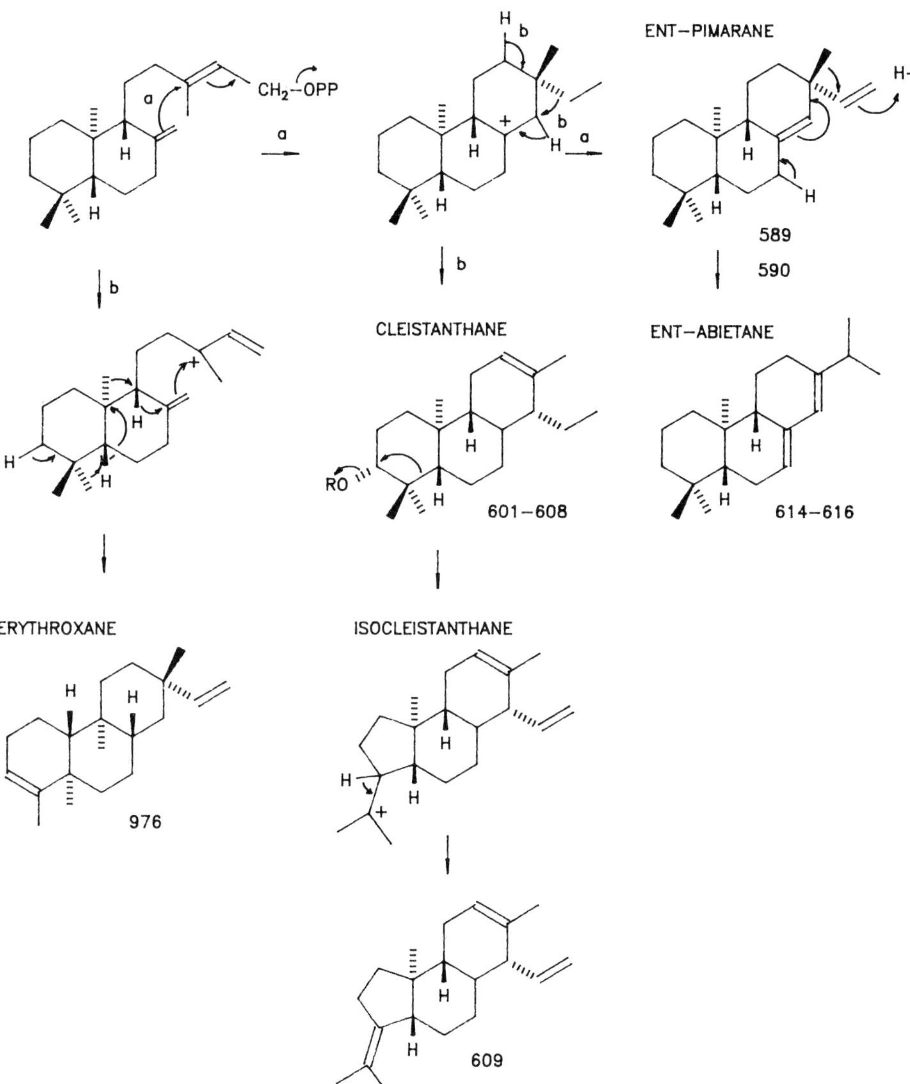

FIGURE 15. Proposed biogenetic routes for tricyclic *ent*-pimarane, *ent*-abietane, erythroxane, cleistanthane and isocleistanthane diterpene skeletons derived from *ent*-labdane precursors.

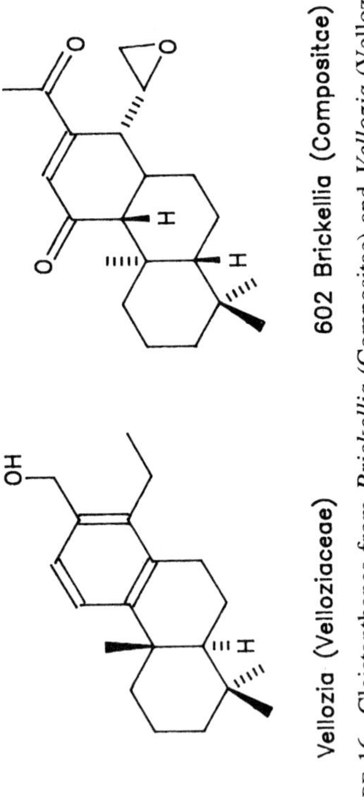

FIGURE 16. Cleistanthanes from *Brickellia* (Compositae) and *Vellozia* (Velloziaceae).

FIGURE 17. Proposed biogenesis of acritoconfertane (**617**) and acritoconfertane-acetal (**1100**) constituents of *Acritopappus confertus* (*184*).

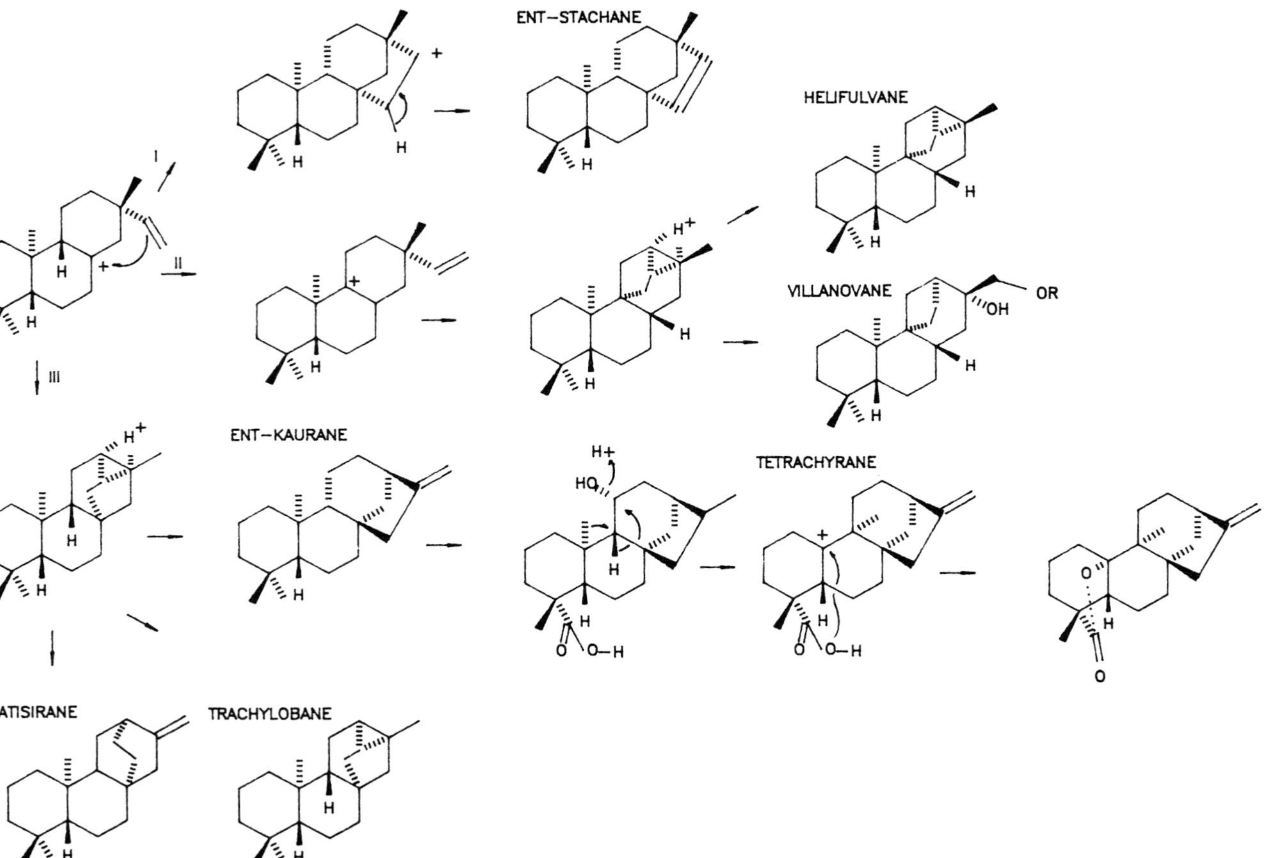

FIGURE 18. Tetracarbocyclic skeletons of *ent*-labdane origin and their proposed biogenetic relationship (in part *208*).

CHAPTER 4

Diterpene Distribution: Compositae

Green plants possess (1) geranylgeraniol, the required precursor for both carotenoid and gibberellin biosynthesis, (2) the enzymatic machinery to cyclize geranylgeraniol (or its isomer) to *ent*-kaurene, and (3) the enzymes for the oxidation of *ent*-kaurene and other diterpene substrates (*420*). Normally, the steady-state concentration of the products of these pathways falls below the level of detectability of all but the most sensitive analytical instruments. By comparison, the compounds that are the focus of this review constitute an "abnormal" biosynthetic excess detectable in the form of resins. Often, general plant metabolism is isolated from these biologically active resins by the comparmentation of resin biosynthesis and storage within glandular trichomes.

The production of copious trichome resin products raises the question of whether or not these resins are adaptive. Various proposals have been advanced regarding the defensive role of such compounds and the link between their structural variability and the coevolutionary process. However, there is little evidence linking these compound's biological function to the fine detail of their structural variability. The extent to which the observed chemical variability is an expression of past selection for specific novel chemical deterrents rather than the product of near-random activation of alternative terpenoid biosynthetic pathways is not known. Thus, in looking at the diterpene variability within the Compositae, care should be exercised in automatically associating a discrete adaptive function with each novel chemistry.

Whatever the cause of the variability, chemical variation patterns correlate with taxonomic boundaries and may prove useful as taxonomic characters. Consequently, chemical differences between taxa may be useful chemical indicators of phylogenetic relationships. As a corollary, our understanding of the factors governing terpenoid biogenesis benefits from superimposing a plant taxonomic framework on the discussion of diterpene distribution and the underlying biosynthetic diversity.

IV.1. Subfamilial Distribution Patterns

The Compositae is so well differentiated from other families that any claims concerning its affinities are difficult to defend. In contrast, its own internal boundaries separating tribes and genera are ill defined, due to complex and overlapping morphological and ecological diversity (*233*). No taxonomic problem better exemplifies these poorly demarcated internal boundaries than the definition of subfamilies. On this subject, Cronquist noted (*233*)...

The classical organization of one subfamily for the Lactuceae, and one for everything else, remains the best such expression of the pattern of diversity within the family. Perhaps three subfamilies could be recognized, one for the Lactuceae, one for the radiate [flowering head type with central flowers displaying regular tubular corollas and marginal flowers in which part of the corolla is prolonged into a strap-shaped ligule] tribes, and one for the discoid [flowering head type in which all flowers have regular tubular corollas] tribes plus the Mutisieae...

Although alternative classifications have been proposed (*226*), the findings of the recent study of subfamilial boundaries by Wagenitz (*488*) were largely compatible with Cronquist's view. In the Wagenitz phylogenetic proposal, the Compositae was divided into two subfamilies, the Asteroideae and Cichorioideae. The Asteroideae was further subdivided into two groups (Table 4 [p. 438]). A major area of uncertainty for Wagenitz was the placement of the Anthemideae in one of these two Asteroideae groups. He opted for group 1, but admitted that with the exception of a pollen anatomical feature, this tribe could be equally well situated in group 2. The most unusual aspect of his scheme was the assignment of the Vernonieae and Eupatorieae to different groups (*488*). Traditionally these two tribes had been associated by their lack of ray-flowers and their purple flower color. Wagenitz considered the Cichorioideae to be more closely related to group 1 than to group 2.

Distributions of three major natural products groups, sesquiterpene lactones (*458*), benzofurans/chromenes(benzopyrans) (*429*) and diterpenes, were superimposed on the Wagenitz arrangement of the Compositae (Table 4). While sesquiterpene lactones appear to be well-distributed across the tribes, both diterpenes and benzofurans and pyrans are much more heavily concentrated in Wagenitz's group 2. Their presence in the group 1 taxa has only rarely been reported. In support of his separation of the Vernonieae and Eupatorieae, the former tribe, despite intensive chemical investigation, has yielded no reports of either diterpenes or bezofurans/pyrans, while the latter is a major producer of both.

A tribe with uncertain affinities based on nonchemical characters, the Anthemideae, displays a closer chemical link with group 1. Intensive chemical analysis of this tribe failed to produce reports of either diterpenes of benzofurans/pyrans. Finally, the chemistry of the Lactuceae (Cichorioideae) more closely resembles that of the group 1 tribes, a finding which is in agreement with Wagenitz.

IV.2. Tribal Affinities

Patterns of diterpene variation are relevant as taxonomic characters to relationships only within the Wagenitz group 2 tribes. Within this group, diterpene patterns distinguish (1) the Heliantheae because of its *ent*-kaurane-based polycyclic profile, (2) the Calenduleae based on its sandaracopimarane chemistry, and (3) the Eupatorieae and Astereae based on their complex labdane/clerodane bicyclic profile (Table 5 [pp. 439–440]).

IV.2.1. Heliantheae

Natural products variability within the Heliantheae usually parallels the extensive variability of morphological features. As noted earlier (*458*), the sesquiterpene lactones display as much variability within the Heliantheae as exists within the entire family. When viewed against this background, the diterpene chemistry of the Heliantheae seems remarkably uniform. Of the approximately 180 Heliantheae taxa analyzed, only 29 yielded reports of non-polycyclic diterpenes. These 29 reports included 16 of linear diterpenes, 9 or normal-labdanes, 1 of *ent*-labdanes, 2 of *trans*-clerodanes, and 2 of *ent*-pimaranes (Table 5). The picture that emerges for this tribe is one of a fairly uniform kaurane-based polycyclic chemistry and a small group of exceptions, the majority of which contain linear homologs of the geranylgeraniol precursor. As yet no significant structural novelty is reported for any diterpene group other than the polycyclics.

III.2.2. Calenduleae

The Calenduleae is the source for most sandaracopimarane (normal-pimarane) reports (Table 5). Of the 24 investigated taxa, only two produced exclusively non-sandaracopimarane chemistries; *Dimorphoteca aurantiaca* and *D. pseudoaurantiaca* gave *ent*-stachanes. A possibly misleading report (*78*) included a series of *Garuleum sonchifolium* structures drawn as *ent*-pimaranes, but referred to as sandaracopimaranes. As two of these structures were described as being previously reported from *G. pinnatifidum*, it was concluded that the structures were drawn incorrectly. Bohlmann and Grenz (*79*) suggested that this tribe was divisible into two groups, a sandaracopimarane-producing group (*Dimorphotecha, Osteosperum, Garuleum* and *Chrysanthemoides*) and a smaller sandaracopimarane-deficient group (*Calenula* and *Castalis*). Investigations of *Calendula* and *Castalis* taxa resulted in the detection of no diterpenes (*79*). *Dimorphotheca* yielded sandaracopimaranes, but as mentioned above two species produced *ent*-stachanes.

IV.2.3. Euaptorieae and Astereae

The similar diterpene profiles linking the Astereae and Eupatorieae are notable because of their relevance to the preliminary outcome of a recent phylogenetic analysis of the Compositae. The preliminary working cladogram (phylogenetic tree) resulting from this analysis indicated that these two tribes were a major monophyletic lineage within the family (K. Bremer, pers. commun.). Parallels in the diterpene chemistries of these tribes adds to the set of characters shared by them.

IV.3. Infrageneric Distribution Patterns

As yet, no systematic study focusing on generic-level diterpenoid distribution patterns has been reported for the Compositae. Consequently, little is known about how diterpenes vary within groups of related species. As a partial remedy for this deficiency, diterpene interspecific variation is described for representative genera of the major diterpene-producing tribes: Eupatorieae (*Brickellia*, *Stevia*), Astereae (*Baccharis*), Inuleae (*Helichrysum*), and Heliantheae (*Helianthus*, *Montanoa*).

Diterpenoid distribution was contrasted to that of other terpenoids for the two genera, *Baccharis* and *Helianthus*. These two taxa were selected for a more detailed examination, because their chemistries represented the two major types of diterpene complements: a bicyclic labdane/clerodane type (*Baccharis*) and a polycyclic type (*Helianthus*).

IV.3.1. *Brickellia*

Most of the approximately ninety *Brickellia* (Eupatorieae: Alomiinae) species inhabit the North American deserts, although three species are South American disjuncts. The diterpene chemistries of the ten investigated taxa (Table 6 [p. 441]; Figure 19 [pp. 457–461]) are remarkably uniform. Normal-labdanes often characterized by novel $2\alpha,3\alpha$-hydroxylation/esterification occur in all diterpene-producing species. In *B. eupatoriedes*, these $2\alpha,3\alpha$-substituted normal-labdanes co-occurred with cleistanthanes and other *ent*-pimarane-derived skeletal types. Production of normal-labdanes together with tri- and polycyclic *ent*-labdane-derived skeletons is an often-repeated theme in the diterpene chemistries of Compositae taxa.

IV.3.2. *Stevia*

The 150 to 300 *Stevia* species (Eupatorieae: Piqueriinae) range from southwestern North America to northern Argentina. Seventy of its approximately ninety North American species occur in Mexico (*464*). Chemical investigation of this genus has been hastened by the discovery of high concentrations of sweet-tasting *ent*-kauranoid glycosides in several of the species, one of which, *Stevia rebaudiana*, is currently used as a commercial source of sweetening agents in Japan and Brazil. Despite the common use of these compounds for sweetening pickled vegetables, sea foods, soft drinks, soy sauce, and so on, questions have been raised about possible health risks associated with the consumption of the diterpene aglycone, steviol (*368*).

Like *Brickellia*, *Stevia* species produces mostly mixtures of normal-labdanes and further-cyclized *ent*-labdane-derived compounds, specifically *ent*-kauranes. In several species, only *ent*-labdanes or their clerodane derivative were reported (Table 7 [pp. 442–443]; Figures 20 [pp. 462–463] and 21 [pp. 464–466]). Three of the *ent*-kaurane-producing species, *S. rebaudiana*, *S. ovata* and *S.

paniculata, produce glycosides in which glycosidic groups of varying complexity are attached at one or more sites, specifically C-13, C-15, or C-19. The distribution of such diterpene glycosides inside and outside this genus is probably underestimated due to the lipophilics-specific extraction methods used in most laboratories.

IV.3.3. *Baccharis*

Centered in South America, the approximately four hundred perennial herbaceous or shrubby *Baccharis* species extend in their range into North America (*292*). Their closest relatives most likely include *Archibaccharis* and *Conyza* (A. Cronquist, pers. commun.). The dominant lipophilic constituents of the leaf extracts from 54 taxa are diterpenes, triterpenes, sesquiterpenes, flavonoids, and other phenolics (Table 8 [pp. 443–446]; Figure 22 [pp. 467–468]). Although *ent*-clerodanes are the most frequently reported diterpene components, labdanes and tetracyclic *ent*-kaurenes and *ent*-stachanes are common (Figure 22).

No clear variation pattern emerges from the examination of the diterpene distribution within this genus. Eighteen of the investigated collections failed to produce any detectable diterpenes, yielding instead various combinations of sesqui- and triterpenoids and phenolics. No clear-cut mutual exclusivity is evident in the distribution of diterpenes and these other terpenoid classes.

The significance of individual negative diterpene reports is difficult to estimate. This difficulty is highlighted by the observation that of the three taxa that have been examined twice, *Baccharis ramoisissima*, *B. tucumanensis*, and *B. alaternoides*, two yielded detectable diterpenes in only one of the two investigations. Given that little is known about (1) factors governing the plant developmental stage at which diterpene production is initiated and (2) factors controlling the rate of production, data for these two taxa suggest that species reports of diterpene absence be treated cautiously.

Looking more closely at the diterpene-producing species, it is apparent that *Baccharis* produces relatively more *ent*-labdanes and related skeletons than were found in *Brickellia* and *Stevia* (Table 9, [pp. 447–449], Figures 23 [p. 469], 24 [pp. 470–473], and 25 [pp. 474–475]). The most common of these skeletons is the 5α,10β-*trans-ent*-clerodane. Among the *ent*-skeletal classes, there is a rough mutual exclusivity of the *trans*-clerodanes and the polycyclic compounds (e.g., *ent*-kaurenes). Normal-labdanes and derivatives occur rarely within this genus.

IV.3.4. *Helichrysum*

The large genus *Helichrysum* (approx. five hundred species; Inuleae: Gnaphaliinae) extends through Eurasia, Africa, Madagascar, Australia and New Zealand. As in *Baccharis*, many investigated *Helichrysum* species failed to yield detectable quantities of diterpenes (*171*). Of the diterpene-producing species, the majority produced *ent*-labdane-derived compounds of the kaurane, trachylobane, helifulvane, and atisirane polycyclic types (Table 10 [pp. 449–

451], Figures 26 [pp. 476–477], 27 [pp. 478–479], and 28 [pp. 480–481]). Other species produced chemistries combining novel unicyclic geranylterpinenes, related acyclic compounds, normal-labdanes, and *ent*-labdanes. In one of two separate reports regarding the same species, *H. albirosulatum*, (*171*), *ent*-labdanes were identified, although in the other (*180*) normal-labdanes were found. The report of a macrocyclic diterpene (isocembrene) from *H. subfalcatum* is the only such report from the Compositae.

IV.3.5. *Helianthus*

The distribution of the New World sunflower genus, *Helianthus* (fifty species; Heliantheae: Helianthinae) is centered in North America. It is very closely related to *Viguiera*, and in fact the generic distinctions between the two genera break down in several species (*316*). Many widespread *Helianthus* taxa are the products of hybridization. A summary of the diterpene variability within *Helianthus* (Table 11 [pp. 452–455], Figure 29 [pp. 482–483]) shows that while diterpenes have not been reported from all investigated taxa, many plants produce resins containing both sesquiterpene lactones and diterpenes. Several species (e.g., *H. glaucophyllus* and *H. microcephalus*) produce resins that are rich in sesquiterpene lactones (*249*) but lack diterpenes, while others (e.g. *H. occidentalis* and *H. rigidus*) produce abundant diterpenoids and no detectable sesquiterpene lactones. Chemical profiles of diterpene-producing *Viguiera* and *Helianthus* species display many similarities (Table 11).

IV.3.6. *Montanoa*

Montanoa (25 species; Heliantheae:Montanoinae) ranges from northern Mexico to Peru, with the greatest species concentration occurring in southern Mexico and Guatemala. Although few diterpenes are reported from this genus, their pharmacological value justifies including a discussion of *Montanoa* diterpenes. The focus of attention has been the novel oxepane diterpenes (Figure 30 [p. 484]) identified as the abortifacient agent in the medicinal tea brewed from *M. tomentosa* (*419, 382*) (See Chapter 5). Oxepanes have been reported from seven of the nine investigated *Montanoa* taxa (Table 12 [p. 456]; Figure 30). The only other reported diterpenes are *ent*-stachanes, kauranes and related structures. Typically, oxepanes are present in the extract in lower concentrations than the co-occurring sesquiterpene lactones.

IV.3.7. Interspecific Variation: An Overview

In *Baccharis*, *Helichrysum*, and other genera detectable levels of diterpenes may typify only a fraction of the species. The absence of detectable quantities of diterpenes, a trait which can characterize an entire tribe (e.g., Vernonieae) or groups of genera within tribes, seems also to appear within groups of related species some other members of which produce diterpenes. As noted in

Baccharis, some collections of a species may yield diterpenes while others do not.

This "spotty" and sometimes unpredictable distribution can have many possible explanations. Any environmental or developmental factor which disrupts the proper functioning of glandular tissues can affect diterpene production. Another important consideration is the central role the diterpene precursor, geranylgeranyl pyrophosphate, plays in terpenoid biogenesis. Its biosynthesis is dependent on a supply of farnesyl pyrophosphate (the C_{15} precursor of both sesqui- and triterpenes) and isopentenyl pyrophosphate (C_5). The existence of competing demands for the C_{15} precursor is suggested by instances in which sesquiterpenes (e.g., hydrocarbons, alcohols or lactones) and/or triterpenes characterize extracts lacking diterpenes. Additionally, geranylgeraniol is the precursor of phytol, a necessary component of chlorophyll, and tetraterpenoids (C_{40}), such as carotenoids. Any of the competing demands for the C_{20}, C_{15}, or C_5 precursors could effect diterpene production.

In summary, several distributional trends characterize the diterpene-producing species of the Compositae genera:

1. Representatives of both the normal- and *ent*-series occur within genera and often within the same species.
2. When normal- and *ent*-series structures co-occur it is usually as a mixture of normal-labdanes and some tri- or polycyclic *ent*-labdane derivatives. Normal-series structures infrequently exist as tri- or polycyclics; the most common of such structures, the normal-pimaranes (sandaracopimaranes), achieve prominence only within the small tribe Calenduleae.
3. Structural novelty (e.g., oxpanes of *Montanoa*, $2\alpha,3\alpha$-dihydroxylation of *Brickellia*) can be diagnostic for groups of related species.
4. Despite early statements regarding the rarity of linear diterpenes in plants (*279*), linear diterpenes and their hetero- and carbocyclic derivatives are hardly rare within genera of the Compositae.

TABLE 4. Distribution of sesquiterpene lactones (SQTL), diterpenes (DIT), chromenes (CHR), and benzofurans (BENZF) in the Compositae tribes organized according to Wagenitz (*488*). Sesquiterpene lactone data are from *458*. Chromene and benzofuran data are from *429*.

	Natural Product Classes			Head Type	
Tribes[1]	SQTL	DIT	CHR & BENZF	DISC	RAD[5]
Asteroideae:					
Group 1:					
Vernonieae	216[3]	-[4]	-[4]	+	
Liabeae	16	-	-		+
Mutisieae	29	1	2	+	
Cynareae[2]	132	1	-	+	
Arctoteae	13	1	-		+
Anthemideae	412	-	-		+
Group 2:					
Eupatorieae	184	118	65	+	
Astereae	7	96	15		+
Heliantheae	892	192	33		+
Senecioneae	744	10	30		+
Calenduleae	1	23	1		+
Inuleae	114	29	19		+
Cichorioideae:					
Lactuceae	30	-	-	ligulate	

[1] Tribes organized according to Wagenitz (<u>488</u>).

[2] Cynareae includes Wagenitz's tribes Cardueae and Echinopeae.

[3] Number of reports of compounds belonging to this chemical class in taxa of this tribe.

[4] No. of taxa reported to produce compounds of this chemical class.

[5] Flowering heads are either discoid or radiate.

TABLE 5. The Number of taxa (organized by tribes) reported to produce major diterpene skeletal types.

Diterpene Skel. No.	VE	EU	AS	IN	HE	AN	SE	CA	AR	CY	MU	LA	LI	AR
Linear														
1.		12		1	5									
2.		8	1		3									
Cyclic														
I. Normal														
16.		29	9	3	9		1							
17.			17											
18.		3	2		1									
19.		2			5									
20.			2											
21.		1	1											
31.		1	7	1										
32.		2	1		3		2	20						
34.								1						
35.		2			1									
37.			2	1										
II. _ent-_														
38.		18	10	2	1									
39.		1												
46.		7	16		1						1			
50.			5											
52.		2			5		2							
55.			4	1										
58.		30	17	17	115		5		1	1				
64.			2	3	9		2							
66.					24									

Representatives of the major diterpene skeletal types characteristic of the Heliantheae, Calenduleae, Astereae and Eupatorieae.

TABLE 5. *(contd.)*.

HELIANTHEAE

ent—Kaurane

CALENDULEAE

Sandaracopimarane

(normal—pimarane)

ASTEREAE

EUPATORIEAE

normal—Labdane

ent—Labdane

normal—Clerodane

ent—Clerodane

TABLE 6. Distribution of diterpenes in *Brickellia* (Eupatorieae) species.

| | Linear | Normal | | | Ent- | | | |
| | A./Uni. | Bi. (L) | Bi. (C) | Tri. | Bi. (L) | Bi. (C) | Tri. | Poly. |
Species								
B. annulosa						511		
B. argyrolepis		101 102 122 123						
B. corymbosa		106 108						
B. diffusa		97 215 216 217 218						
B. eupatoriedes		78 98 99 103 104 105 126 127 128					601 602 603 604 605 606 607 608 609 610	
B. paniculata		126						
B. squarrosa		106 107 108 109						
B. vernicosa		991 992 993 994 995						
B. veronicaefolia		124 125 129 220						
B. veronicaefolia A. Gray var. typica		129						

TABLE 7. Distribution of diterpenes in *Stevia* species.

Species	Linear A./Uni.	Normal Bi. (L)	Normal Bi. (C)	Normal Tri.	Ent- Bi. (L)	Ent- Bi. (C)	Ent- Tri.	Poly.
S. berlan-diera		145 1021 1022 1023						
S. bupthal-miflora								649 650 651 652
S. eupatoria								641 998
S. jaliscensis		118						
S. lemmonia	1020							
S. lucida Lag. var. Bipontini					415			
S. monar-daefolia		143 153						625 649
S. myriadenia	3 23				421	493 494		625 650 651 774
S. origa-noides					428 429 808 809 810 811			
S. ovata								895 896 897 898 899
S. paniculata								895 896 897 898 899
S. poly-cephala						570		
S. rebaudiana		195			428 429			682 683 684 685

TABLE 7. (*contd.*).

Species	Linear A./Uni.	Bi. (L)	Bi. (C)	Tri.	Bi. (L)	Bi. (C)	Tri.	Poly.
								686
								687
								688
								689
S. sali- cifolia		121 1024 1025 1026			154[abs. ster. unknown] 429			
S. setifera								625

Stevia Diterpene Cyclic Classes — Normal / Ent-

TABLE 8. Major natural products reported from *Baccharis* (Astereae).

Species	C20	C30	C15	Flavonoid	Phenolic	Acetylenic
B. alaternoides H.B.K. (1) Peru	6	11	14	16		
B. alaternoides H.B.K. (2) Peru				16		
B. articulata (Lam.)Pers. Arg.	5,6					
B. caespitosa (Ruiz and Pavón) Pers. Ecua., Peru, Bol.						
B. calvescens DC. Braz.	6	10,13	14			
B. cassinaefolia DC. Braz.	4	11,12,13	14			20
B. chilco H.B.K. (1) Col.	4,6	10				
B. chilco H.B.K. (2) Col., Peru		13	14			
B. concinna Barroso	8	12,13	14	16		

TABLE 8. (*contd.*).

Species	Terpenoid			Flavonoid	Phenolic	Acetylenic
	C20	C30	C15			
B. conferta H.B.K. Mex.	6					
B. eggersii Hieron. Peru	3	11	14			
B. genistelloides (Lam.)Pers. Col., Ecua., Peru	1,6		14			
B. grandicapitulata Hieron. Peru	6	11	14			
B. halimifolia		11				
B. helichryisoides DC. Braz.					19	
B. heterophylla H.B.K.		10				
B. hutchisonii Cuatr. Peru	6	11	14			
B. intermixta Gardn. Braz.	2,8		14			
B. juncea (Lehrn.)Desf. Braz.					18	
B. kingii Cuatr. Peru	6		14	17		
B. latifolia (Ruiz & Pavón) Pers. Bol., Ecua.		11,13	14,15			
B. leptocephala DC. Braz.		11	14	16		
B. microcephala (Less.)DC. Para.	6		14			
B. minutiflora Braz.	8	12	14			

TABLE 8. (*contd.*).

Species	Terpenoid			Flavonoid	Phenolic	Acetylenic
	C20	C30	C15			
B. nitida (Ruiz & Pavón)Pers. Peru	4	10	14			
B. oxydonta DC. Braz.	2	13	14	16		
B. pedunculata (Miller)Cabrera Ven., Peru		11	14			20
B. polyphylla Gardn. Braz.	8		14			
B. pylicoides H.B.K.	8	10	14			
B. quitensis H.B.K.	8	12	14			
B. ramoisissima Gardn. (1) Braz.		12	14	16		
B. ramoisissima Gardn. (2) Braz.	8				18	
B. reticularia DC. Braz.		11,12,13	14	16		
B. salicifolia (Ruiz & Pavón) Pers. Col., Arg., Chile	6	10,11		17		
B. salzmannii DC. Braz.	8	12	14			
B. sarothroides A. Gray	6					
B. scandens R. & P. Peru			14			
B. scoparia (L.)Sw. Jamaica	3,5	10,11	14			
B. serrulata (Lam.)Pers. Braz.		13	14	16		

TABLE 8. (*contd.*).

Species	Terpenoid			Flavonoid	Phenolic	Acetylenic
	C20	C30	C15			
B. sternbergiana Steud. Peru	1,2		14	16	18	
B. subdentata DC. Braz.	4	10	14			
B. tola Chile	2,9	10				
B. tricuneata var. lineata Cuatr.	6	10				
B. tricuneata var. ruiziana Cuatr. Col, Ven.					21	
B.tricuneata (L.f)Pers. var.tricuneata Col., Ven.	4,5,7			17	19	
B. trimera Braz.	6			17		
B. trinervis (Lam.)Pers. Braz., Mex.		12,13	14			20
B. trinervis var. rhexoides (H.B.K)Baker Bol., Ecua.				17		20
B. truncata	8	12	14	16		
B. tucumanensis H. & A. (1) Arg.	6					
B. tucumanensis H. & A. (2) Arg.	2 or 3			17		
B. ulcina H. & A. Bol., Ecua.			14			
B. vaccinoides H.B.K.	6	10				
B. varians Gardn. Braz.				16		

TABLE 9. Distribution of diterpenes reported from *Baccharis* (Astereae).

| | Linear | Normal | | | Ent- | | | |
Species	A./Uni.	Bi. (L)	Bi. (C)	Tri.	Bi. (L)	Bi. (C)	Tri.	Poly.
B. alaternoides						484 569		
B. articulata						510 527		
B. calvescens						518		
B. cassinaefolia						531 532 533		
B. chilco						529		
B. concinna (concinnia)								646 649 625
B. conferta						521		
B. eggersii		965 966						
B. genistelloides						561		
B. grandicapitulata						963		
B. hutchisonii						507 967		
B. intermixta					363 364			625 646
B. kingii						846		
B. microcephala						561		
B. minutiflora								618 619 621 624 625 697 698 713 720

Baccharis Diterpene Cyclic Classes

TABLE 9. (*contd.*).

| | Linear | Normal | | | Ent- | | | |
Species	A./Uni.	Bi. (L)	Bi. (C)	Tri.	Bi. (L)	Bi. (C)	Tri.	Poly.
								729
								730
								731
								732
								733
								734
								735
								736
								737
								738
								739
B. nitida						968		
B. oxydonta					442			
					443			
					444			
					445			
					446			
B. polyphylla						569		625
								646
								649
								661
B. quitensis								699
								740
B. ramosissima								624
								625
B. salicifolia						518[1033]		
						904		
						1070		
						1093		
B. salzmannii								625
								650
B. sarothroides						518[1033]		
						1034		
B. scoparia	964					969		
						970		
						1061		
B. stern- bergiana					331			
					332			
					333			
					334			
					357			
					359			

TABLE 9. (*contd.*).

| | Baccharis Diterpene Cyclic Classes | | | | | | | |
| | Linear | Normal | | | Ent- | | | |
Species	A./Uni.	Bi. (L)	Bi. (C)	Tri.	Bi. (L)	Bi. (C)	Tri.	Poly.
B. subdentata						529		
B. tola					455			765 766 771 772 779 780 782
B. tricuneata (L. f.)Pers. var. tricuneata						509 527 528 529 530		
B. tricuneata var. lineata						508 518 520		
B. trimera						559 562 563		
B. truncata (tricuneata?)						565		624 625 648 650
B. tucumanensis	110					962		
B. vaccinoides						518[1033]		

TABLE 10. Distribution of diterpenes reported from *Helichrysum* (Inuleae).

| | Helichrysum Diterpene Cyclic Classes | | | | | | | |
| | Linear | Normal | | | Ent- | | | |
Species	A./Uni.	Bi. (L)	Bi. (C)	Tri.	Bi. (L)	Bi. (C)	Tri.	Poly.
H. acutatum	69 70 71 72 1101							

TABLE 10. (*contd.*).

| | Linear | Normal | | | Ent- | | | |
Species	A./Uni.	Bi. (L)	Bi. (C)	Tri.	Bi. (L)	Bi. (C)	Tri.	Poly.
H. albirosulatum		1105 1106 1107			447 448 449			
H. argentissimum								625
H. aureum (Houtt.) Merile var. monocephalum								625 629 640 690
H. bellum								625
H. calliconum	65 66 67							
H. chiono-sphaerum							614	625 783 796 797
H. confertum		132 137 996 1102 1103 1104						
H. cooperi								625 629 640
H. dendroideum								619 623 626 698 764 765 767 770 771
H. diosmifolium								728
H. fulvum								625 774 791 798 799

The header spans: "Helichrysum Diterpene Cyclic Classes"

TABLE 10. (*contd.*).

Species	Linear A./Uni.	Normal Bi. (L)	Bi. (C)	Tri.	Ent- Bi. (L)	Bi. (C)	Tri.	Poly.
H. heterolasium								619 624 629 640 648
H. Krebsianum	32 70							
H. miconiifolium								624 625
H. mundii	23							
H. nudifolium	68							
H. odoratissimum	68							
H. pallidum								624 625 640 720 722
H. pinifolium				447				
H. platypterum								690
H. refluxum								976 625 774 791
H. ruderale								640 690
H. subfacultatum	23							
H. subfalcatum	73							
H. suterlandii				430				
H. vernum				447				625

Table header spanning: Helichrysum Diterpene Cyclic Classes

TABLE 11. Major di- and sesquiterpenoids of *Helianthus* (Heliantheae).

<table>
<thead>
<tr><th rowspan="3">Species</th><th rowspan="3">KA[1]</th><th colspan="11">Terpenoids</th></tr>
<tr><th colspan="8">C20</th><th colspan="3">C15</th></tr>
<tr><th>LA</th><th>PI</th><th>ST</th><th>TE</th><th>IS</th><th>TR</th><th>AT</th><th>VI</th><th>GE</th><th>HE</th><th>FU</th></tr>
</thead>
<tbody>
<tr><td>HELIANTHUS</td><td></td><td></td><td></td><td></td><td></td><td></td><td></td><td></td><td></td><td></td><td></td><td></td></tr>
<tr><td>1.HELIANTHUS</td><td></td><td></td><td></td><td></td><td></td><td></td><td></td><td></td><td></td><td></td><td></td><td></td></tr>
<tr><td>H. annuus</td><td>15,18[2] 19,21-23 28,29</td><td></td><td></td><td></td><td></td><td>35</td><td>38-40</td><td>42,43</td><td></td><td></td><td>47</td><td>49</td></tr>
<tr><td>H. anomalus</td><td></td><td></td><td></td><td></td><td></td><td></td><td></td><td></td><td></td><td></td><td></td><td></td></tr>
<tr><td>H. argophyllus</td><td>18,33</td><td></td><td></td><td></td><td></td><td></td><td>39</td><td></td><td></td><td></td><td>47</td><td></td></tr>
<tr><td>H. bolanderi</td><td></td><td></td><td></td><td></td><td></td><td></td><td></td><td></td><td></td><td></td><td></td><td></td></tr>
<tr><td>H. debilis subsp. cucumerifolius</td><td>23,28,29</td><td></td><td></td><td></td><td></td><td>35-37</td><td></td><td></td><td></td><td></td><td></td><td></td></tr>
<tr><td>H. debilis subsp. debilis</td><td>23,28,29</td><td></td><td></td><td></td><td></td><td>34</td><td>38</td><td></td><td></td><td></td><td></td><td></td></tr>
<tr><td>H. deserticola</td><td></td><td></td><td></td><td></td><td></td><td></td><td></td><td></td><td></td><td></td><td></td><td></td></tr>
<tr><td>H. neglectus</td><td></td><td></td><td></td><td></td><td></td><td></td><td></td><td></td><td></td><td></td><td></td><td></td></tr>
<tr><td>H. niveus subsp. canescens</td><td>23,28</td><td></td><td></td><td></td><td></td><td></td><td>39</td><td></td><td></td><td></td><td>47</td><td></td></tr>
<tr><td>H. niveus subsp. niveus</td><td></td><td></td><td></td><td></td><td></td><td></td><td></td><td></td><td></td><td></td><td>47</td><td>48</td></tr>
<tr><td>H. paradoxus</td><td></td><td></td><td></td><td></td><td></td><td></td><td></td><td></td><td></td><td></td><td></td><td></td></tr>
<tr><td>H. agrestis</td><td></td><td></td><td></td><td></td><td></td><td></td><td></td><td></td><td></td><td></td><td></td><td></td></tr>
<tr><td>H. arizonensis</td><td>33</td><td></td><td></td><td></td><td></td><td></td><td>39</td><td></td><td></td><td></td><td></td><td></td></tr>
<tr><td>H. ciliaris</td><td>18,21,33</td><td></td><td></td><td></td><td></td><td></td><td>39</td><td></td><td></td><td></td><td></td><td>49</td></tr>
<tr><td>H. laciniatus</td><td>18,21, 23,33</td><td></td><td></td><td>16, 17</td><td></td><td></td><td>39</td><td>42</td><td></td><td></td><td></td><td></td></tr>
<tr><td>H. cusickii</td><td>28</td><td></td><td></td><td></td><td></td><td></td><td>39</td><td></td><td></td><td></td><td></td><td></td></tr>
</tbody>
</table>

TABLE 11. (*contd.*).

| | | Terpenoids | | | | | | | | | | |
| | | C20 | | | | | | | | C15 | | |
Species	KA[1]	LA	PI	ST	TE	IS	TR	AT	VI	GE	HE	FU
H. gracilentus										47	48	49
H. pumilus										47		
H. californicus							39			47		
H. decapetalus	23,26	1-3					38	45		47		
H. divaricatus										47		
H. eggertii												
H. giganteus	23,24,26 28,29	5					38			47		
H. grosseserratus	24,28					35	39			47		49
H. hirsutus	23,26		11, 12				38			47		49
H. maximiliani	24	3, 5								47		49 (50)
H. mollis										47		
H. nuttallii	23,28					35	39					49
H. resinosus										47		
H. salicifolius	18,33						39			47		
H. schweinitzii												49
H. strumosus			7,8									49
H. tuberosus	23,24 26,29	3,5									48	
H. glaucophyllus												(50)
H. laevigatus												
H. microcephalus												(50)

TABLE 11. (*contd.*).

| | | Terpenoids | | | | | | | | | | |
| | | C20 | | | | | | | | | C15 | |
Species	KA[1]	LA	PI	ST	TE	IS	TR	AT	VI	GE	HE	FU
H. porteri												
H. smithii												
H. atrorubens												
H. occidentalis	21,23,28,29-31	3,5				35-37	38,39	44				
H. rigidus	23,26,33						38,39					
H. silphioides												
H. angustifolius	23,24,28,33	3,5										49
H. carnosus											48	
H. floridanus												
H. heterophyllus	23,24,25,27										48	49 (50)
H. longifolius												
H. radula	19-21,23,26,29,31-33					35,36	41					
H. simulans	23,26,28,29,31					35,36					48	
V. bishopii	18,23,24	4,6		15			38					
V. cordifolia	23											
V. dentata	23,24,28,29	6					38					
V. excelsa	23-25,27,31											49
V. grammatoglossa	23			13-15								
V. insignis	23,24			15				43				

TABLE 11. (*contd.*).

Species	KA[1]	Terpenoids										
		C20								C15		
		LA	PI	ST	TE	IS	TR	AT	VI	GE	HE	FU
V. lanceo-lata							38					
V. line-aris											48	49
V. macu-lata	23,28,34											
V. oaxa-cana	23											
V. pazen-sis	29						38		46			
V. porteri	23,28,29					35						
V. procum-bens	23,24						38					
V. quin-queradiata	29											
V. steno-loba	23,29				34						48	

[1] KA = Kaurane
  LA = *ent*-Labdane
  PI = *ent*-Pimarane
  ST = *ent*-Stachane
  TE = Tetrachyrane
  IS = Isokaurane
  TR = Trachylobane
  AT = Atisirane
  VI = Villanovane

  GE = Germacrolide
  HE = Heliangolide
  FU = Furanoheliangolide

[2] See Figure **19** for representative structures

TABLE 12. Distribution of diterpenes reported from *Montanoa* (Heliantheae).

| | | Montanoa Diterpene Cyclic Classes | | | | | | | |
| | Linear | | Normal | | | Ent- | | | |
Species	A./Uni.	Bi. (L)	Bi. (C)	Tri.	Bi. (L)	Bi. (C)	Tri.	Poly.
M. atriplicifolia								625
M. frutescens	56[homolog]							
M. leucantha subsp. arborescens	55							
M. leucantha subsp. leucantha	55							
M. mollissima	55							
M. pteropoda								618 619 624 625 649 652 655 690 694 773
M. tomentosa subsp. microcephala	55							
M. tomentosa subsp. tomentosa	55 57 1094 1095 1096 1097 1098 1099							
M. tomentosa subsp. xanthiifolia	56							800

SCHEME A

SCHEME B

101 R = Ang
102 R = Tig

122 R = Ang
123 R = Tig

119 X = α−OH,H
129 X = O

991 R = CO₂H
992 R = CO₂Me
993 R = CH₂OH
994 R = CH₂O2−Mebut

220

995

FIGURE 19. Structures of normal-labdanes reported from *Brickellia* (Eupatorieae): compounds with common substitutional patterns (Scheme A) and with unusual 2α,3α-dihydroxy-pattern (Scheme B). Structures of *Brickellia ent*-clerodanes [**511**], cleistanthanes, etc. [**601–609**] (Scheme C).

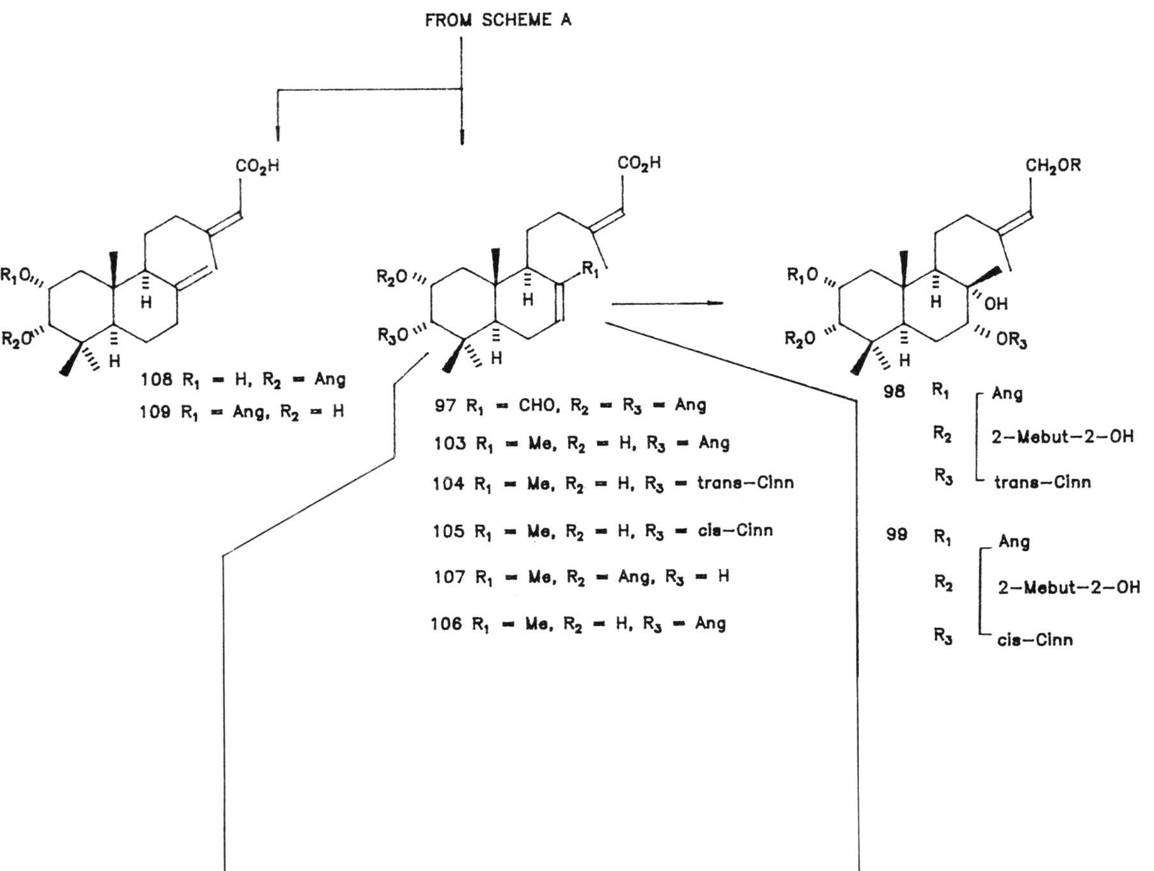
SCHEME B

FROM SCHEME A

CO₂H
R₁O
R₂O
H
H
108 R₁ = H, R₂ = Ang
109 R₁ = Ang, R₂ = H

CO₂H
R₂O
R₃O
R₁
H
H
97   R₁ = CHO, R₂ = R₃ = Ang
103 R₁ = Me, R₂ = H, R₃ = Ang
104 R₁ = Me, R₂ = H, R₃ = trans-Cinn
105 R₁ = Me, R₂ = H, R₃ = cis-Cinn
107 R₁ = Me, R₂ = Ang, R₃ = H
106 R₁ = Me, R₂ = H, R₃ = Ang

CH₂OR
R₁O
R₂O
H
H
OH
OR₃

98 R₁ — Ang
    R₂ — 2-Mebut-2-OH
    R₃ — trans-Cinn

99 R₁ — Ang
    R₂ — 2-Mebut-2-OH
    R₃ — cis-Cinn

215 $R_1 = R_2 = $ Ang, $R_3 = $ H
216 $R_1 = R_2 = $ Ang, $R_3 = \alpha$-OH
217 $R_1 = R_2 = $ Ang, $R_3 = \beta$-OH

218 $R_1 = $ H, $R_2 = $ Ang

124 $R_1 = R_2 = $ H
125 $R_1 = $ H, $R_2 = $ 2-Mebut-2-OH
126 $R_1 = $ H, $R_2 = $ Ang

C-13 Orientivity Undefined:
127 $R_1 = $ H, $R_2 = $ trans-Clnn
128 $R_1 = $ H, $R_2 = $ cis-Clnn

FIGURE 19. (*contd.*).

SCHEME C

608

511

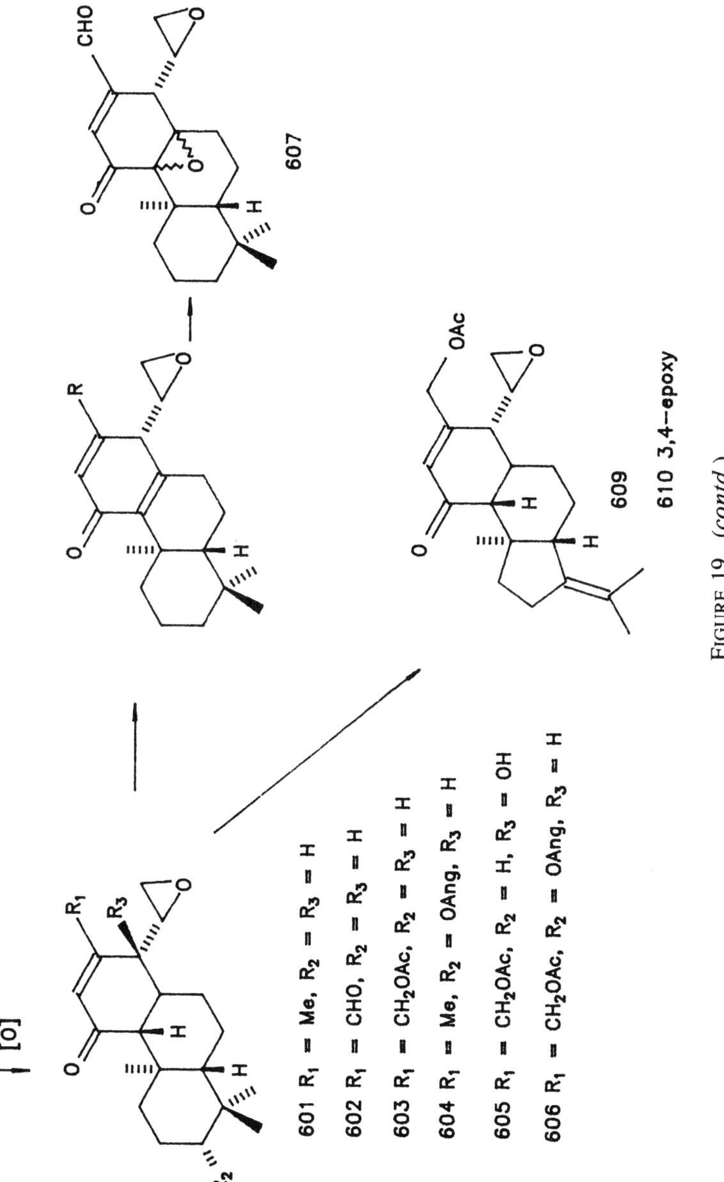

FIGURE 19. (*contd.*).

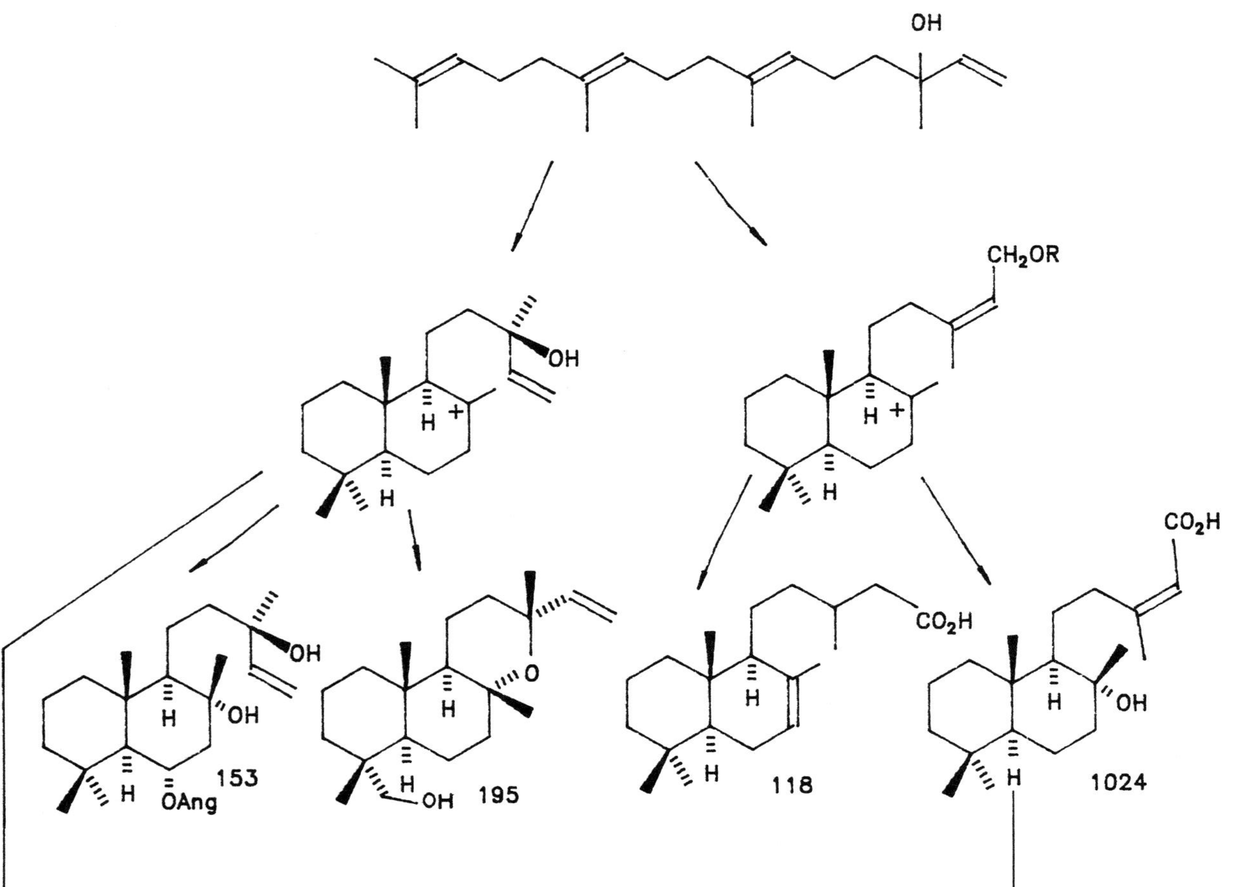

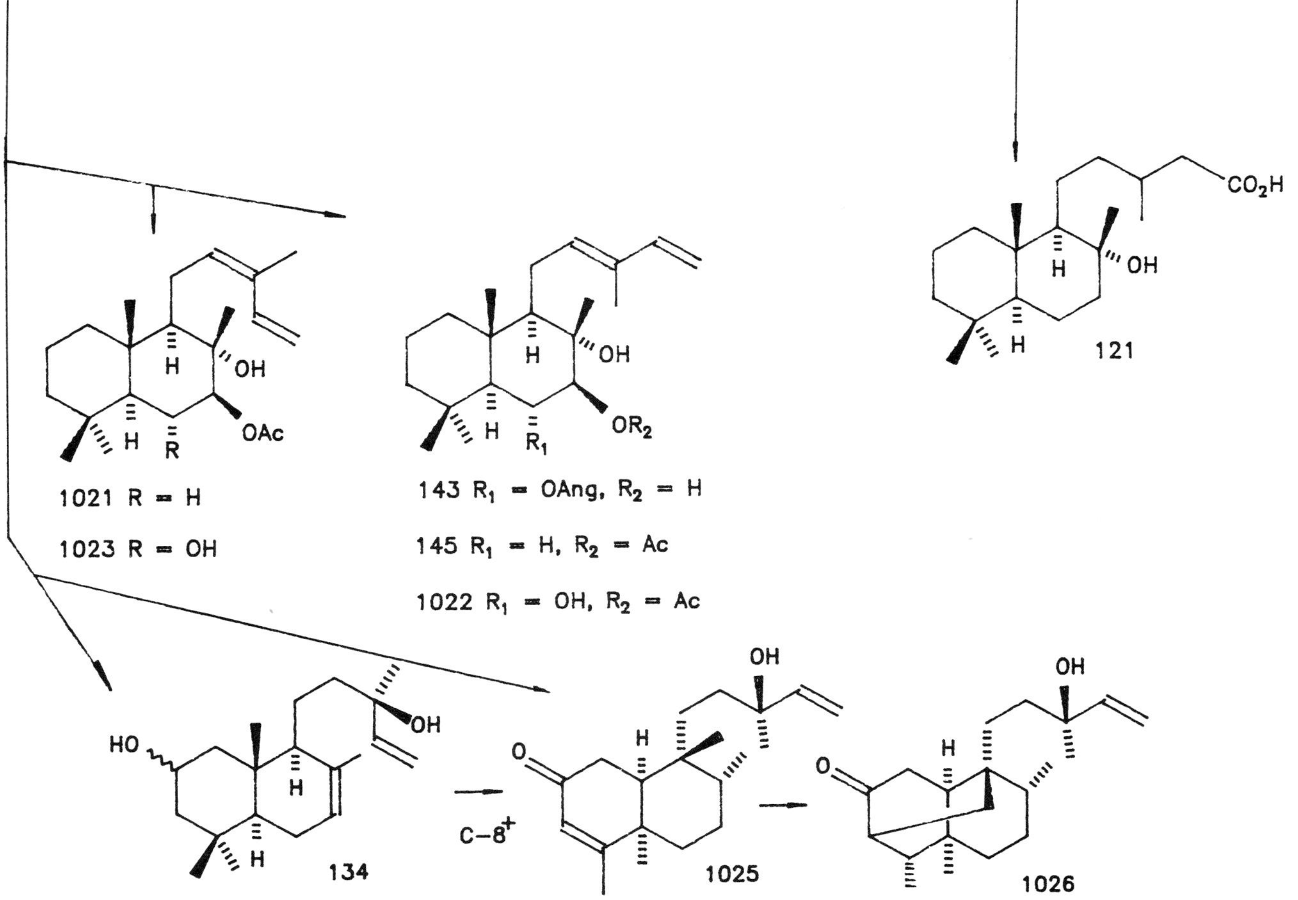

FIGURE 20. Structures of normal-labdane and related diterpenes reported from *Stevia* (Euaptorieae).

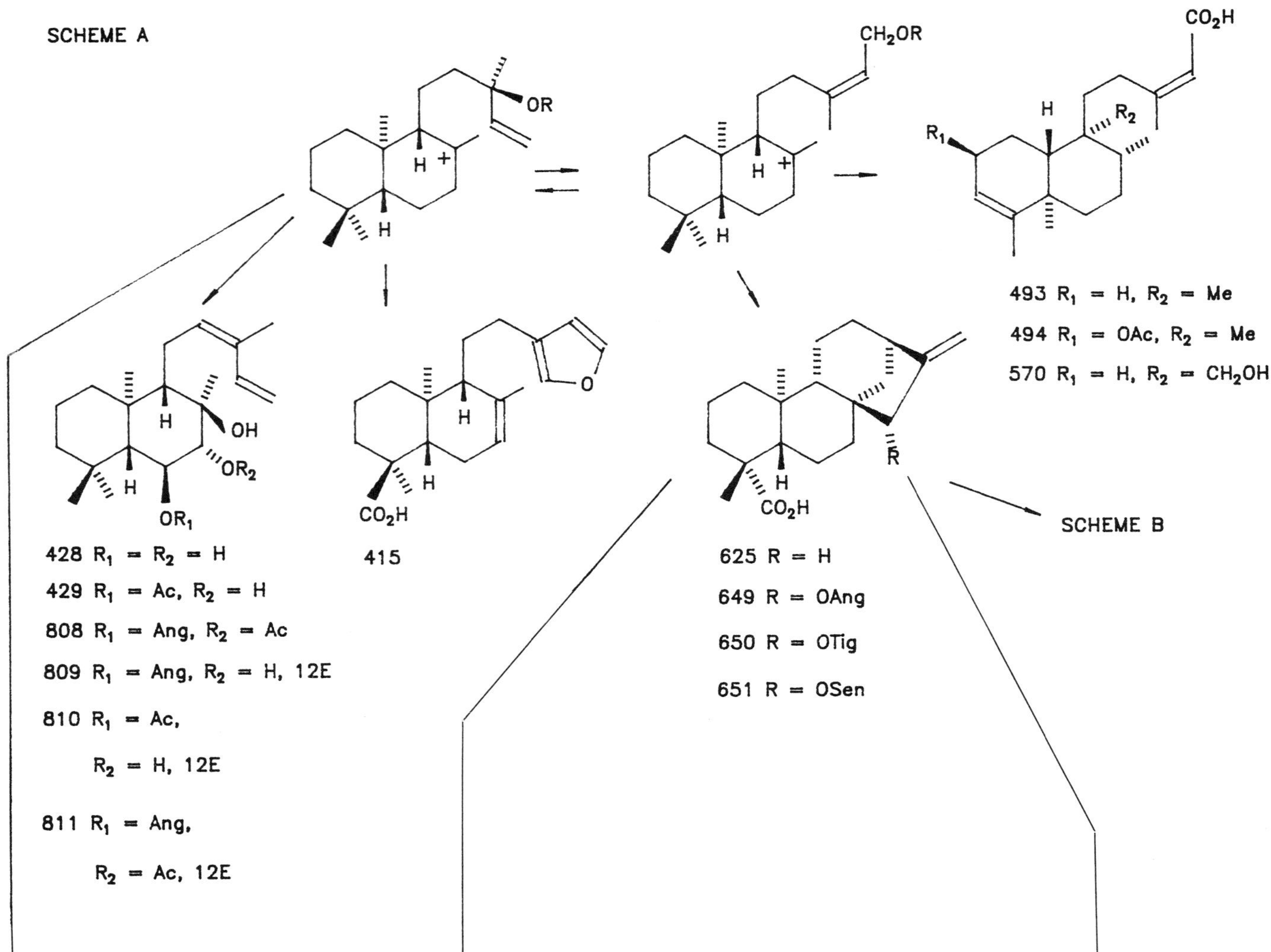

SCHEME A
CH₂OR
OR
CO₂H
493 R₁ = H, R₂ = Me
494 R₁ = OAc, R₂ = Me
570 R₁ = H, R₂ = CH₂OH
OH
OR₂
OR₁
428 R₁ = R₂ = H
429 R₁ = Ac, R₂ = H
808 R₁ = Ang, R₂ = Ac
809 R₁ = Ang, R₂ = H, 12E
810 R₁ = Ac,
    R₂ = H, 12E
811 R₁ = Ang,
    R₂ = Ac, 12E
CO₂H
415
CO₂H
625 R = H
649 R = OAng
650 R = OTig
651 R = OSen
SCHEME B

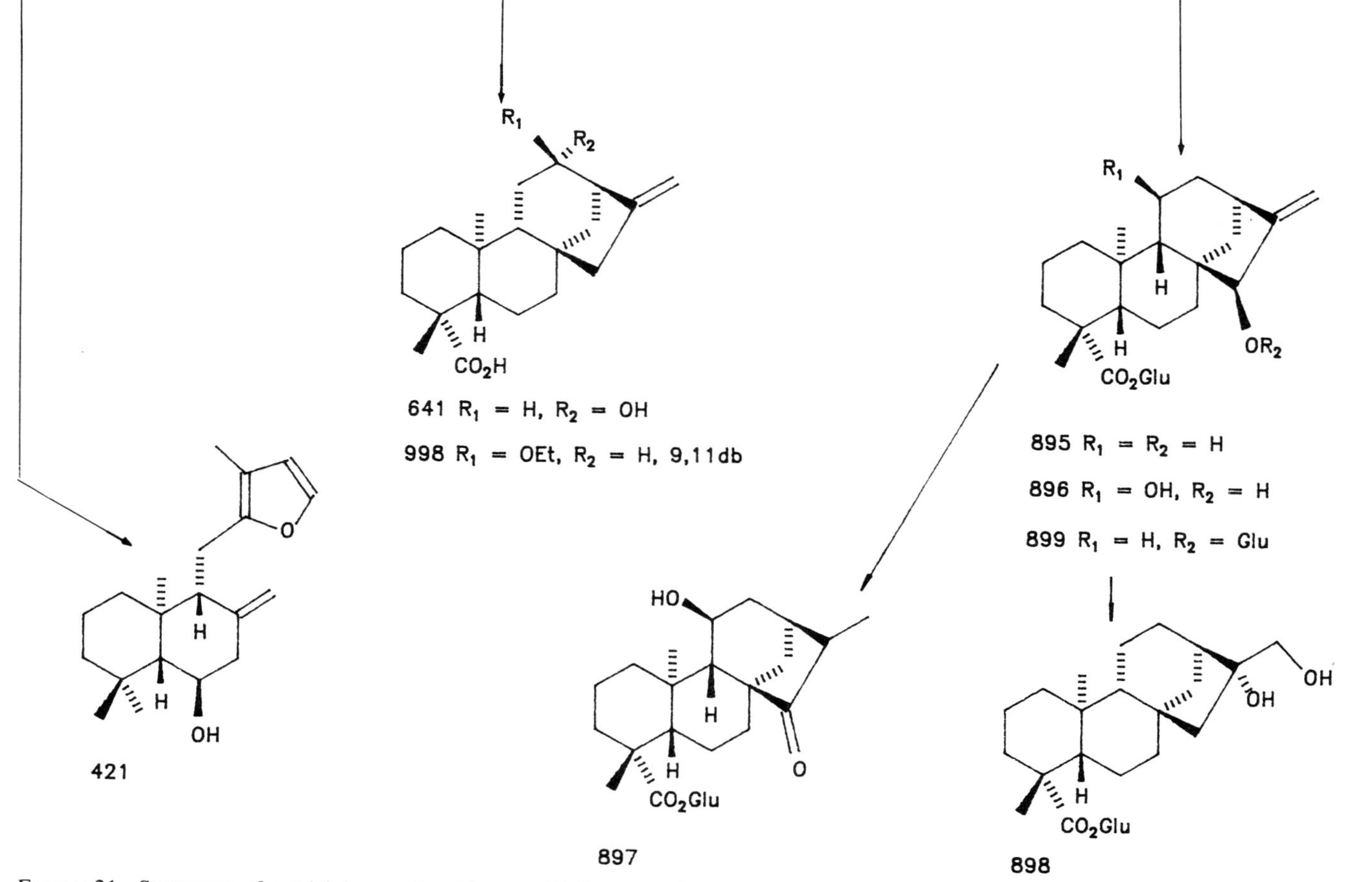

FIGURE 21. Structures of *ent*-labdane and *ent*-kauranoid diterpenes isolated from *Stevia* (Eupatorieae): *ent*-labdanes and kaurene mono- and di-glycosides (Scheme A); C-13-kaurene-glycosides (Scheme B).

SCHEME B

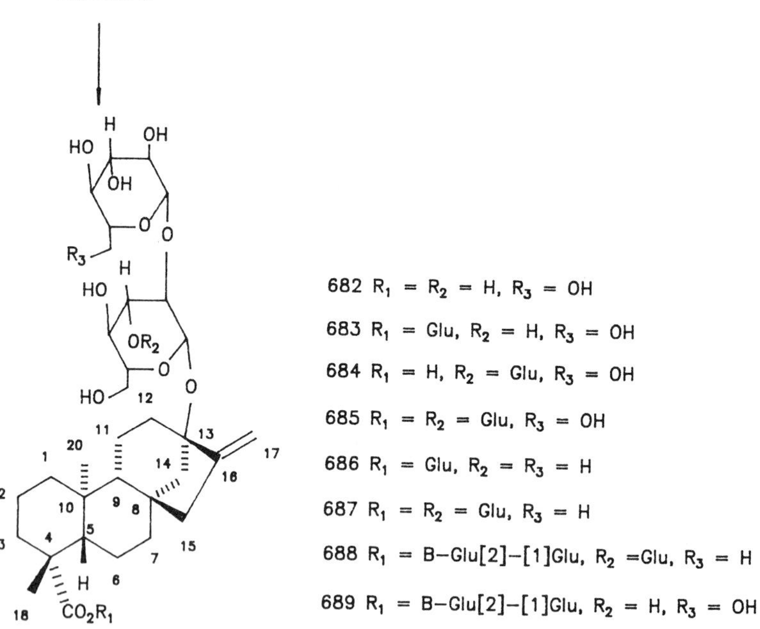

FIGURE 21. (*contd.*).

FIGURE 22. Representative structures of di-, and tri-, and sesquiterpenes, flavonoids, acetophenones, coumarins, and polyacetylenes reported from *Baccharis* (Astereae). The numbers assigned to the structures correspond to the numbers listed in Table 8.

TRITERPENES:

10· Oleanolic Acid—type   11·

12· Lupeol—type   13· Squalene

SESQUITERPENES:

14· Germacrene—D—type   15· Eudesmane—type

FLAVONOIDS:

16· Flavonone—type   17· Flavone—type

FIGURE 22. (*contd.*).

ACETOPHENONE:

18·

COUMARINS: P−COUMARIC ACIDS

19·   **21·**

POLYACETYLENES:

FIGURE 22. (*contd.*).

110   **964**   **965 R = CH$_2$OH**

                                                966 R = CHO

FIGURE 23. Structures of normal-labdanes reported from *Baccharis* (Astereae).

FIGURE 24. Structures of *ent*-labdanes (Scheme A) and related *ent*-clerodanes (Schemes B, C, and D) reported from *Baccharis* (Astereae).

FIGURE 24. (*contd.*).

SCHEME C

SCHEME B

FIGURE 24. (*contd.*).

SCHEME D

FIGURE 24. (*contd.*).

765 R$_1$ = Me, R$_2$ = CH$_2$OH, R$_3$ = Me

766 R$_1$ = CH$_2$OH, R$_2$ = R$_3$ = Me

771 R$_1$ = Me, R$_2$ = R$_3$ = CH$_2$OH

772 R$_1$ = R$_2$ = CH$_2$OH, R$_3$ = Me

782 R$_1$ = Me, R$_2$ = OH, R$_3$ = Me

779 R$_1$ = CH$_2$OH, R$_2$ = Me

780 R$_1$ = Me, R$_2$ = CH$_2$OH

720 R$_1$ = R$_2$ = Me, R$_3$ = OH

729 R$_1$ = Me, R$_2$ = H, R$_3$ = CHO

730 R$_1$ = CH$_2$OH, R$_2$ = H, R$_3$ = CHO

731 R$_1$ = Me, R$_2$ = H, R$_3$ = CO$_2$H

732 R$_1$ = CH$_2$OH, R$_2$ = H, R$_3$ = CO$_2$H

736 R$_1$ = CH$_2$OH, R$_2$ = H, R$_3$ = COMe

737 R$_1$ = CHO, R$_2$ = H, R$_3$ = CHO

738 R$_1$ = CO$_2$H, R$_2$ = H, R$_3$ = CH$_2$OH

FIGURE 25. Structures of *ent*-stachane and *ent*-kauranoid diterpenes isolated from *Baccharis* (Astereae).

618 R_1 = Me, R_2 = H

619 R_1 = CH_2OH, R_2 = H

621 R_1 = CH_2OCH_2OCHO, R_2 = H

622 R_1 = CH_2O–p–OH–H–Cinn, R_2 = H

624 R_1 = CHO, R_2 = H

625 R_1 = CO_2H, R_2 = H

646 R_1 = CO_2H, R_2 = OH

648 R_1 = CO_2H, R_2 = OAc

649 R_1 = CO_2H, R_2 = OAng

650 R_1 = CO_2H, R_2 = OTig

651 R_1 = CO_2H, R_2 = OSen

661 R_1 = CO_2Me, R_2 = OH

733 R = Me

734 R = CH_2OH

735 R = CHO

713 R = CO_2H

740

697 R = H

698 R = OH

699 R = O–p–OH–H–Cinn

OR
23 R = H, X = H,H
32 R = H, X = O
65 R = CH_2OH
66 R = CHO
67
X
R
HO_2C

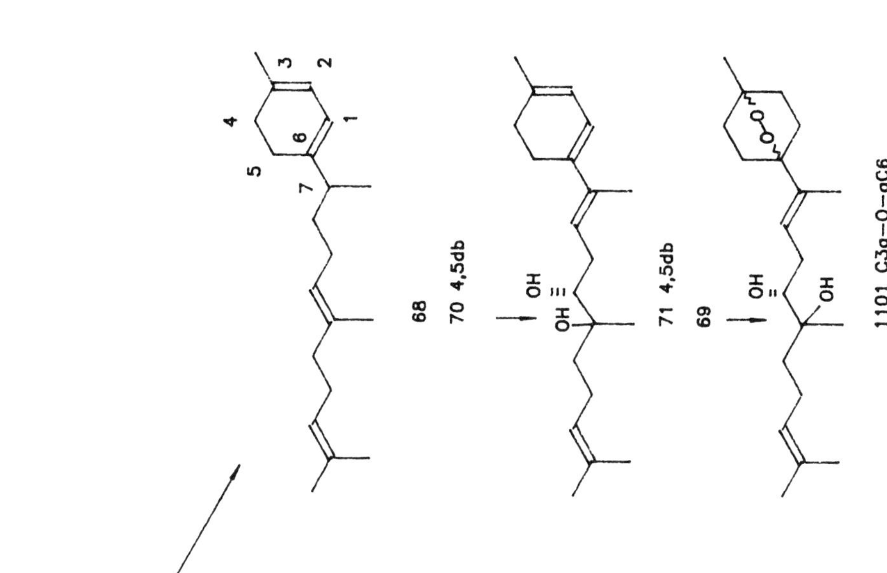

FIGURE 26. Structures of linear (**23**, **32**), unicyclic (**65–72**, **1101**), and macrocyclic (isocembrene, **73**) diterpenes reported from *Helichrysum* (Inuleae).

23 R = H

996

FIGURE 27. Structures of normal-labdane diterpenes isolated from *Helichrysum* (Inuleae).

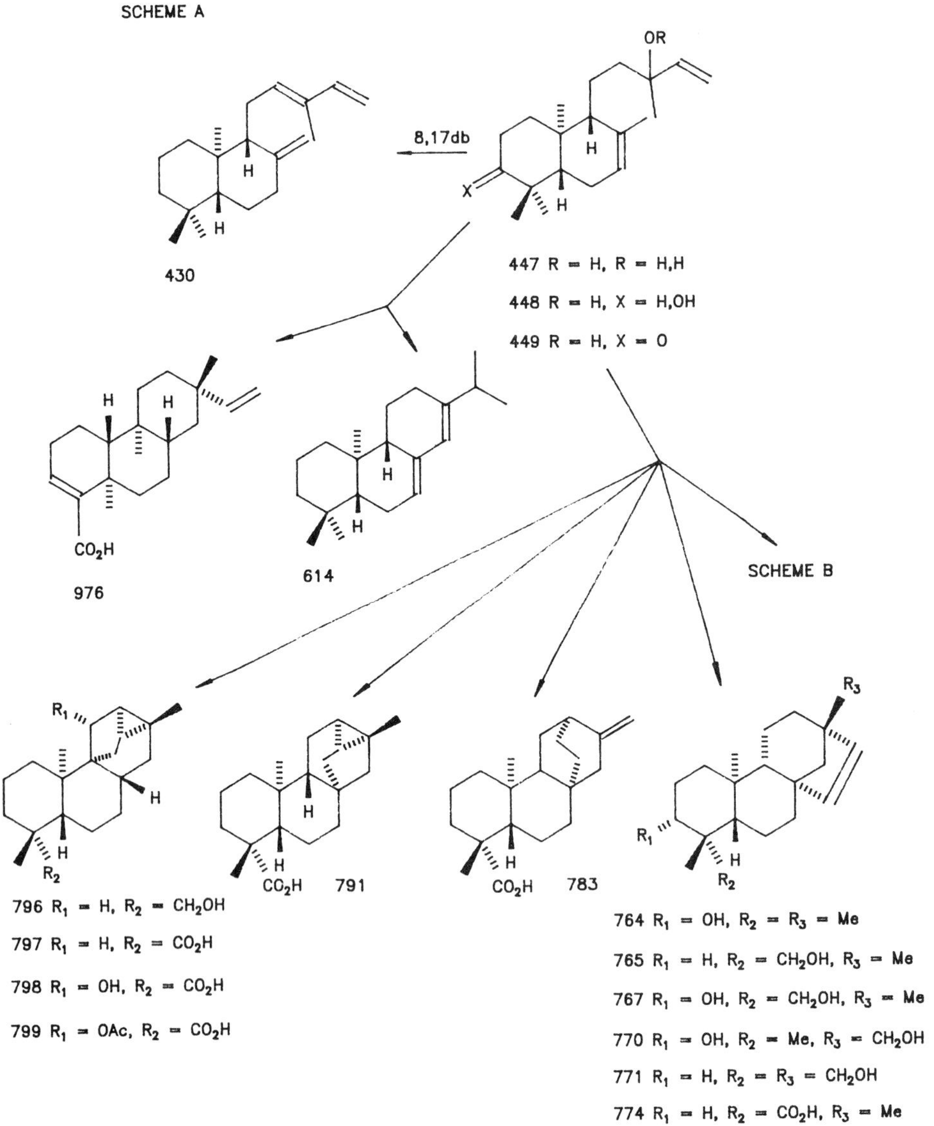

FIGURE 28. Structures of *ent*-labdane (**430, 447–449**), *ent*-pimarane (**976**), *ent*-abietane (**614**), *ent*-helifulvane (**796–799**), *ent*-trachylobane (**791**), *ent*-atisirane (**783**), *ent*-stachane (**764, 765, 767, 770, 771, 774**) (Scheme A), and *ent*-kauranoid (Scheme B) diterpenes reported from *Helichrysum* (Inuleae).

SCHEME B

720 R = Me

722 R = CH₂OAc

619 R₁ = H, R₂ = CH₂OH

623 R₁ = a−OH, R₂ = CH₂OH

624 R₁ = H, R₂ = CHO

625 R₁ = H, R₂ = CO₂H

626 R₁ = B−OH, R₂ = CH₂OH

629 R₁ = a−OAc, R₂ = CO₂H

690 R₁ = H, R₂ = CO₂H, 9,11db

728

698

640 R₁ = CO₂H, R₂ = H, R₃ = OAc

648 R₁ = CO₂H, R₂ = OAc, R₃ = H

FIGURE 28. (contd.).

ENT–LABDANES

1· R = CH₃

2· R = CHO

3· R = CO₂H

4· R = CH₃

5· R = CO₂H

6·

ENT–PIMARANES

7· 7(8) d·b·

8· 8(9) d·b·

9· 9(11) d·b·

11·

12·

10·

ENT–STACHANES

13· R = CH₂OH

14· R = CHO

15· R = CO₂H

16· R = H

17· R = OH

FIGURE 29. Major diterpenes isolated from *Helianthus* (Heliantheae). The numbers assigned to the compounds correspond to the numbers listed in Table 11.

ENT—KAURANES

18· R_1 = a—OH, R_2 = CH_3

19· R_1 = a—OH, R_2 = CH_2OH

20· R_1 = a—OH, R_2 = CO_2H

21· R_1 = B—OH, R_2 = CH_3

22· R_1 = B—OH, R_2 = CO_2H

23· R = H

24· 9(11)d·b·, R = H

25· 9(11)d·b·, R = B—OH

26· 9(11)d·b·, R = a—OH

27· 9(11)d·b·, R = =O

28· R_1 = H, R_2 = OH

29· R_1 = H, R_2 = OAng,OTig, etc·

30· R_1 = R_2 = OH

31·

32·

33·

TETRACHYRANE

34·

ISOKAURANE

35· R = CH_2OH

36· R = CHO

37· R = CO_2H

38· R_1 = R_2 = R_3 = H

39· R_1 = OH, R_2 = R_3 = H

40· R_1 = H, R_2 = OH, R_3 = H

41· R_1 = R_2 = H, R_3 = =O

42· R = a—OH

43· R = B—OH

44· R_1 = OH, R_2 = H

45· R_1 = H, R_2 = OH

FIGURE 29. (*contd.*).

FIGURE 30. Structures of oxepane diterpenes (**55–57, 1097–1099**) and their linear precursor-like compounds (**1094–1096**) reported from *Montanoa* (Heliantheae).

CHAPTER 5

Biological Activity of Diterpenes

V.1. Introduction

Various biological activities (Table 13 [pp. 489–490]) have been reported for plant diterpenes, principally from members of the Ericaceae, Euphorbiaceae, Lamiaceae, and Compositae. Each of the major diterpenoid classes from this last family will be discussed below in terms of its biological properties.

V.2. Linear Diterpenes

Linear diterpenes derived from geranylnerol, geranylgeraniol, and geranyllinalool occur commonly within the Compositae. The most diverse group of linear diterpenes is derived from geranylnerol and includes lactonic, oxepane, and furanoid members as well as unicyclic and macrocyclic derivatives. In addition to the previously mentioned *Montanoa* oxepane diterpene series, oxepanes and lactones have been detected in *Acanthosperum*, *Ichthyothere*, *Melampodium*, and other genera mostly within the Heliantheae.

For hundreds of years, a tea brewed from the leaves of *Montanoa tomentosa* has been used medicinally to induce labor and as an antifertility agent (*419, 382*). The *M. tomentosa* extract possessed abortifacient, labor induction, and menstruation induction effects (*372, 297*). Zoapatanol (55) appears to be the tea component responsible for most of the activity (*368*). In addition to zoapatanol, several other oxepane diterpenes (**56, 57, 1097, 1098, 1099**) and their uncyclized precursors (**1094, 1095, 1096**) were isolated from this species. Commercial interest in these compounds has led to the synthesis of analogous structures sharing similar pharmacological properties (*353*).

V.3. Bicyclic Diterpenes

Grindelane normal-labdanes **185** and **188** isolated from *Grindelia*, *Chrysothamnus*, *Solidago*, and *Haplopappus* displayed significant antifeedant activity against the Colorado potato beetle (*440*) and the aphid *Schizaphis graminum*

(*441*). The anatomical distribution of grindelanes was examined in *Grindelia camporum* and the development of anatomically defined chemistries was traced over time. This work led to the observations that most diterpene resin accumulated on the plant surface and the quantity of accumulated cuticular resin was positively correlated with the number of multicellular epidermal resin glands (*340*).

Surface resin production in *G. camporum* and other resin-producing xerophytic plants (e.g., *Chrysothamnus*, *Gutierrezia* and *Xanthocephalum*) may serve at least three (not mutually-exclusive) functions: (1) when coating the leaf surface, the resin makes the cuticle less permeable to water and thus reduces water loss, (2) the resin coating increases the reflectance of the leaf surface thereby reducing the amount of solar radiation penetrating the surface and, consequently, lowering the internal temperature, and (3) the resin deters herbivores and/or plant pathogens (*340*).

Although the plant source, *Nicotiana glutinosa*, is a member of the Solanaceae not the Compositae, the major leaf-surface-resin normal-labdane constitutents, sclareol (**1102**) and 13-epi-sclareol, either have been isolated from the Compositae (sclareol) or resemble other normal-labdanes isolated from this family. When tested against a variety of fungi, these *Nicotiana* compounds significantly inhibited the radial growth rate of common fungal colonies. Radial growth was apparently retarded by the effect of diterpenes on the growth-regulatory processes controlling fungal branching (*24*).

Solidago clerodanes are frequently referred to as the "bitter principal" responsible for the characteristic taste of the root tissue. For example, solidagonic acid (**500**) was described as a bitter principle from the roots of *Solidago altissima* (*365*).

V.4. Tricyclic Diterpenes

When each of the diterpene resin acids, pimaric acid, dihydropimaric acid, sandaracopimaric acid, isopimaric acid, neoabietic acid, levopimaric acid, palustric acid, and abietic acid (Figure 31 [p. 491]) was included in the artificial diet of the pink bollworm, *Pectinophora gossypiella*, larval growth was inhibited. In order to ascertain the importance of the compounds' C-19 acid function in growth inhibition, methyl isopimarate and isopimarol, (Figure 31) were tested under the same conditions and found to be substantially less active than the free acid. Isopimarol had virtually no effect on larval growth.

Resemblance of the tricyclic diterpene skeleton to the ABC-ring system of common steroids prompted the suggestion that the diterpenes interfered with the pink bollworm's steroid metabolism. Insect hormones are produced from dietary phytosterols which are absorbed, transported, and biochemically altered. Interference by the diterpene resin acids at any stage of this process could significantly reduce larval viability. To test this hypothesis, feeding studies were conducted using a levopimaric acid test diet with added cholesterol (addition of cholesterol diminishes the insect's requirement for phytosterols). In comparison to the control diet that contained 0.2 percent levopimaric acid and no cholesterol, test diets containing levopimaric acid and steadily incremented cholesterol concentrations from 0.05 to 1.0 percent produced corresponding increments (increases) in larval growth. By 1.0 percent cholesterol, the larval weight had reached 86 percent of larvae grown on an artificial diet containing no levopimaric acid, suggesting that added cholesterol canceled the deleterious effect of the diterpene. It was concluded that reduced larval growth resulted from the antagonism between resin acids and steroids and that the insect's hormonal system was affected (*258*).

V.5. Tetracyclic Diterpenes

Feeding tests using kaur-16-en-19-oic acid (**625**) and trachyloban-19-oic acid (**791**) from the florets of sunflower (*Helianthus annuus*) varieties resistant to the sunflower moth, *Homoeosoma electellum*, produced significant reduction in this insect's larval growth (*491*). Resistance of these varieties to seed predation by the sunflower moth was attributed to the relatively high concentration of these compounds in the sunflower's seed-bearing florets. Further studies (*332*) showed that when other *Helianthus* diterpenes, ciliaric acid (**792**) and angelylgrandifloric acid (**649**), were incorporated (one percent) in the diet of sunflower moth larvae, the larvae experienced a significantly higher level of mortality than those larvae feeding on the diet lacking the diterpenes. The compounds did not affect the length of the development period but did appreciably reduce the size of seven-day-old larvae.

Kaur-16-en-19-oic acid (**625**) obtained from *Mikania monagasensis* inhibited the growth of *Staphylococcus aureus* and *Candida albicans* (*384*). In the latter case, this diterpene was appreciably less active than the co-occurring sesquiterpene lactone, mikanolide.

Diterpenes of *Wedelia* (Heliantheae) were responsible for this plant's toxicity to sheep and cattle. The first of two kaurenoid glycosides, wedeloside (**679**) and atractyloside (**681**), caused the severe loss of sheep in Queensland, Australia, due to grazing on *Wedelia asperrima* (*408*), while the second proved fatal to South American cattle grazing on *W. glauca* (*456*). Investigation of *W. buphthalmiflora*, another South American shrub toxic to cattle, (using extraction methods unlikely to isolate kaurenoid glycosides), revealed a complex chemistry consisting of kaur-9(11),16-dien-19-oic acid (**690**), kaur-16-en-19-oic acid (**625**), and a variety of grandifloric acid (**646**) esters (*455*). One of these esters, 15α-tiglinoyloxy-kaur-16-en-19-oic acid (**650**), isolated from *W. scaberrima* was miracidal (toxic to the miracidia stage of *Schistosoma mansoni*). Kaur-16-en-19-oic acid was molluscicidal against *Biomphalaria glabata* (*476*).

The sulfated glycoside, carboxyatractyloside (**678**), was reported from two sources, *Xanthium strumarium* (*232*) and *Atractylis gummifera* (*239*). This hypoglycemic compound strongly inhibits translocation of adenine nucleotides across the mitochondrial membrane and shows an *in vivo* toxicity ten times greater than that of atractyloside (**681**, which co-occurs in *A. gummifera*) (*239*). Carboxyatractyloside was responsible for cocklebur poisoning of pigs in the southern United States (*238*). Later work on the inhibitors responsible for the dormancy properties of *Xanthium* seeds identified carboxyatractyloside as a potent plant growth inhibitor (*238*).

Sweet-tasting *ent*-kaurene glycosides, stevioside (**683**), rebaudiosides A (**685**), B (**684**), C, D (**688**) and E (**689**), dulcoside A (**686**), and steviolbioside (**682**) (See Figure 21), have been isolated from the New World genus *Stevia* (*359, 464*). Eighteen sweet-tasting *Stevia* taxa have been identified from the total of 121 tested taxa (*464*). Stevioside, which is estimated to be three hundred-fold sweeter than sucrose, is produced commercially as a sweetening agent for desserts, soft drinks, sea foods, pickled vegetables, and so on. Although no evidence indicates that stevioside is harmful to humans, the likely hydrolysis product, steviol (Figure 32 [p. 492]), is metabolized to mutagenic structures by human enzymes and appears to covalently interact with DNA. When steviol was incubated in the presence of rat-liver microsomes (S-9), the principal metabolic product was 15α-hydroxy-steviol. Testing of this latter compound did not reveal any mutagenicity. However, 15-oxo-steviol, a closely-related derivative (Figure 32) was a direct-acting mutagen and a bactericide (*425*).

The role of diterpenes in the resistance of four *Solidago* (Astereae) species to the beetle *Trirhabda* (Coleoptera) is under investigation (*368*). Preliminary results indicate that three of the four kauranoids isolated from the leaves of *S. nemoralis* and *S. altissima*, *ent*-kauran-16α-ol (**720**), 15α-hydroxy-*ent*-kaur-16-en-19-oic acid (**646**), and 17-hydroxy-*ent*-kaur-15-en-19-oic acid reduced *Trirhabda* larval growth by 40 percent and adult growth by 32 to 49 percent. The fourth, *ent*-kaur-16-en-19-oic acid (**625**), had no deterrent effect on the larvae and actually stimulated feeding by adults.

The diterpene alkaloids, lycoctonine (**1009**) and anthranoyllycoctonine (**1010**) appear to be responsible for the blood pressure depressant, smooth muscle relaxant and curareform properties of the source plant, *Inula royleana* (*34*).

TABLE 13. Plant diterpene biological activities.

Activity Compound	Source	Family	Reference
Antifeedant			
ajugarin I-III	Ajuga remota	Lamiaceae	305,306
grindelanes	Chrysothamnus nauseosus	Compositae	340
	Grindelia (7 spp.)	"	"
	Haplopappus sp.	"	"
ciliaric acid	Helianthus sp.	"	306
kalmitoxins	Kalmia latifolia	Ericaceae	259
grayanotoxins	" "	"	"
kaurenoic acids	Helianthus annuus	Compositae	258,306
trachylobanic acids	" "	"	" "
kaurenoid glycosides	Wedelia spp.	Compositae	408,455,456
	Xanthium strumarium	"	238
	Atractylis gummifera	"	239
teucjaponin A	Teucrium japonicum	Lamiaceae	304
Antifungal			
casbene	Ricinus communis	Euphorbiaceae	261
pseudolaric acid	Pseudolarix kaempferi	Pinaceae	306
sclareol,etc.	Nicotiana glutinosa	Solanaceae	24
Fish poison			
eremone	Emerocarpus setigerus		304
hautriwaic acid	"	"	"
Insecticidal			
ajugarin IV	Ajuga remota	Lamiaceae	305
Molluscicidal, etc.			
kaurenoic acid esters	Wedelia scaberria	Compositae	476
Antibacterial			
longikaurins	Rabdosia longituba	Lamiaceae	306
kaurenoic acid	Mikania monagasensis	Compositae	384

TABLE 13. (*contd.*).

Activity Compound	Source	Family	Reference
Antibiotic			
przewaquinone	_Salvia przewalskii_	Lamiaceae	304
sonderianol	_Croton sonderianus_	Euphorbiaceae	305
Antimicrobial			
7,18-dihydroxy-sandaracopimaradiene	_Iboza riparia_	Lamiaceae	304
Curareform			
lycoctonine,etc.	_Inula royleana_	Compositae	34
Antiviral			
bacchotricuneatin	_Baccharis tricuneata_	"	490
Oxytocic			
zoapatlin, etc.	_Montanoa tomentosa_	"	297,372
Plant growth inhibitor			
podolactone	Coniferae spp.	Coniferae	304
carboxyatrac-tyloside	_Xanthium strumarium_	Compositae	238
Sweetening agent			
kaurenoid glycosides	_Stevia_ (18 spp.)	"	464

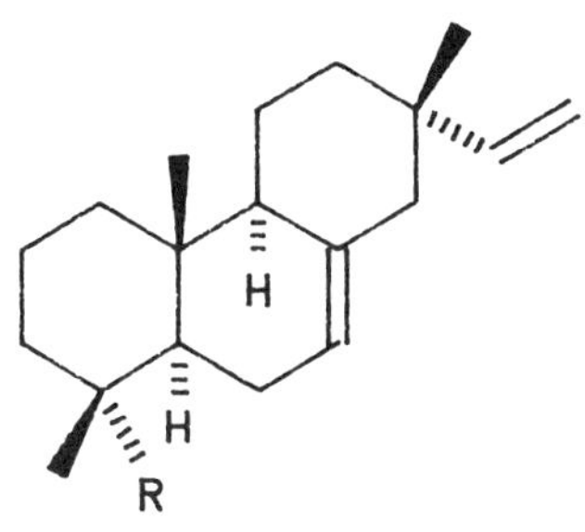

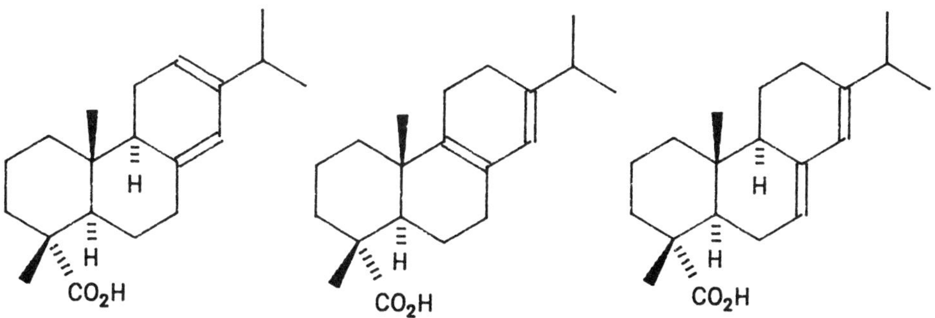

FIGURE 31. Diterpene resin acids tested for biological activity against pink bollworm (*Pectinophora gossypiella*) larvae.

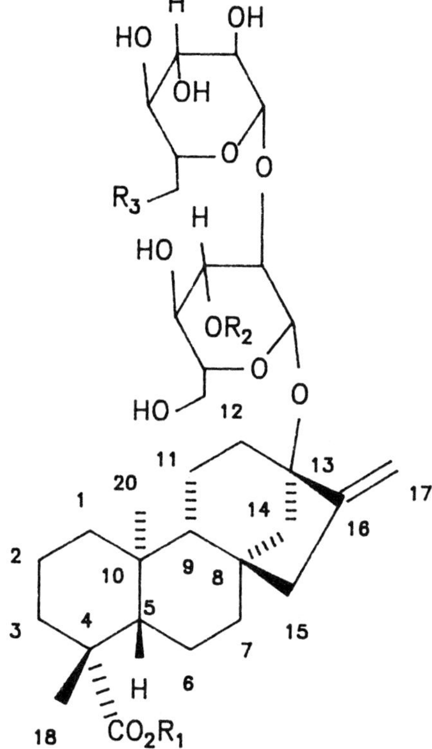

Stevioside R_1 = Glu, R_2 = H, R_3 = OH

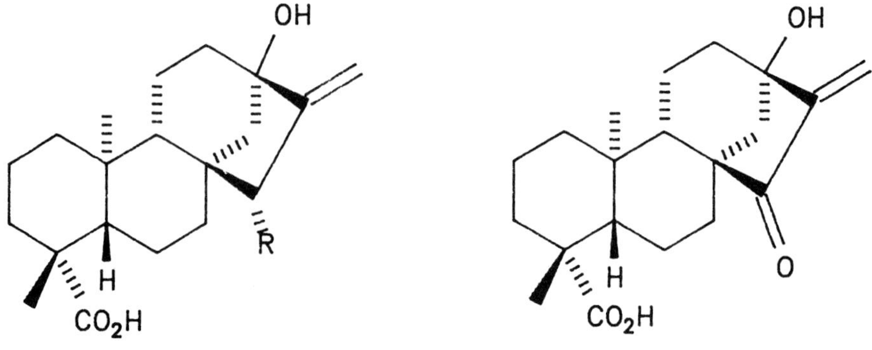

Steviol R = H

15-Oxo-steviol

15α-Hydroxy-steviol R = OH

FIGURE 32. *Stevia* diterpene, stevioside, its aglycone, steviol, and a mutagenic derivative, 15-oxo-steviol.

CHAPTER 6

Diterpene Analysis with Emphasis on Clerodanes

VI.1. Clerodanes

VI.1.1. Major Clerodane Types

VI.1.1.1. trans-ent-*Clerodanes (5α,10β*-trans-*Clerodanes)*

Early work on Compositae diterpenes focused on *Solidago* and the normal-labdane, solidagenone (Figure 33 [pp. 513–514]; Structure **158**) (*9, 16, 17*). Later, *ent*-clerodanes (kolavanes) were found to co-occur with the labdanes (*364, 365*). The first identified novel clerodane, solidagonic acid (Figure 33; Structure **500**), was shown by chemical interconversion (*364, 365*; Figure 33) to share the orientivity of the major chiral centers, C-5, C-8, C-9, and C10, with the co-occurring clerodanes of known stereochemistry, kolavenic acid (**495**) and kolavenol (Figure 33; Structure **475**).

Structural correlation of solidagonic acid with **495** and **475** was made feasible by the recent assignment by Misra and coworkers (*391*) of absolute stereochemistry to the chiral centers of the *Hardwickia pinnata* clerodanes, (-)-hardwickiic acid (**511**), kolavenic acid, and kolavenol (Figure 34 [p. 515]). Their pivotal work on clerodane stereochemistry proceeded as described below:

1. Chemical degradation of 511 led to a series of products that permitted assignment of the ring fusion configuration. The CD spectrum of derivative **511c** produced a strong Cotton effect which was almost a mirror-image of that produced by the steroid, 3β-hydroxy-5α-androstan-17-one. This suggested that the two rings of **511c** were *trans*-fused, but with an absolute stereochemistry opposite to that of the steroid's CD rings.
2. Orientivity at C-9 was established from the **511d** ^{1}H-NMR data which failed to indicate a shielding effect of the C-9 carbomethoxy group on the C-5 methyl. An α-oriented carbomethoxy group would have a 1,3-diaxial relationship with the C-5 methyl which would then be subject to deshielding. The absence of deshielding indicated a β-oriented sidechain and an α-oriented methyl at C-9.
3. The paramagnetic ^{1}H-NMR chemical shift ($\delta 1.19$) of the **511f** C-8 methyl signal (H-17) relative to the shift observed for the parent compound and related structures (c. $\delta 0.83$) was attributed to the deshielding influence of the

keto group. Misra argued that since a carbonyl group can deshield only those protons that lie in its plane, the C-8 methyl must be equatorial (α-oriented). The absolute stereochemistry of kolavenic acid and kolavenol follow from their relationship with (-)-hardwickiic acid (*391*).

Unfortunately, the early assumption of (-)-hardwickiic acid-based biogenetic uniformity among diterpenes of related *Solidago* species led to several incorrect stereochemical assignments: Anthonsen and McCrindle (*18*) isolated from *Solidago elongata* a series of clerodanes that co-occurred with kolavenol, kolavelool (**479**), kolavenic acid plus several angelate and acetate ester derivatives (**480, 495, 498**). Based on ¹H-NMR data that paralleled those of the kolavane series, this new series of five clerodanes, elongatolides A–E, were also assigned *trans-ent*-clerodane absolute stereochemistries (Figure 35 [pp. 516–517]). Two other clerodanes from *S. shortii* (**556** and **557**) were assigned the same stereochemistry (*9*).

Diterpenoids from *S. altissima*, first identified as a series of *trans-ent*-clerodanes, solidagolactones I–VI (*413*), were in some instances identical to members of the previously reported elongatolide series (Figure 35). Solidagolactone (Figure 35; Structure **543**), the only member of this series lacking a C-6 oxygen function, was shown via reduction/methylation to be identical to a kolavenic acid derivative. Consequently, the absolute stereochemistry of the kolavanes was assigned to **543** as well as the other six *S. altissima* clerodanes, solidagolactones II–VII.

Co-occurrence with (-)-hardwickiic acid of known absolute stereochemistry (*391*) also led to the assignment of its stereochemistry to the clerodane constituents of *S. juncea* (*318*).

VI.1.1.2. cis-normal-*Clerodanes (5α,10α-*cis-*Clerodanes)*

The first non-*trans*-fused clerodane (*235*) from the Compositae was identified by X-ray as a C5α,C10α-*cis*-clerodane, gutierolide (**270**). Perhaps prompted by this finding and the growing number of such *cis*-clerodanes reported from other plant families, the absolute stereochemistries of the next reported pair of *Solidago* clerodanes (*15*), solidagoic acids A (**239**), and B (**244**), and a related series of neutral clerodanes (*317*) were studied exhaustively. These structures were assigned a 5α,10α-*cis*-ring junction indicating a normal-labdane origin. Initially, an attempt was made to correlate structure **239** with the established stereochemistry of (-)-hardwickiic acid (**511**; Figure 36A [p. 518]). Chemical conversion of **239** to **239a** and the corresponding conversion of **511** to **507** permitted the comparison of the properties of the two constitutionally equivalent derivatives, **239a** and **507**. Significant differences in their ¹H-NMR chemical shifts proved the nonequivalence of the two parent structures, **239** and (-)-hardwickiic acid (Figure 36A).

Attempts were then made to correlate **239** with a *cis*-clerodane of known absolute stereochemistry [revised as per *15*], plathyterpenone (**1075**). Comparison of the CD results for **239b** (a derivative of **239**) and **1075** suggested a *cis*-AB ring-fusion at C-5 and C-10 identical to the revised **1075** (Figure 36B [p. 518]).

The *Solidago shortii* clerodanes reported initially as *trans-ent*-clerodanes (See above) were reinvestigated (*386*) as part of the *S. arguta* study. Preparation of the 3,4-β-epoxide of 5α,10α-*cis*-fused solidagoic acid A produced a structure with ^{1}H-NMR properties strikingly similar to those of a *S. shortii* compound, **556**. It was then suggested that this and other *S. shortii* structures be revised to the corresponding *cis*-normal-clerodanes.

Reinvestigation of the above-mentioned *Solidago elongata* and *S. altissima* C-6 oxygenated clerodanes was also prompted by the identification of tricyclosolidagolactone (Figure 37 [p. 519]; Structure **947**), a *cis*-normal-clerodane (5α,10α-*cis*-fused) from *S. altissima* (*500*). This tricyclic clerodane contained a cis-fused AB ring system. Niwa and Yamamura proposed that the biogenesis of **947** (Figure 37) involved a precursor that was constitutionally identical to the first-proposed structure for solidagolactone V (Figure 35; Structure **262**) but belonging to the *cis*-normal-clerodane class (*402*). They reinvestigated the *S. altissima* clerodanes in order to resolve this inconsistency. Consequently, the structures of solidagolactones II–VII and the corresponding elongatolides were revised to *cis*-normal-clerodanes (Figure 35). Confirmation of the absolute stereochemistry of the solidagolactone series was provided by X-ray analysis of the C-6 p-bromobenzoate derivative of desacylsolidagolactone VIII (*401*).

VI.1.1.3. cis-ent-*Clerodanes (5β,10β-cis-Clerodanes)*

A *Solidago arguta* report identified a mixture of seven clerodanes, one *trans-ent*-clerodane (**507**) and six novel 5β,10β-*cis-ent*-clerodanes, **578–583** (*7*). The absolute stereochemistry assigned these structures was confirmed by X-ray analysis (*263*) and by chemical interconversion (*386*). First, a nonheavy atom X-ray analysis of **583** established its relative stereochemistry (Figure 38 [p. 519]). The lactone **583** was transformed through reduction and acetylation into the co-occurring compounds **581** and **582**, respectively. Then, a study of optical data for *S. arguta* compound **578** (Figure 39 [p. 520]) confirmed the absolute stereochemistry for these structures as well as other compounds of similar properties. Haplopappic acid (**576**), isolated from *Haplopappus angustifolius* and *H. foliosus* (*462*), and the series of cistodioic acid-related structures from *Cistus monspeliensis* (Cistaceae; *33, 469*) were correlated through a series of derivatives with these *Solidago* structures of established stereochemistry (Figure 39).

CD (circular dichroism) and other optical data indicated that haplopappic acid, the *S. arguta* clerodanes and the cistodioic acid-related structure series had a common relative configuration that was largely opposite that of solidagoic acid A (**239**) and B (**244**). The two enones, **576c** and **578b**, derived as indicated in Figure 39 have positive Cotton effects for the n → π* transition of almost identical molecular amplitudes (α + 65 and + 64, respectively). The two enones derived from solidagoic acid A had an opposite and weaker Cotton effect (α − 46 and − 33) compared to those observed for **576c**, **578b** and related structures.

Despite opposite orientations at C-5, C-9, and C-10, $5\beta,10\beta$- and $5\alpha,10\alpha$-*cis*-clerodanes are epimeric rather than enantiomeric because both have α-methyls at C-8.

VI.1.1.4. trans-*normal-Clerodanes ($5\beta,10\alpha$-trans-*Clerodanes)*

Of the four possible AB-ring junction orientations, three are well-documented among the clerodanes of the Compositae. Identification of the fourth, a $5\beta,10\alpha$-*trans*-normal-clerodane, is at best tentative. In the first report of *trans*-normal clerodanes, structures **236** and **237** from *Bedfordia salicina* were illustrated (*122*) without specifying orientation of the methyl groups (Figure 40 [p. 521]). Although constitutionally identical to kolavenol (**475**), **237** differed in its spectral properties. In particular, the optical rotations for **237** and **475** were opposite ($[\alpha]_D$ = $+7.9$ and $[\alpha]_D$ = -57.1, respectively, *365*). The authors suggested that the data for **236** and **237** were consistent with structures bearing C-5β-, C-8β-, and C-9β-oriented methyls. Nevertheless, the positive optical rotation and methyl chemical shifts (e.g., **237**: H-17,δ0.79; H-18,δ1.69; H-19,δ1.05; H-20,0.82) for **236** and **237** are also consistent with the *cis-ent*-clerodanes (Table 14 [pp. 507–510]).

Koanophyllon admantium compounds **245** and **250** were depicted as *trans*-normal-clerodanes with β-oriented C-8 methyls (Figure 40; *42*). However, the structures were accompanied by the statement that the absolute configuration of these diterpenes was not determined, raising the possibility that **245** and **250** were (-)-hardwickiic acid and its 1,2-dehydro-derivative, respectively. The *Koanophyllon* clerodanes shared with (-)-hardwickiic acid a negative optical rotation. The confusion regarding these structures was heightened by the report of these two structures from *Centipeda* (*128*) and **251**, the methylester of **250**, from *Nidorella* (*72*) in which the structures were displayed with C-8α-methyls (The C-8α-methyl epimers of **245** and **250** are treated here as **249** and **248**, respectively). Although **251** was reported as a methylester (nidoreseda acid methyl ester) from *Nidorella*, when this structure report was cited in the *Centipeda* article the structure was described and drawn as the free acid.

The *Centipeda* compound **249** assignment was based on correlation with the *Solidago* (*286*) *trans-ent*-clerodane, (-)-hardwickiic acid. The discrepancy between the structure illustrated for **249** and that of (-)-hardwickiic acid was not discussed. Possibly, the above-described confusion results from incorrectly illustrated structures for **248**, **249**, and **251**. Because **249** was described as being previously reported by Misra et al. (*391*), it is assumed that this structure is actually (-)-hardwickiic acid and is probably identical to **245**. The other structures, **248** and **250**, are presumably the same compound and identical to 1,2-dehydro-(-)-hardwickiic acid. Very likely, **248** and **249** are the same as **250** and **245**, respectively.

A previously characterized clerodane, **238**, and two novel clerodanes, **253** and **254**, were reported as $5\beta,10\alpha$-*trans*-clerodanes from *Nidorella* (*72*). The citation for the known compound was a *Solidago* article (*317*) in which the structure was drawn differently (as **1064**, a $5\alpha,10\alpha$-*cis*-clerodane). Structures **238** and **1064** differ in that their C-5 hydroxymethylenes are β- and α-oriented, respectively (Figure 40). The authors noted that the absolute stereochemistry of these compounds was not determined and that the displayed structures were only tentative. Possibly, the authors intended all three of these structures (**238**, **253**, and **254**) to be shown with the more common configuration of **1064**.

Based on the above-described reports, there is no well-established example of $5\beta,10\alpha$-*trans*-clerodanes in the Compositae.

The previously mentioned mix-up regarding the *Nidorella* compounds has exacerbated an already confused situation regarding the name(s) and structure(s) proposed for the 5,10-seco-clerodane structures. The first report of a compound with these properties from *Conyza stricta* (*460*) incorrectly assigned the structure of Δ^1-polyalthic acid (Figure 40) to the compound, conyzic acid. A second compound, secoreseda acid, from *Nidorella resedifolia* (*72*) (Figure 40) was assigned a relative (but not absolute) stereochemistry, based on analogy with co-occurring compounds. Concurrently, a diterpene isolated from *Conyza stricta*, strictic acid, (*470*) was assigned a structure epimeric at C-9 to that first proposed for secoreseda acid. Comparison of the properties of these three compounds led to the subsequent proposal that conyzic acid, secoreseda acid, and strictic acid were identical (*376*) and that the probable structure was that of secoreseda acid (Figure 41 [p. 522]; Structure 2). The C-9 methyl was assigned a β-orientation because secoreseda acid co-occurred with compounds with β-oriented C-9 methyls. Later, when secoreseda acid was reported from *Conyza scabrida*, the structure was drawn as a *trans-ent*-clerodane derivative (Figure 41; Structure 3; *83*). The assigned stereochemistry was at least in part based on the co-occurrence with biogenetically closely related *trans-ent*-clerodanes (see Figure 11). This revision of the secoreseda acid structure was published without comment by the authors (*83*). However, this revised version has the same absolute stereochemistry as that originally assigned to strictic acid (*470*) and is currently the best guess at the structure of conyzic acid/strictic acid/secoreseda acid.

VI.1.2. ^{1}H-NMR Spectrometry of Clerodanes

As suggested by the preceding discussion, assignment of orientivity at the four clerodane asymmetric centers, C-5, C-8, C-9, and C-10, is likely the most problematic aspect of structure elucidation. The co-occurrence in the same species of clerodanes representing different ring-junction types erodes the reliability of biogenetic arguments in assigning stereochemistry based on stereochemical properties of compounds previously isolated from the same species or from related species. Given the range of biogenetically feasible structure candidates, assignment of stereochemistry may require choosing between pairs of enantiomeric structures or between constitutionally identical structures that are epimeric at one or more centers. Such a problem greatly complicates the assignment of absolute stereochemistry.

VI.1.2.1. General Properties of Methyl Chemical Shifts

A reference point for the evaluation of the ^{1}H-NMR chemical shifts of the clerodanes is the study initiated by Zürcher on the steroid ^{1}H-NMR properties. Zürcher examined the effects of individual substituents on the chemical shifts of the C-10 and C-13 angular methyls of the four androstane skeletal types (*504, 505*; Figure 42A [p. 523]). His findings became codified as the "Zürcher Rules." Chemical shifts for the four reference compounds (Figure 42A) demonstrate that angular methyl shift values depend in part on the *cis* or *trans* nature of the ring junction.

Other results indicate that the largest deshielding effect on angular methyls results from substituents that have a 1,3-*syn*-diaxial relationship with the methyl group. Comparable substituents that are equatorial at these positions have little influence on the methyl chemical shifts. In general, substituents on carbons that are vicinal to the methyl-bearing ring carbon display the second greatest deshielding influence.

Similar patterns can be observed in the constitutionally equivalent representatives of the three clerodane types (Figure 43 [p. 524]):

1. The $5\alpha,10\beta$-*trans*-representative (Figure 43; Structure C-1) with its *trans*-AB ring junction and the syn-diaxial arrangement of the C-9 and C-5 methyls is comparable to the 5α-14α-androstane structure (Figure 42A; Structure 1): Its relatively upfield C-5 and C-9 methyl chemical shifts are largely a consequence of the methyls's axial relationship to the ring system.
2. The introduction of the A-B and/or C-D *cis*-ring junction into the last three representative androstane skeletons results in an appreciable downfield shift of the angular methyl situated at the *cis*-junction. Relative to the 5α-10β-*trans*-clerodane structure, C-5 methyls of both the $5\alpha,10\alpha$-, and $5\beta,10\beta$-*cis*-representatives display similar downfield shift effects (Figure 43; $5\alpha,10\beta:\delta1.00$; $5\alpha,10\alpha:\delta1.17$; $5\beta,10\beta:\delta1.06$).
3. In both the $5\alpha,10\beta$-*trans*- and $5\beta,10\beta$-*cis*-clerodanes, the C-9 methyl adopts an axial orientation that is *trans* to H-10. Both produce chemical shifts ($\delta0.74$ and 0.84, respectively) that are upfield from the shift of the equatorial C-9 methyl of the $5\alpha,10\alpha$-*cis*-structure ($\delta1.08$). Despite the flexibility of the $5\alpha,10\alpha$-*cis*-decalin system, evidence (discussed below) suggests that for this system the conformation with the C-9 equatorial methyl is generally favored.

The usefulness of ^{1}H-NMR chemical shifts in differentiating the two major clerodane isomeric types ($5\alpha,10\beta$-*trans*- and $5\alpha,10\alpha$-*cis*-fused) was first noted in the comparison of the chemical shifts of the solidagoic acid A (**239**) derivative (**239a**) and the (-)-hardwickiic acid (**511**) derivative (**507**) (Figure 36A). Analysis of additional structures of the two isomeric types demonstrated that the most useful ^{1}H-NMR parameters for assigning stereochemistry were the C-5-, C-8-, and C-9-methyl chemical shifts. Niwa and Yamamura (*402*) observed that the C-8-methyl doublet ($\delta0.90$) of **239a** appeared upfield from the C-9 methyl singlet ($\delta1.08$), while the C-9 methyl singlet ($\delta0.74$) of **507** occurred upfield from the C-8 methyl doublet ($\delta 0.86$). These differences were attributed to the presence of an equatorial C-9 methyl in **239a** and an axial C-9 methyl in **507**.

Based on examination of compiled C-4, C-5, C-8, and C-9 methyl chemical shifts (Table 14), with very few exceptions the relative positions of C-8 and C-9 methyl chemical shifts observed by Niwa and Yamamura characterize all listed compounds of the two major clerodane types.

VI.1.2.2. ^{1}H-NMR Properties of cis-normal-Clerodanes (5α,10α-cis-)

VI.1.2.2.1. C-6 Oxygenation

The C-9 methyl (H-20) chemical shift generalization made by Niwa and Yamamura (*402*) for the 5α,10α-*cis*-clerodanes seems to be consistent for all compounds listed in Table 14. The compounds with the most divergent H-20 chemical shifts, **262** ($\delta 1.21$), **265** ($\delta 1.17$), and the synthetic product **262a** ($\delta 1.24$) (*402*), represent all the listed structures with C-6 keto functions. While no H-20 deshielding is evident for C-6 hydroxyl- or ester-containing structures, such deshielding is pronounced for the C-6 carbonyl-bearing compounds (Figure 44 [p. 525]).

VI.1.2.2.2. Conformational Variability

^{1}H-NMR data support the view that two basic conformations, I and II, (or some minor variant of I or II in which, for example, the A-ring exists as a flattened half-boat form or some twist analog) occur within the 5α-10α-*cis*-clerodanes (Figures 45A [p. 526] and 46 [p. 527]).

Individual circumstances tend to favor one conformation or the other. For instance, the presumed conformation for hemiacetal or lactone ring-containing compounds with C-18,19-oxido linkages (Figure 45B [p. 526]) is conformation I because the closure of the hemiacetal or lactone ring imposes less constraint on the decalin system in this conformation. Compound **1068** (Figure 45B) is characterized by a H-3 ^{1}H-NMR signal that is relatively narrow ($w_{1/2} = 3$Hz), in agreement with the proposed conformation (*317*). However, as will be described below not all such C-18,C-19-heterocyclic structures agree with this interpretation.

The assignment of conformation was more problematic for compounds lacking C-18,19-oxido linkages. Based on a variety of data (*15*), Anthonsen and coworkers proposed that conformation II with the A-ring flattened into a half-boat was more reasonable than conformation I. They argued that their proposal was supported by the C-9 methyl (H-20) chemical shift for compounds with the A-ring substituents shown in Structure 4 of Figure 45A. As drawn, the C-9 methyl is directed over the enone function and is shielded by it. In compounds lacking the enone, the methyl produces a signal at approximately $\delta 1.05$. However, when the enone is present the methyl signal appears at about $\delta 0.85$ (*15*).

These results, in particular those for structure **239b**, suggest that the shielding zone of the ketone (in the absence of α,β-unsaturation) alone is sufficient to cause the observed H-20 chemical shift. Other examples of this shielding effect include structures **239d**, **239b**, and **239c** (Table 14; Figure 60 [pp. 541–544]).

A derivative, **263a**, prepared from a variety of sources (solidagolactones II and III and **263**) by Goswami and coworkers, displayed methyl chemical shifts (Table 14) that were inconsistent with the above-described shielding effect (*291*). In depicting the conformation of the related structures **268** and **269**, Goswami and associates (*291*) opted for the conformation that was most consistent with their results (Figure 45C [p. 527]). As in conformation I, the C-9 methyl was equatorial, and the C-6 ester was axial.

The results reported by Anthonsen and coworkers (*15*) and Goswami and associates (*291*) appear on the surface to be contradictory. However, these and other results (described below) are made compatible by assigning the favored conformation according to "rules" based on several alternative A and B ring substitutional patterns. If R_1 (Figure 46) is a keto function and R_2 is a pair of hydrogens, then conformation II is favored. However, the results suggest that if R_2 is an oxygen substituent the favored conformation is I regardless of the R_1 substituent. Given that the majority of $5\alpha,10\alpha$-*cis*-clerodanes have C-6 oxygen substituents, the most common conformation is expected to be I. If R_2 is a keto function, maximum deshielding of the C-9 methyl is expected in conformation I (or some minor variant of this conformation). This effect is pronounced in the C-9 methyl chemical shifts of **262**, **262a**, and **265** (δ1.21, 1.24, and 1.17, respectively).

These "rules" are consistent with the data reported by Goswami and coworkers (*291*) and with data reported for the two structures, **1066** and **244a**, (Figure 47 [p. 528]). However, not all results are consistent with the generalization that conformation I should be favored in structures bearing a C-18,C-19-heterocyclic ring. The comparison of the methyl chemical shifts for **1066** and **244a** indicate a sizable difference in the chemical shifts of H-20 (δ1.03 and 0.84, respectively, Figure 47A). Partly, this difference can be attributed to the occurrence of the tetrahydrofuryl moiety in **244a** and the furyl moiety in **1066**. However, comparison of similar structures differing in this manner (Figure 47B) reveals that the deshielding effect of the furyl moiety on H-20 is approximately δ0.07–0.08 greater than that of the tetrahydrofuryl moiety. The difference in H-20 chemical shift between **1066** and **244a**, δ0.19, suggests that some factor other than the furyl moiety is contributing to this difference. The only other structural difference is the C-2 keto function. If the C-2 keto group is responsible for this evident shielding effect of H-20, then conformation II is plausibly assigned to **244a**. This result is consistent with the proposed "rule" that if $R_1 = O$ and $R_2 = H,H$ (Figure 46) then conformation II is favored.

The "rules" for the interpretation of conformational variability are further supported by the data for the two **1025** conformations detected at $-50°C$ (Figure 48 [p. 529]; *170*). While the room temperature $CDCl_3$ spectrum was difficult to interpret, two distinct sets of signals were obtained at $-50°C$. The first (**1025a**; Figure 48) was characterized by a B-ring conformation in which the C-6 acetate adopted an axial position and the C-9 methyl was equatorially oriented. This corresponds most closely to conformation I described above. The C-9 methyl chemical shift ($\delta 0.93$) is in agreement with this equatorial position. In the second conformation (**1025b**; Figure 48), the C-6 acetate adopts an equatorial orientation and the C-9 methyl is axially oriented (conformation II). The upfield position of the C-9 methyl signal ($\delta 0.73$) is consistent with the shielding effect of the C-2 keto on the axial methyl. These assignments are corroborated by the H-6 coupling constants. For **1025a**, the H-6 signal appears as a broad singlet, while the H-6 signal for **1025b** is a doublet (J = 12Hz), values which are consistent with the proposed conformations. In compliance with the above "rules," this structure which possesses a C-6 β-oxygen substituent adopts conformation I as the dominant conformation at room temperature (Figure 48).

Several other lines of evidence were developed for the assignment of conformation I to C-6 oxygen-substituent-bearing structures. In the first, Eu(dpm)$_3$-induced shift results for C-6 hydroxy-representatives of $5\beta,10\beta$- and 5α-10α-clerodanes supported the two conformations indicated in Figure 49 (p. 530). The normalized ratio for the $5\beta,10\beta$-fused structure (Figure 49; *386*) was interpreted as support for a steroidlike conformation. The C-19 value for the $5\alpha,10\alpha$-fused structure (5.3) was indicative of an *anti-trans* relation of C-19 to the axial C-6 hydroxyl and of a nonsteroidal conformation (*378*).

The second line of evidence was based on the chemical transformations of **545** to yield two products, **545a** and **545b** (Figure 50 [p. 531]). The formation of a compound with an ether bond between C-3 and C-6 (**545b**) was considered proof of a $5\alpha,10\alpha$-*cis*-ring junction and a nonsteroidal conformation (*378*).

VI.1.2.2.3. 3,4-Epoxides

The 3,4-epoxide functions of clerodanes can adopt either an α- or β-oriented *cis* configuration. As this represents one more complication in the already difficult task of assigning absolute stereochemistry, it is not surprising that epoxide orientation within the elongatolide (solidagolactone) series of $5\alpha,10\alpha$-*cis*-clerodanes has been misassigned. Elongatolides D (solidagolactone VI, *413*; **553**) and E (solidagolactone VII, *413*; **260**) were the first reported 3,4-epoxide-containing clerodanes of this type (currently accepted structures shown in Figure 51A [p. 532]). Although they were initially described as *trans*-clerodanes with unspecified epoxide orientation (Figure 35; *18*), Niwa and Yamamura (*402*) revised these structures to $5\alpha,10\alpha$-*cis*-clerodanes with α-epoxides (Figure 51). Their assignment was based on a series of transformations involving the furyl derivative of **260** (Fig. 51B [p. 532]). They argued that the formation of **260b** and **260c** suggested that the C-4 tertiary hydroxyl group must be axially oriented which would be compatible with the original presence of an α-epoxide ring in **260**. However, the conformational flexibility of the $5\alpha,10\alpha$-clerodane skeleton provides a variety of reasonable conformations: In the instance of the C-4 α-hydroxyl, conformations with either the axial or equatorial configuration of this functionality can reasonably be proposed. Likewise for the C-4 β-hydroxyl option, there are two possible orientations (axial or equatorial).

The structural questions relating to **553** and **260** were finally resolved by Nishino et al. (*401*). Epoxidation of solidagolactone IV (**545**) yielded (in order of abundance) two products, the β- and α-epoxides (Figure 52B [p. 534]). The observed ratio of products was in agreement with the usual mechanism for the epoxidation of homoallylic alcohols: In the presence of a C-6 β-OH group, the epoxide of the major product should have the same configuration as the hydroxyl group. Data for the major product (β-epoxide) included an IR absorption at 3470 cm^{-1} (unaffected by dilution) indicative of hydrogen bonding and no absorption at 3600 cm^{-1} which, if present, would suggest a free OH. The IR spectrum of the minor product (α-epoxide) included an absorption at 3600 cm^{-1} and only a weak absorption at 3475 cm^{-1} which was affected by dilution. Further evidence for the hydrogen bonding of the epoxide oxygen and the C-6 hydroxyl was the presence in the ^{1}H-NMR spectrum of the β-epoxide-containing structure of a sharp singlet at $\delta 5.41$ (OH proton), whereas the OH proton of the α-epoxide occurred as a multiplet at approximately $\delta 1.8$. Because this intramolecular hydrogen bonding is only possible if the 3,4-epoxide is β-oriented, the major product was assigned this configuration. The β-epoxide was correlated with solidagolactone VII (**260**) and VIII (**988**) by a set of reactions (Figure 51C [p. 533]) in which **260** and **988** and the p-bromobenzoate derivative (Figure 51C; Structure 5), were transformed to the corresponding furyl derivatives and through hydrolysis to the common product (Figure 51C; Structure 4). More importantly, the other compounds were correlated with the p-bromobenzoate derivative (Figure 51C; Structure 5) for which X-ray analysis provided the absolute configuration. Consequently, the α-epoxide structure proposed for solidagolactone VII (and presumably solidagolactone VI) was revised to a β-epoxide.

An unexpected result of the study by Nashino and his coworkers was the ratio of products in the solidagolactone IV epoxidation reaction (Figure 52B), 1.7(β-epoxide):1.0(α-epoxide). Given that the homoallylic hydroxyl system should favor an attack on the same side of the molecule as the hydroxyl function (β), the surprisingly nonskewed observed ratio suggests that in the absence of this C-6 hydroxyl, attack from the α-face of the A-ring is favored. Support for this proposal came from an apparently unlikely source (*291*) in which just the opposite assumption was made. Solidagolactones VII (**260**) and VIII (**988**) were isolated by Goswami and associates from *Solidago virgaurea* and correctly identified as $5\alpha,10\alpha$-*cis*-clerodanes. The epoxide was assigned an α-orientation in agreement with the previously published findings of Niwa and Yamamura (*402*). Treatment of the solidagolactone II (**258**) and III (**259**) mixture with m-chloroperbenzoic acid yielded an angelate/tiglate pair of epoxide products that were isomeric to **260** and **988** (Figure 52C [p. 535]). The authors assumed that this angelate/tiglate pair possessed a β-epoxide because the H-3's chemical shift ($\delta 3.04$) and coupling constant (d, J = 4Hz) were distinct from those ($\delta 2.74$, broad singlet) of the naturally occurring pair of structures (**260** and **988**) to which they had previously assigned an α-epoxide. However, the absolute configuration of solidagolactone VII was subsequently established and revised to a β-epoxide (*361*). Given that Goswami and coworkers originally misassigned the epoxide of **260** and **988**, then the epoxide of the isomeric transformation product (Figure 35C) must have just the opposite configuration (α) to the one originally assigned (β). The similarity of the ^{1}H-NMR data for the H-3 proton of the epoxidation product (d, J = 4Hz) and the α-epoxide synthesized by Nishino et al. (d, J = 5.1Hz) supports this conclusion. The observation that this α-epoxide was the major product (no other products were identified) suggests that in the absence of the C-6 β-hydroxyl, epoxidation occurs most favorably from the α-face. A similar epoxidation reaction (Figure 52D [p. 535]) involving the C-6 acetate of solidagolactone IV (**545**, *413*) yielded one detectable product (probably an α-epoxide) characterized by a doublet H-3 signal with J = 4.5Hz.

In another epoxidation reaction (Figure 52A), a transformation product of solidagoic acid, **239**, (*15*) yielded two epimeric epoxides (*386*). Differentiating the two epimers was attempted by the further transformation of the β-epoxide (Figure 53 [p. 535]) (*386*). ^{1}H-NMR evidence of an axial (β) hydroxyl function at C-3 in the product of this Lewis acid treatment was used as evidence in the assignment of spectroscopic properties to each epoxide. Two aspects of the assignments appear contradictory to previously reported findings. First, the ratio of products, 3(β-epoxide):1(α-epoxide), indicates that in this instance attack from the β-face of the 5α-10α-*cis*-clerodane is favored. Second, the H-3 coupling constants reported for the two epimers are opposite those assigned by Nishino et al. (Figure 52B). Accordingly, it appears that the two epoxide products (Figure 52A) should be revised such that the major product is the α-epoxide. However, caution should be exercised in concluding that the reaction-A products are misassigned. There are several reasons for caution. The absence of a C-6 oxygen function may affect the conformation of the products of reaction A, with the result that coupling constants differ from those of C-6 oxygenated compounds. Further, the effect of the absence of the C-6 function on the likelihood of β-face attack during epoxidation is not well understood.

A recent definitive study of the stereochemistry of C-6 substituted 5α-10α-*cis*-clerodanes (*483*) confirmed the above interpretation of the results for the epoxidation of the C-6 β-acetate and formally revised the epoxide configurations (described above) originally assigned by Niwa and Yamamura (*402*) and Goswami et al. (*291*). The authors also used α- and β-epoxide analysis results to determine the feasibility of applying Tori equation calculations to the prediction of H-3 coupling constants (Figure 54 [p. 536]). Dihedral bond angles for H3-C3-C2-Hβ-2 and H3-C3-C2-Hα-2 for the β-epoxide were known and those for the α-epoxide were estimated based on an anticipated change of 18° in this angle during epoxidation of the olefin. In each instance the calculated coupling constants closely approximated the observed values.

VI.1.2.3. ^{1}*H-NMR Properties of trans-ent*-Clerodanes

The axial configuration of the C-9 methyl (H-20) results in a relatively high-field ^{1}H-NMR chemical shift for this signal. Departures from the typical value of c. δ0.75 for H-20 can usually be attributed to the deshielding effect of an oxygen function at one of several possible positions:

VI.1.2.3.1. C-7α-OR

The ^{1}H-NMR spectra of C-7α-hydroxyl-bearing compounds **509**, **559**, and **562** and C-7α-acetoxyl-bearing **500** and **509a** produce H-20 singlets at δ1.02, 0.92, 0.86, 0.97, and 1.02, respectively (Figure 55 [p. 537]). Spectra of compounds **509**, **500**, and **509a** display H-19 singlets at δ1.36, 1.17, and 1.34, respectively. The spectra of the C-7 keto derivatives of **500** and **509** are characterized by H-19 and H-20 singlets that are δ0.2–0.3 upfield from the parent compounds (Figure 55). These observed chemical shifts are attributable to the 1,3,5-triaxial relationship of the C-5 methyl, the C-9 methyl, and the C-7α-hydroxyl or ester function. The interaction between these substituents is shown in Figure 56 (p. 538) from data supplied by Ohsuka et al. (*412*).

VI.1.2.3.2. C-12-OR

The C-12 hydroxyl function is the probable cause of the relatively downfield position of the H-20 singlet (δ0.94) in the spectrum of **1061**.

VI.1.2.3.3. C-19-OR and Aldehyde

Introduction of a hydroxyl function at C-19 has minimal effect on the chemical shift of H-20 (Table 14 and Figure 57 [p. 539]). However, attachment of increasingly bulky ester functions (e.g., angelate) at C-19 appears to result in modest deshielding of H-20. Structure **904** which differs from the related compounds **900**, **902**, **903**, and **1033** (hautriwaic acid) (Figure 57) by the further oxidation of C-19 to an aldehyde is characterized by an H-20 singlet at δ0.60. The considerable shielding of this methyl is attributed to the aldehyde (C-19) which is 1,3 diaxially oriented with respect to the C-10 methyl (H-20).

VI.1.2.3.4. C-17 Aldehyde

Oxidation of the C-8 methyl (H-17) in compound **967** appears to have an appreciable deshielding effect on H-20 (δ1.00).

VI.1.2.4. cis-ent-*Clerodanes*

McCrindle and coworkers (*386*) have proposed a conformation for the 5β,10β-*cis*-clerodanes (Figure 58 [p. 539]). The H-20 chemical shifts for the members of this series (Table 14) are consistent with the depicted axial arrangement for this methyl. A comparison of the values reported for the cistodioic acid series of compounds (Figure 59 [p. 540]; *33, 469*) and those of the Compositae (Table 14) suggest that the range of observed C-9 methyl chemical shifts results largely from differences in the sidechain moiety. A saturated sidechain is correlated with the relatively upfield values, while the furyl sidechain is associated with the most downfield.

VI.1.3. ^{13}C-NMR Spectrometry of Clerodanes

Reviews of diterpene ^{13}C-NMR data have focused in one instance on all reported diterpenes (*494*), in the second instance on the *trans*-clerodanes (*375*), and in the third (*378*), on the use of these data to differentiate between 5α,10α-*cis*- and *trans*-clerodanes. As this third review deals specifically with the solidagolactone problem, it will be discussed at length here.

Pendant methyl groups of clerodanes display some of the same properties as those previously elucidated for steroids (Figure 42; *504* and *505*). Principally, the configuration of the ring junction influences the chemical shifts of the associated methyls. Comparison of the C-17, C-19, and C-20 chemical shifts of a series of C6-substituted 5α,10α-*cis*-clerodanes with those of several *trans*-clerodanes reveals differences consistent with those observed for *cis*- and *trans*-decalin (Figure 42B) and steroid systems (Table 15 [pp. 510–512]; Figure 61 [pp. 545–549]). In the *cis*-clerodanes, the C-19 signal appears in an area centered on δ25, while in the *trans*-clerodanes it occurs between δ17 and 19. Similarly, the C-20 signal resonates at a lower field (δ21 to 29) than that of the C-20 of the *trans*-clerodanes (δ17 to 19). The C-17 signals of both *cis*- and *trans*-fused compounds display relatively constant chemical shifts near δ15.

The association between ring junction and methyl chemical shift was attributed to the shielding effect from carbons with a τ-*gauche* relationship to the resonating carbon. In *trans*-clerodanes, the number of carbon atoms having a τ-*gauche* interaction with C-19 is larger than the number in *cis*-clerodanes. In *trans*-clerodanes, C-20 has an axial orientation, while in *cis*-clerodanes, this methyl has an equatorial orientation. The relatively constant chemical shift of C-17 in *trans*- and *cis*-clerodanes is attributable to its equatorial orientation in both ring junction types (*378*). Evidently, C-19 and C-20 ^{13}C-NMR chemical shifts are useful in distinguishing *cis*- and *trans*-clerodanes (Figure 61). Just how useful these data will be depends on how many exceptions (if any) will be found to the data in Table 15.

VI.1.4. Conformational Analysis (X-ray)

Quantitative conformational analysis of *trans*-clerodanes (*256*) and 5α,10α-*cis*-clerodanes (*378*) has been conducted on X-ray diffraction results.

VI.2. Other Bicarbocyclics

^{13}C-NMR studies of labdanes have focused in one instance on the normal-labdanes, manoyl oxide (**272**), manool (**996**), sclareol, and marrubiin (*4*), and in a second on the C-2β- and C-3α-hydroxylated *ent*-labdanes (*424*) reported from a non-Composite, *Phlogacanthus thyrsiflorus*. The results of the first study and other unpublished ^{13}C-NMR findings for normal-labdanes were discussed in detail by Wehrli and Nishida (*494*).

VI.3. Tricarbocyclics

Early ^{1}H-NMR work on the resin acids, sandaracopimaric, pimaric, isopimaric acids and their dehydro- and tetrahydro- derivatives, applied chemical shift calculation methods to the prediction of C-17, C-18, and C-20 methyl chemical shifts (*19, 20*). ^{13}C-NMR results for several other normal-pimaranes were recently reviewed (*494*).

Investigation of ^{1}H-NMR properties of the sandaracopimaranes isolated from *Garuleum* and *Osteosperum* (*156*) led to the proposal of a conformation for these structures in which the A-ring adopted a twist form (Figure 62 [p. 550]). Evidence for this conformation was provided by the comparison of the C-4β-methyl chemical shifts for sandaracopimar-15-en-8β-ol and its C-6β-hydroxy and acetoxy derivatives (Structures 1, 3 and 4, respectively; Figure 62; Table 16 [p. 512]). The C-6β-hydroxyl group had a strong deshielding effect on the C-4β-methyl, an effect which was diminished by introduction of an acetyl group. This effect agrees with the proposed 1,3-*syn*-diaxial relationship between the C-6 hydroxyl and the C-4β-methyl. The deshielding effect of ths C-6 hydroxyl on the C-10 methyl suggests that the same 1,3-*syn*-diaxial relationship exists for these substituents. The introduction of a C-11β-hydroxyl group (Structures 11 and 12; Figure 62) displayed the expected deshielding effect on the C-10 and C-13 methyls (Table 16). However, equatorially oriented hydroxyl groups at C-11 and C-12 (e.g., Structures 5, 7, and 10; Figure 62) had much less effect on neighboring protons.

Detailed ^{13}C-NMR results for ferruginol (**319**) and other aromatic abietane compounds were used to indicate the stereochemistry at C-4 for compounds possessing either an equatorial or axial carbinol group (*494*).

The only review of ^{13}C-NMR results for the cassane series focused on a group of compounds from a non-Compositae source (*Caesalpinia bonducella*) that were substitutionally distinct from those reported from *Osteospermum* (*231*).

VI.4. Tetracarbocyclics and Related Skeletons

Pre-1978 ^{13}C-NMR analyses of *ent*-kauranes (*357, 310, 309, 449*) have been reviewed (*494*). Another report (*502*) included complete data for the *Stevia ent*-kaurene glucosides, stevioside (**683**), rebaudioside A (**685**), and paniculosides I-III (**895–897**). Recently, a detailed study of kaurenoic acid derivatives (*345*) sought to compare the ^{13}C-NMR chemical shifts of *ent*-kaur-16-en-19-oic acid (**625**) with those of hydroxy derivatives at C-6, C-7, C-11, C-12, and C-15 and with kaurenolide. Another study (*265*) reported the results of complexation of *ent*-kaurane polyols with boric acid. Compounds containing 1,2- or/and 1,3-glycol systems displayed boric acid-induced shifts in their ^{13}C-NMR spectra together with broadening of signals of carbons attached to or adjacent to the glycol function.

ent-Beyerane and beyerene ^{13}C-NMR data have been reported for a series of C-1- and C-12-substituted structures (*486, 494*) and for a series of C-2- and C-3-substituted compounds (*308*).

The ^{13}C-NMR parameters for trachylobane and C-19 derivatives were consistent with predicted values derived from 9-methyl-*trans*-decalin and tricyclo[2.2.1.0^{2,6}]octane (*21*).

TABLE 14. ^{1}H-NMR Chemical Shifts (δ) for the C-17, C-18, C-19, and C-20 Methyls of Clerodanes.

Compound No.	H-17	H-18	H-19	H-20
5α,10α-cis				
1. 1025(RT)[1]	0.93	1.95	1.32	0.93
2. 1025b	1.04		1.18	0.73
3. 1025a	0.84		1.29	0.93
4. 239[2]	0.89	1.52	COOH	0.98
5. 239a[3]	0.92	1.62	1.17	1.08
6. 310	0.94	(4.24,3.96)	(3.83,3.50)	1.09
7. 258	0.86	1.56	1.28	1.08
8. 259	0.85	1.56	1.28	1.08
9. 545	0.88	1.70	1.15	1.03
10. 262	0.83	1.50	1.17	1.21
11. 263	0.89	1.64	1.26	1.08
12. 265	0.84	1.57	1.30	1.17
13. 268(Ang)/ 269(Tig)	0.86	0.98	1.33	1.10
14. 261	0.88	1.29	1.21	1.03
15. 260	0.88	1.31	1.22	1.03
16. 260a	1.03	(4.93-5.01)	1.28	1.18
17. 262a	0.87	1.54	1.24	1.24
18. 260b	0.87	1.08	1.16	1.16
19. 260c	0.86	1.23	1.08*	1.00*
20. 260d	0.85	1.26	1.20*	1.03*
21. 263a	0.97		1.37	1.04
22. 1091a	0.93		COOH	0.91
23. 1091	0.92		COOH	0.89
24. 1085	0.98	(4.47,4.95)	COOH	0.98
25. 1062(CCl_4)	0.78	1.42	CHO	0.78
26. 1063(")	0.85	CHO	CHO	0.96
27. 1064(")	0.90	1.67	CH_2OH	1.08

TABLE 14. (*contd.*).

Compound No.	H-17	H-18	H-19	H-20
28. 1065(")	0.85	(4.16)	1.18*	1.03*
29. 1066(")	0.90	COOR	(4.44,3.80)	1.03
30. 1067(")	0.85	(4.21)	(5.42)	0.97
31. 1068(")	0.81	(4.89)	-CHOH-	0.88
32. 1026	0.96	0.87	0.96	$-CH_2-$(1.58,1.69)
33. 556	0.88	1.23	1.11	0.94
34. 239b	0.87	0.97	1.22	0.90
35. 239c	0.96	1.94	1.28	0.83
36. 239d	1.02	1.95	1.27	0.90
37. 244a	0.82	COOR	(4.43,3.88)	0.84

5α,10β-<u>trans</u>

Compound No.	H-17	H-18	H-19	H-20
38. 1030	1.03	1.86	1.15	0.83
39. 475	0.79	1.55	0.97	0.71
40. 479	0.75	1.55	0.97	0.70
41. 480	0.82	1.53	1.21	0.97
42. 495	0.85	1.60	1.02	0.76
43. 494(C_6D_6)	1.00	1.52	0.82	0.67
44. 1032	1.06	1.80	1.04	0.80
45. 1031	0.75	1.55	1.21	0.83
46. 1027	0.81	1.63	1.20	0.82
47. 1028	0.83	1.55	1.24	0.84
48. 1029	0.82	1.56	1.19	0.81
49. 962	0.84	1.29	1.10	0.74
50. 507	0.86	1.64(CCl_4)	1.00	0.74
51. 1034a	0.86	COOH	(4.55,4.31)	0.79
52. 509	1.04	(4.12)	1.36	1.02
53. 509a	1.05	4.5	1.34	1.02
54. 509b	0.97	(4.49)	1.06	0.75

TABLE 14. (*contd.*).

Compound No.	H-17	H-18	H-19	H-20
55. 846	0.85	(4.23,3.67)	(3.85,3.98)	0.78
56. 1033a	0.86	COOH	(4.59,4.32)	0.80
57. 1034	0.88	COOH	(4.22)	0.76
58. 1034a	0.86	COOH	(4.55)	0.79
59. 1034b	0.86	COOMe	(4.60)	0.81
60. 967	CHO	1.60	1.07	1.00
61. 551b	0.82	1.22	1.02	0.67
62. 1061		COOR	(4.15,4.09)	0.94
63. 543	0.84	1.57	1.01	0.77
64. 563	0.82	COOR	(4.28,3.89)	0.58
65. 562	1.03	COOR	(5.31,3.9)	0.86
66. 559	1.06	COOR	(5.33,3.92)	0.92
67. 561	0.99	COOR	(3.92,4.01)	0.64
68. 562	1.04	COOR	(3.91,5.30)	0.87
69. 969		COOR	(4.09,4.05)	0.69
70. 970		COOR	(4.00,4.10)	0.87
71. 525		COOR	(4.15,4.18)	1.10
72. 526		COOR	(4.29,4.27)	1.13
73. 526a		COOR	(4.13,4.45)	1.16
74. 900	0.89	COOMe	(4.36,4.60)	0.84
75. 902	0.89	COOMe	(4.62,4.54)	0.89
76. 903	0.89	COOMe	(4.62,4.37)	0.85
77. 904	0.86	COOMe	CHO	0.60
78. 500	0.92	1.59	1.17	0.97
79. 500a			0.96	0.72
80. 500b	1.00	1.59	1.25	0.98
81. 500c			1.27	1.01
82. 500d	0.89		0.95	0.71
83. 960	0.81	2.02	0.92	0.85

TABLE 14. (*contd.*).

Compound No.	H-17	H-18	H-19	H-20
5β,10β-<u>cis</u>				
84. 1057	0.74	(4.23,4.11)	1.12	0.76
85. 578(CCl$_4$)	0.83	1.72	1.06	0.84
86. 580(")	0.82	(4.54)	1.20	0.80
87. 1056a(C$_6$D$_6$)	0.70		1.16	0.76
88. 1055a	0.79	(4.58)	1.12	0.80
89. 579(CCl$_4$)	0.80	(4.24)	1.12	0.78
90. 1058	0.76	1.70	(3.36,3.23)	0.77
91. 582(CCl$_4$)	0.83	(4.82)	1.28	0.80
92. 582a(")	0.87	(4.70)	1.15	0.83
93. 581	0.85	(4.22)	1.35	0.80
94. 581a	0.87	1.85	1.21	0.82
95. 585	0.80	COOR	1.20	0.69
96. 584	0.85	COOR	1.26	0.77
97. 577	0.77	COOR	1.24	0.79
98. Cistodioic Acid	0.71	COOR	1.20	0.75
99. Cistodiol- Acetate	0.73		1.07	0.75

[1] See Figure 48 for conformations of compounds **1025a** and **1035b**.

[2] See Figure 1 for structures of non-derivative compounds.

[3] See Figure 60 for derivative compounds and naturally occurring compounds not included in Figure 1.

TABLE 15. ^{13}C NMR Chemical Shifts of Methyl Carbons, C-17, C-19, and C-20, of 5α,10α- and 5α,10β-clerodanes.

Structure	C-17	C-19	C-20	Reference
I. 5α,10α-<u>cis</u>-Clerodanes				
1. [545][1]	15.40	24.18	28.34	<u>483</u>
2. [262]	15.17	21.55	24.24	<u>483</u>
3. [260]	15.75	26.64	27.46	<u>483</u>

TABLE 15. (*contd.*).

Structure	C-17	C-19	C-20	Reference
4. [988]	15.75	26.53	27.46	483
5.	15.75	26.53	28.05	483
6.	15.81	26.47	28.05	483
7.	15.58	25.24	28.58	483
8.	15.52	25.24	28.34	483
9.	15.28	22.37	28.87	483
10.	15.28	24.54	28.11	483
11.	15.05	21.90	28.69	483
12.	15.46	24.13	28.58	483
13.	14.29	21.73	29.28	483
14.	15.23	28.46	26.66	483
15. [1083]	15.5	31.0	17.6	180
16. [1084]	15.7	64.8	17.2	180
17. [1085]	18.3	77.0	22.7	180
18. [1086]	17.7	101.1	22.3	180
19. [1087]	16.7	97.8	21.8	180
20. [1088]	16.4	101.3	21.4	180
21. [1089]	17.0	102.6	22.2	180
22. [1090]	17.0	104.1	23.6	180
23. [1091]	16.8	176.5	21.0	180
II. 5α,10β-_trans_-Clerodanes				
24. [961]	15.8	19.5	17.8	483
25.	15.5	*	17.6	483
26.	15.6	*	17.6	483
27.	15.5	*	17.3	483
28.	15.3	*	17.3	483
29.	15.4	*	17.8	483
30. [1033]	15.9	65.5	18.6	464
31. [1033 Ac]	15.6	71.8	18.1	464
32. [1034 Ac]	15.6	70.1	17.8	464
33. [1034Me,Ac]	15.6	70.1	17.8	464

TABLE 15. (*contd.*).

Structure	C-17	C-19	C-20	Reference
34. [**1070** Me]	19.9	60.6	15.5	468
35. [**1070**Me,Ac]	19.9	62.4	15.6	468
36. [**1093**Me,Ac]	18.9	62.2	15.9	468
37.	15.3	68.1	17.2	341
38.	12.0	72.6	19.0	341
39.	11.9	72.7	19.1	341
40.	14.4	64.2	16.3	341
41.	15.6	70.2	17.2	341
42.	15.4	69.1	16.4	341
43.	15.3	68.1	17.6	341
44.	*	73.4	18.8	341

[1] See Figure 61 for structures of these compounds.

TABLE 16. ^{1}H-NMR Chemical Shifts for the Sandaracopimaranes (Structures 1–12, Figure 62) Isolated from *Garuleum* and *Osteosperum* (*156*).

	[4α-Me] H-19	[4β-Me] H-18	[10-Me] H-20	[13-Me] H-17	H-15	H-16(c)	H-16(t)
1.*	0.85	0.85	1.00	1.22	5.73	4.77	4.82
2.*	0.83	0.85	0.95	1.01	5.64	4.76	4.79
3.	0.97	1.23	1.31	1.23	5.74	4.83	4.87
4.*	0.96	0.98	1.16	1.27	5.65	4.74	4.79
5.	0.87	0.87	1.18	1.26	5.73	4.81	4.86
6.	0.85	0.88	1.00	1.31	5.67	4.90	4.92
7(1)	0.88	0.88	1.20	1.28	5.75	5.00	5.02
(2)*	0.87	0.87	1.16	1.22	5.77	4.93	4.94
8.	0.89	0.89	1.20	1.34	5.67	4.93	4.94
9.	0.86	0.88	1.11	1.27	5.83	4.96	4.97
10.	0.86	0.88	1.02	1.26	5.72	5.04	5.05
11.	0.96	1.25	1.65	1.40	5.70	4.82	4.84
12.	0.89	0.91	1.42	1.42	5.78	5.13	5.01

* Solvent:CCl$_4$

A·

158 Solidagenone

500 Solidagonic Acid

B·

$500 \xrightarrow{\text{LiAlH}}$

500a R = H

500b R = Ac

$500a \xrightarrow[\text{(Ac}_2)\text{O}]{\text{Pyridine}} 500b \xrightarrow{\text{Jones Reagent}}$

500c

$500c \xrightarrow[\text{Hydrazine hydrate}]{\text{KOH}}$

475 Kolavenol

FIGURE 33. Chemical interconversion of solidagonic acid (**500**) to a compound of established absolute stereochemistry, kolavenol (**475**) (*364, 365*).

CO₂H
CO₂CH₃
CH₃O₂C
511e

O
CH₃O₂C
511f

R
CO₂CH₃
CH₃O₂C
H
511a R = CO₂CH₃
511b R = COCH₃

CO₂CH₃
H
O
511c

O
H
CO₂H
511

OH
H
H
A
B
C
D
O
3B—Hydroxy—5a—androstan—17—one

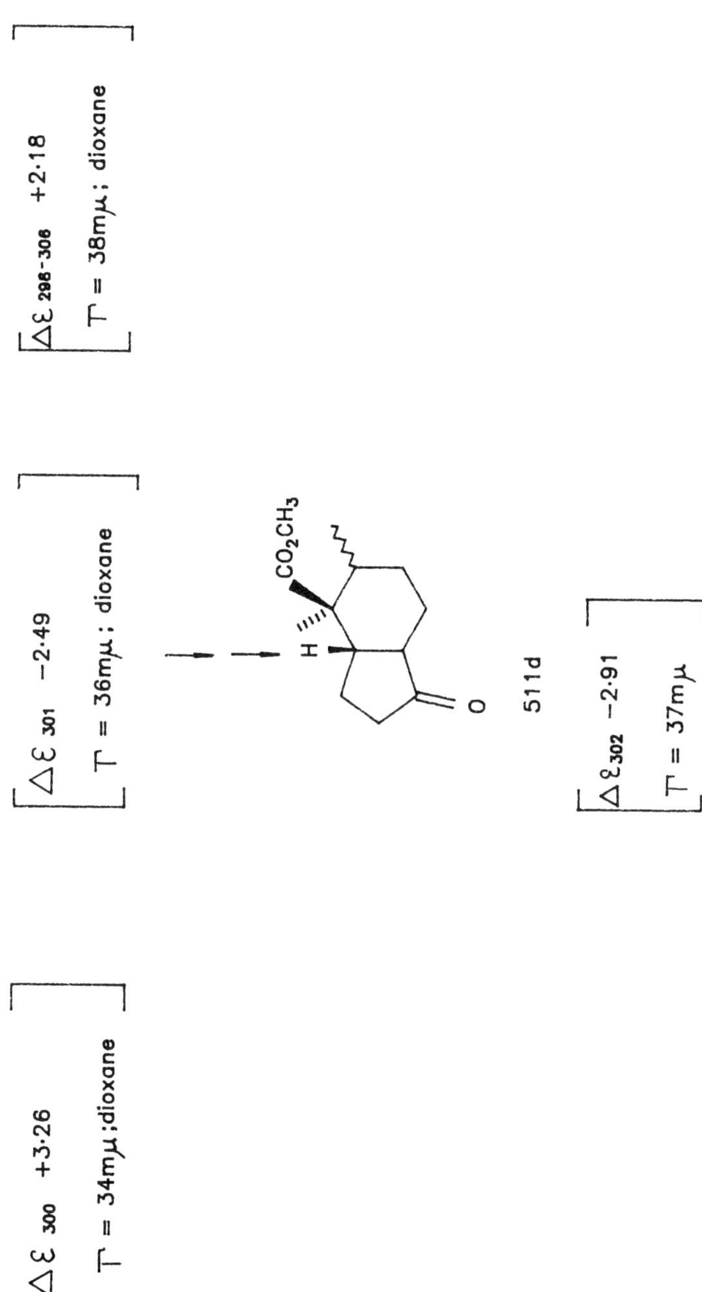

FIGURE 34. Chemical degradation of (-)-hardwickiic acid (**511**) to products permitting the assignment of ring-fusion configuration (*391*).

ORIGINAL   REVISED

258 Elongatolide C (R = Ang)
    (Solidagolactone II)

545 Elongatolide A (R = H)
    (Solidagolactone IV)

546 Elongatolide B (R = Ac)

259 Solidagolactone III (R = Tig)

262 Solidagolactone V

FIGURE 35. Original *trans-ent*-clerodane structures assigned as the elongatolide series from *Solidago elongata* (*18*) and the solidagolactone series from *S. altissima* (*413*) and the revised 5α,10α-*cis*-normal-clerodane structures (*401*, *402*).

FIGURE 35. (*contd.*).

A.

239 ⟶   **239a**

507 ⟵ 511

B.

239 ⟶   **239b**

1075 (revised)

FIGURE 36. Attempts to correlate the structure of solidagoic acid A (**239**) with those of known *cis*-clerodanes: A. comparison of the methyl ^{1}H-NMR chemical shifts (δ) of a solidagoic acid A derivative, **239a**, to those of a (-)-hardwickiic acid derivative, **507**; B. enone derivative (**239b**) of solidagoic acid A and the revised structure of plathyterpenone used for comparison of CD optical properties.

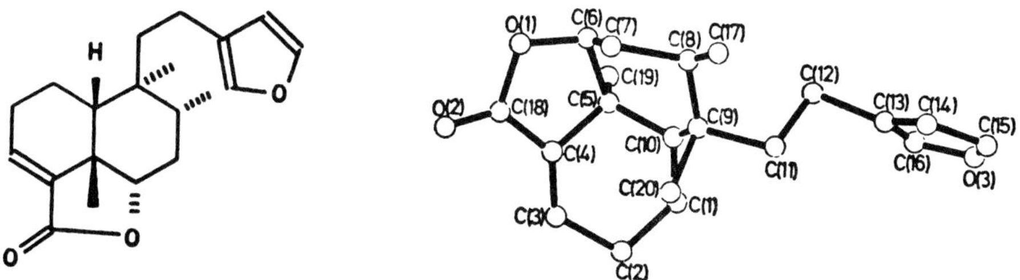

FIGURE 37. Tricyclosolidagolactone (**947**) and possible precursor, solidagolactone V (**262**).

FIGURE 38. Relative stereochemistry of the *cis-ent*-clerodane, **583**, as determined from X-ray analysis.

FIGURE 39. Comparison of the optical data obtained from $5\beta,10\beta$-*cis-ent*-clerodane derivatives of **576** and **578** to that obtained from derivatives of $5\alpha,10\alpha$-*cis*-normal clerodane **239** (*237*).

A· Originally Proposed Structures

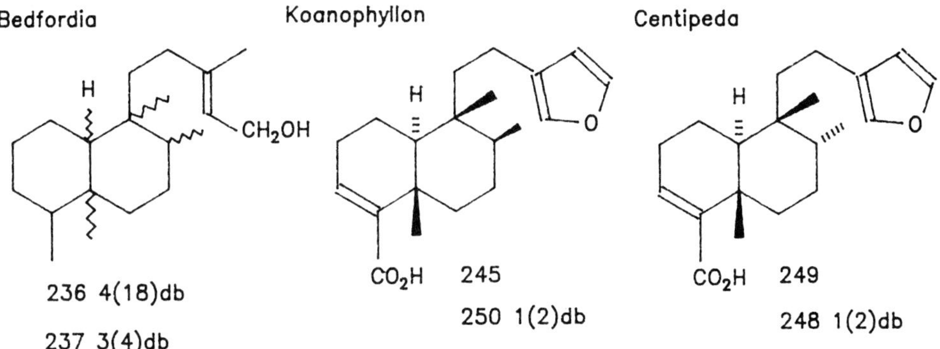

B· Originally Proposed Nidorella Structures

FIGURE 40. Originally proposed *trans*-normal-clerodane structures for compounds isolated from *Bedfordia*, *Koanophyllon*, *Centipeda* and *Nidorella*.

1·   1–Polyalthic acid

2· Secoreseda acid   (1)

3· Secoreseda acid (2)

FIGURE 41. Proposed structures for the 5,10-seco-clerdanes: (1.) conyzic acid was first assigned the structure of Δ^1-polyalthic acid (*411*); (2.) first, proposed structure of secoreseda acid (*61*) with C-9 orientivity opposite to that first assigned to strictic acid (*420*); (3.) second proposed structure of secoreseda acid (*72*) and by extension to strictic (**573**) and conyzic acids.

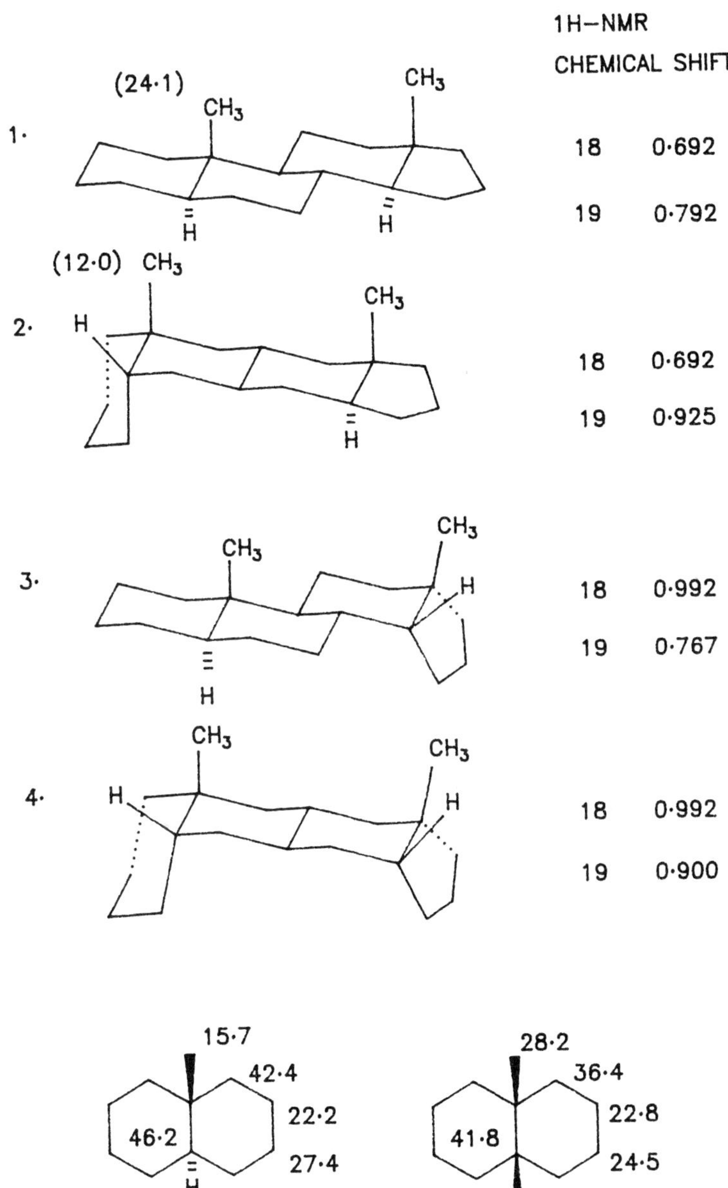

FIGURE 42. The effect of ring junction configuration on the ^{1}H- and ^{13}C-NMR chemical shifts of methyls attached at the ring junction in parallel instances: A. androstane steroid skeletal types 1-4; B. *cis*- and *trans*-methyldecalin.

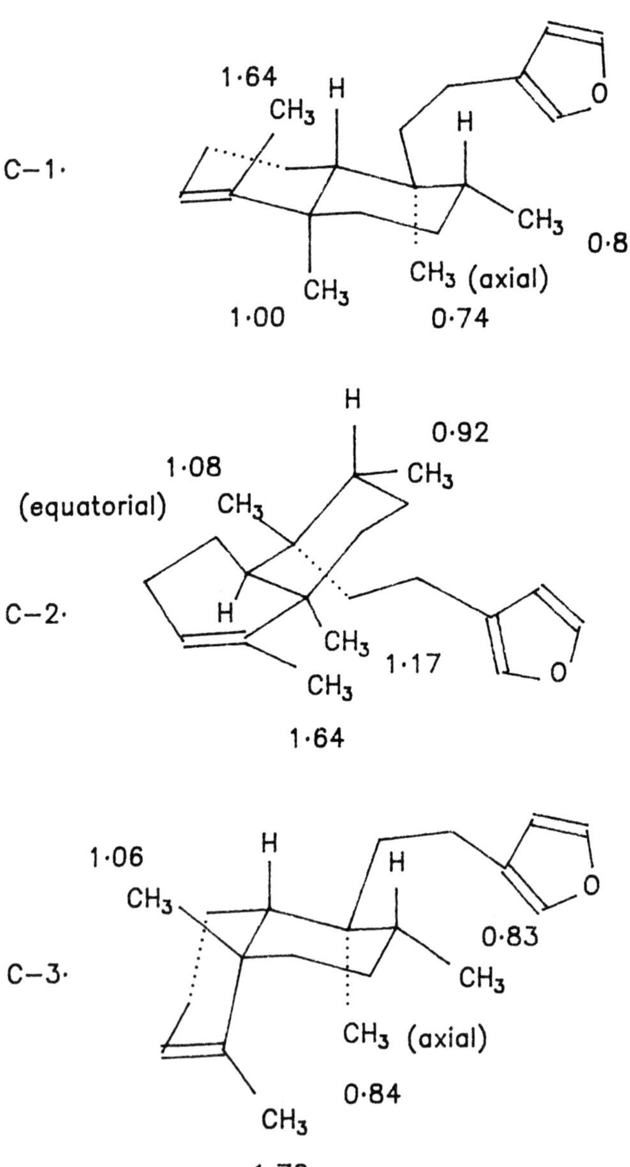

FIGURE 43. ^{1}H-NMR chemical shifts (δ, CDCl$_3$) of the methyls associated with three constitutionally equivalent representatives of the different clerodane ring-junction types, 5α-10β-*trans-ent*-clerodane (C-1), 5α,10α-*cis*-normal-clerodane (C-2) and 5β,10β-*cis-ent*-clerodane (C-3).

^{1}**H-NMR** **Chemical Shift** (δ, CDCl$_3$)

Structure	**H-17**	**H-18**	**H-19**	**H-20**
1. R = H	0.92	1.64	1.17	1.08
2. R = OH	0.86	1.68	1.21	1.05
3. R = OAng	0.86	1.56	1.28	1.08
4. R = =O	0.87	1.54	1.24	1.24

FIGURE 44. The effect of C-6 substitution on the methyl ^{1}H-NMR chemical shifts of $5\alpha,10\alpha$-*cis*-normal-clerodanes (Structures 1 from *15*; Structures 2, 3 and 4 from *402*).

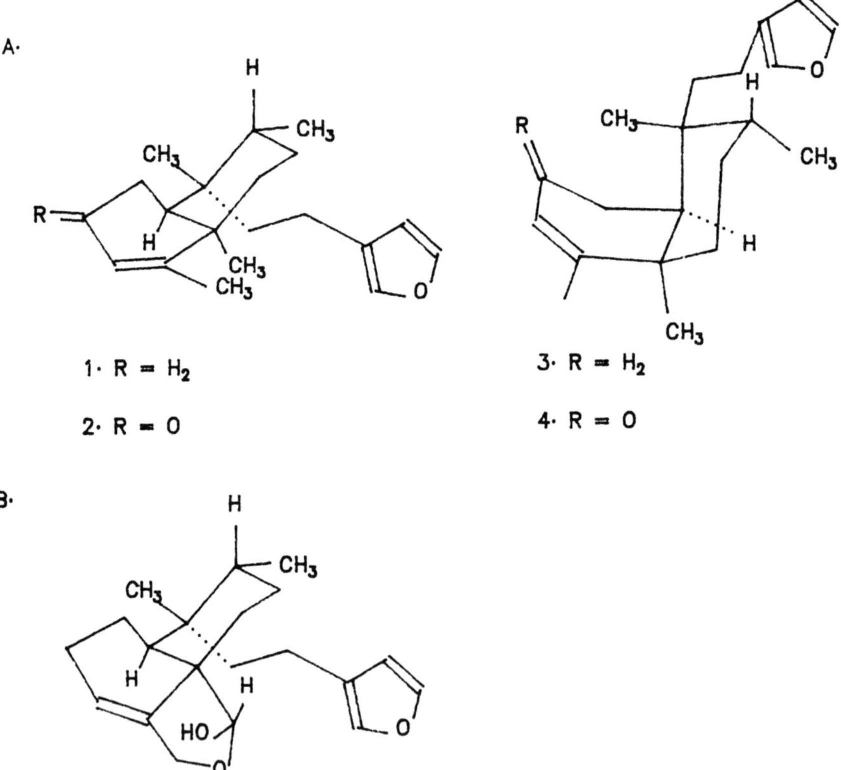

FIGURE 45. Conformational variation attributed to the 5α-10α-*cis*-normal-clerodanes: A. two favored conformational possibilities (*15*); B. probable conformation of compounds containing a 18,19-oxido linkage; C. the effect of the shielding zone of the C-2 ketone on the ¹H-NMR chemical shift of the C-9 methyl (H-20) of *cis*-clerodanes.

FIGURE 45. (*contd.*).

268 R = Ang

269 R = Tig

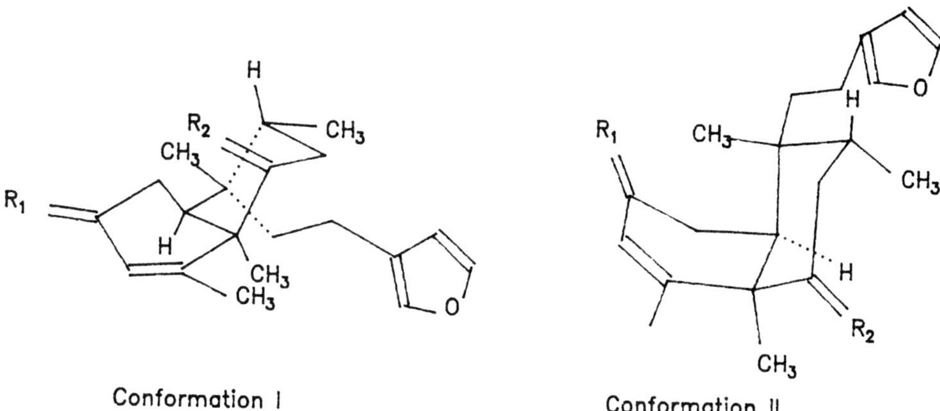

FIGURE 46. Two conformations assignable to $5\alpha,10\alpha$-*cis*-normal-clerodanes based on the nature of the R_1 and R_2 substituents.

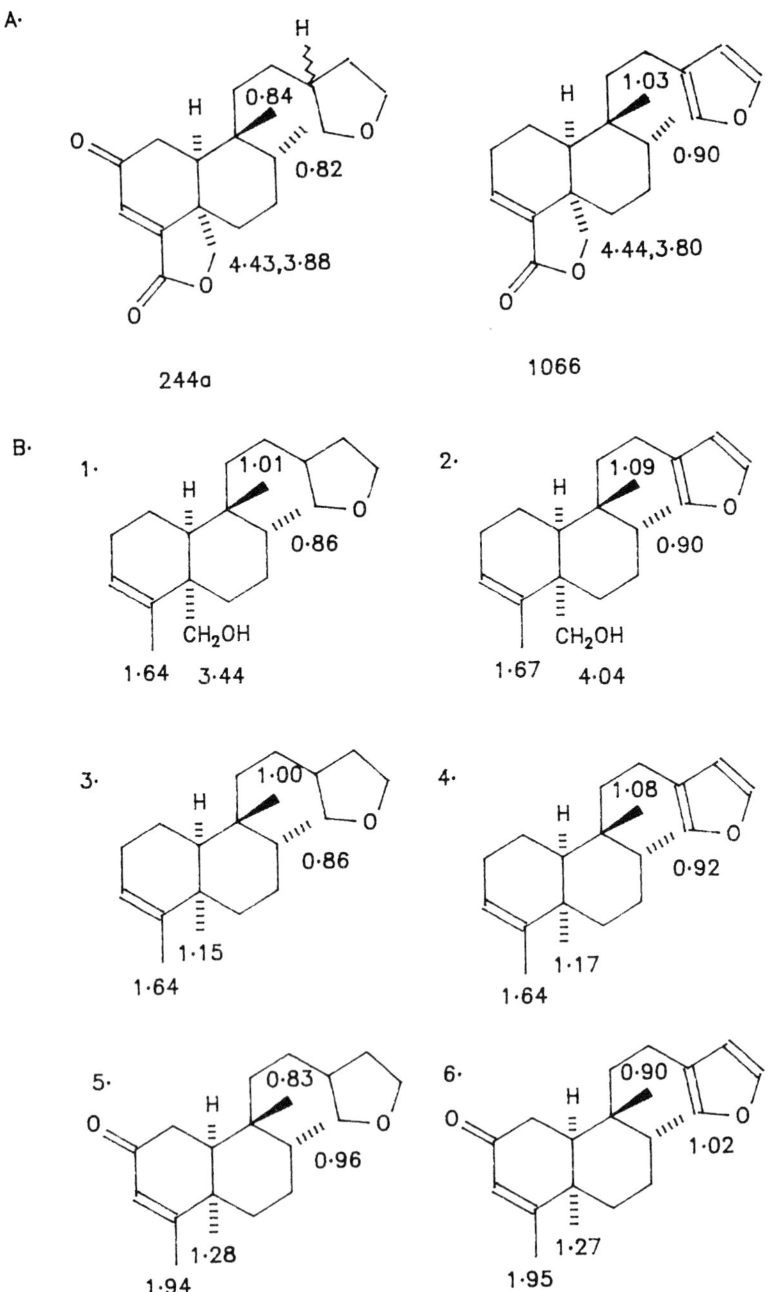

FIGURE 47. The effect of 18,19-oxido linkage on the probability of either conformation I or II: A. comparison of the methyl ^{1}H-NMR chemical shifts (δ, CDCl$_3$) of **244a** and **1066**; B. comparison of the methyl chemical shifts of tetrahydrofuryl-bearing *cis*-clerodanes (Structures 1, 3, and 5) and the corresponding furyl compounds (Structures 2, 4, and 6).

A· Conformations at -50° C

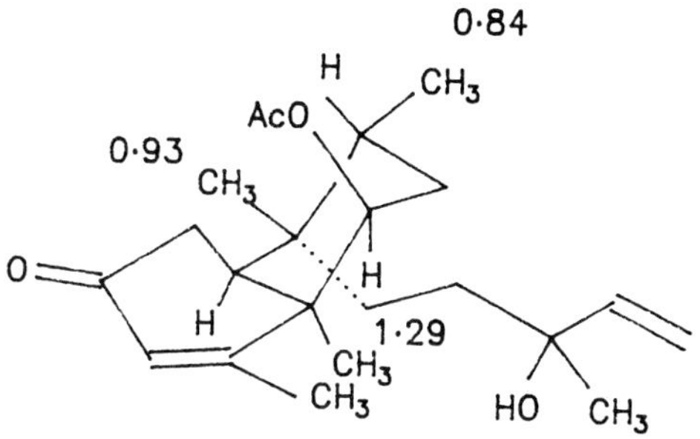

1025a Conformation I (C–6β–acetate: axial)

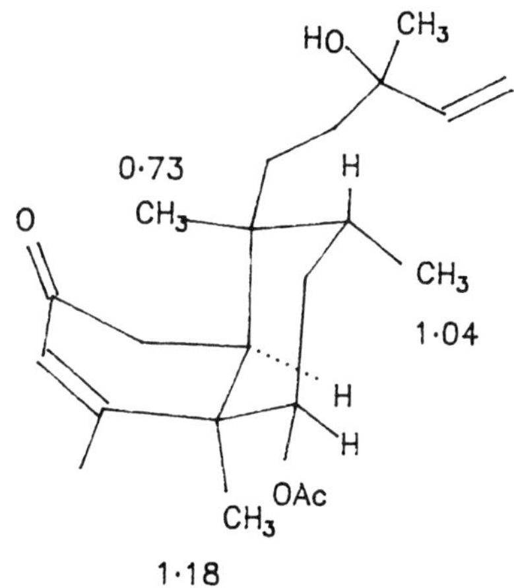

1025b Conformation II (C–6β–acetate: equatorial)

B· Room–Temperature Chemical Shifts:

H–17   0·93

H–19   1·32

H–20   0·93

FIGURE 48. Methyl ^{1}H-NMR chemical shifts (δ, CDCl$_3$) for the two conformations (**1025a** and **1025b**) of 1025 observed at $-50°$C (*170*).

Eu(pdm)$_3$ Addition-Induced Shift

Methyl Proton	1.	2.
H-18	10.0	3.9
H-19	9.6	5.3
H-17	1.7	2.0
H-20	4.2	3.3

FIGURE 49. Comparison of the Eu(pdm)$_3$ addition-induced ^{1}H-NMR chemical shift results for representative compounds of 5α-10α-*cis*-normal- and 5β,10β-*cis-ent*-clerodanes (*386, 378*).

FIGURE 50. The transformation of solidagolactone IV (**545**) to a product with a C-3-C-6 ether function (**545b**) as partial proof of a *cis*-ring junction (*378*).

O
OAc
H
O
553 Elongatolide D
(Solidagolactone VI)
O
OAng
H
O
260 Elongatolide E
(Solidagolactone VII)
A.
O
OAng
H
HO
260b
+ 31%
O
OAng
H
HO
260c
36%
+
OAng
H
O
280
B.

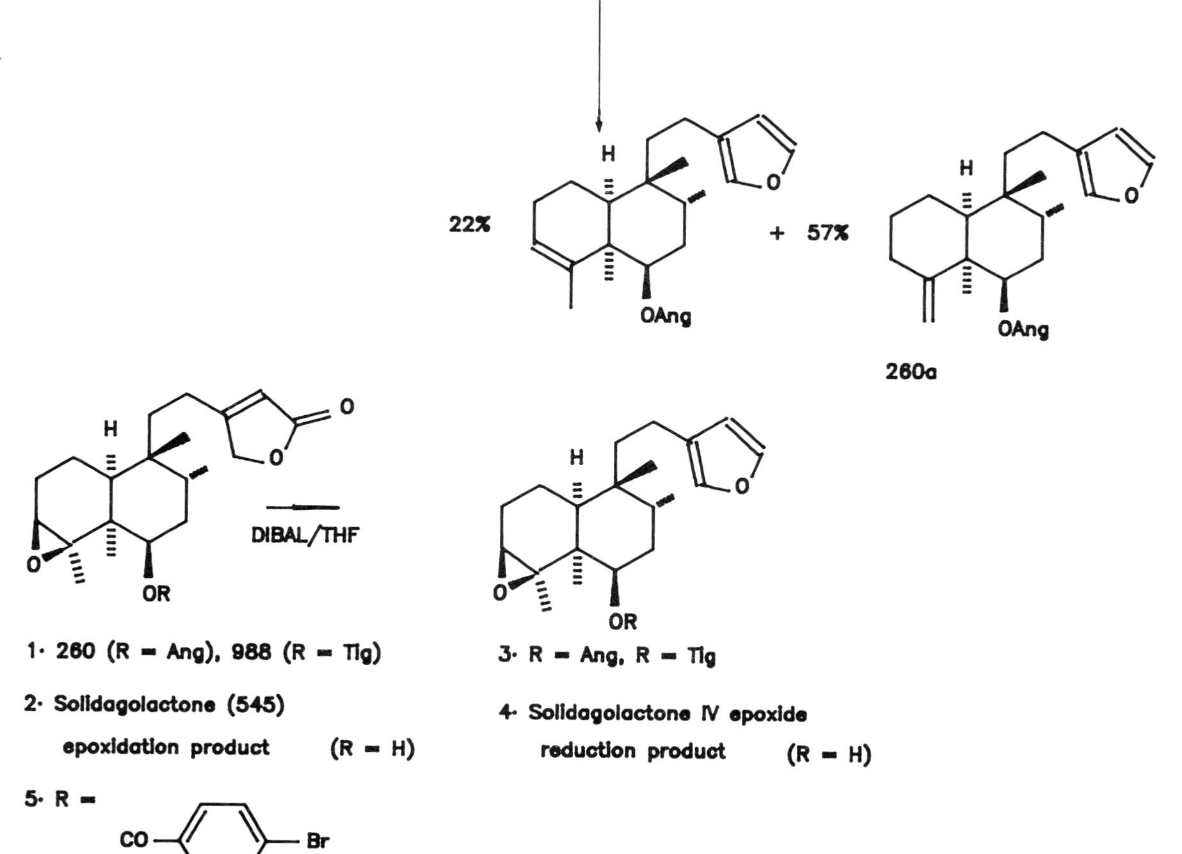

FIGURE 51. Structure determination of elongatolides D (**553**) and E (**260**): A. correct structures of **553** and **260** (*18, 413, 401*); B. incorrect assignment of structure to the products, **260b** and **260c**, of the epoxidation of elongatolide E (**260**) furyl derivative (*402*); C. correlation of the major product of the epoxidation of **545** (Figure 35B.) (Structure 2, R = H) with related butenolide-containing compounds, **260**, **988** and the p-bromobenzoate (Structure 5) (*401*).

FIGURE 52. The effects of C-6 substitution on the products of 3,4-epoxidation of 5α-10α-*cis*-normal clerodanes: A. epoxidation of **239** (*386*); B. epoxidation of **545** (*401*); C. epoxidation of a **258** and **259** mixture (*291*); D. epoxidation of **546** (*413*).

FIGURE 52. (*contd.*).

FIGURE 53. Transformation of solidagoic acid (**239**) epoxy-derivative to the 3β-hydroxy product (*386*).

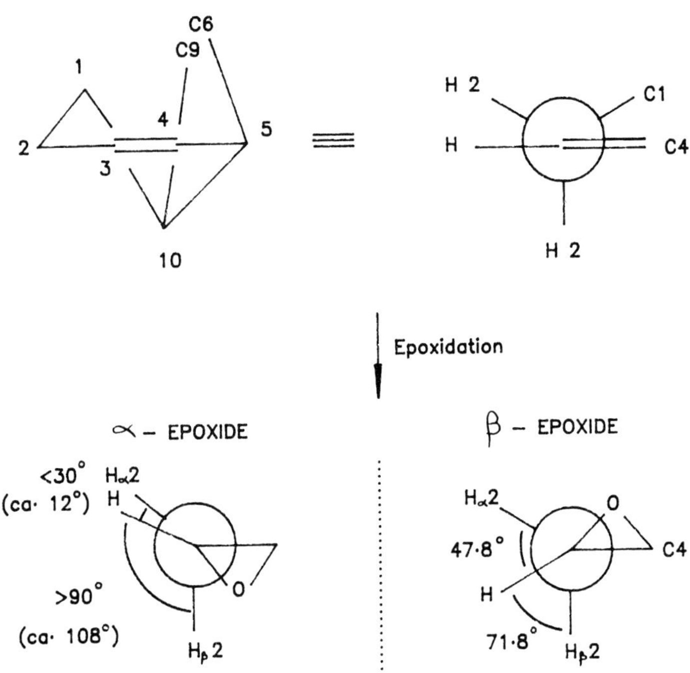

<u>Coupling Constants</u>

Tori equation
calculation: $J_{3,2\alpha}$ = 4.88Hz (12°) $J_{3,2\alpha}$ = 0.50Hz (47.8°)

$J_{3,2\beta}$ = 0.49Hz (108°) $J_{3,2\beta}$ = 2.30Hz (71.8°)

Observed: $J_{3,2\alpha}$ = 5.1Hz $J_{3,2\alpha}$ = 0.0Hz

$J_{3,2\beta}$ = 0.0Hz $J_{3,2\beta}$ = 1.5Hz

FIGURE 54. The application of Tori equation calculations to the prediction of H-3 coupling constants of 5α-10α-*cis*-normal-clerodane 3,4-α- and β-epoxides (*378*).

509 R = H
509a R = Ac

509b

500

500a R = CH$_2$OAc

500b R = CO$_2$CH$_3$

^{1}H-NMR Chemical Shift (δ, CDCl$_3$)

Structure	H-19	H-20
509	1.36	1.02
509a	1.34	1.02
500	1.17	0.97
509b	1.06	0.75
500a	0.95	0.71
500b	0.96	0.72

FIGURE 55. The effect of different C-7 substituents on the H-19 and H-20 ^{1}H-NMR chemical shifts (δ) of 5α,10β-*trans-ent*-clerodanes.

FIGURE 56. The effects of C-7 α- and β-hydroxylation on the ^{1}H-NMR properties of $5\alpha,10\beta$-*trans-ent*-clerodanes (*412*).

Compound	C19 Substitution	H−20 Chem· Shift
900	CH_2OAc	0·84
902	CH_2OAng	0·89
903	CH_2Oi−Val	0·85
1033	CH_2OH	0·79
904	CHO	0·60

FIGURE 57. The effect of different $5\alpha,10\beta$-*trans-ent*-clerodane C-19 substituents on the ^{1}H-NMR chemical shift (δ) of H-20.

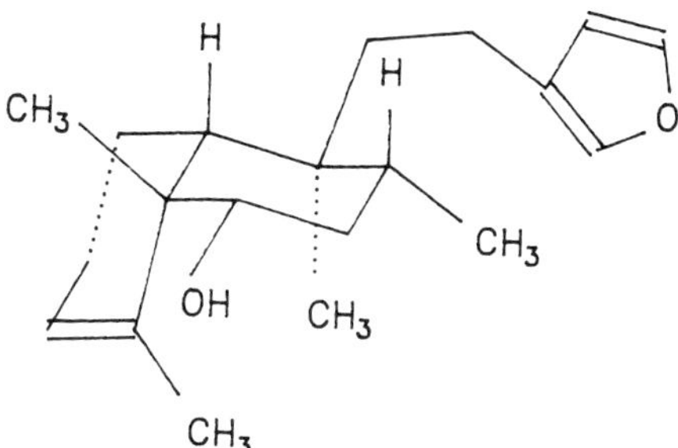

FIGURE 58. Conformation assigned by McCrindle et al. (*386*) to the 5β-10β-*cis-ent*-clerodanes.

16
15
R_1
H
20
17
19
R_2
18

^{1}H-NMR Chemical Shift($CDCl_3$)

Structure	H-17	H-18	H-19	H-20
1. R_1 = CO_2H, R_2 = CH_2OH	0.74	–	1.12	0.76
2. R^1 = CO_2H, R_2 = CH_2OAc	0.74	–	1.11	0.69
3. R_1 = R_2 = CH_2OH	0.75	–	1.12	0.76
4. R_1 = CO_2CH_3, R_2 = CH_2OH	0.74	–	1.12	0.76
5. R_1 = CH_2OH, R_2 = CO_2CH_3	0.75	–	1.22	0.77

FIGURE 59. ^{1}H-NMR chemical shifts (δ, $CDCl_3$) of the methyl protons on cistodioic acid (R_1 = R_2 = CO_2H) derivatives (*33*, *469*).

FIGURE 60. Structures of clerodane diterpenes (included in Table 14 compilation of ^{1}H-NMR data) that are derivatives of the natural products listed in Table 2.

FIGURE 60. (contd.).

FIGURE 60. (*contd.*).

FIGURE 60. (*contd.*).

FIGURE 61. Structures of clerodane diterpenes included in the compilation of ^{13}C-NMR data (Table 15).

FIGURE 61. (*contd.*).

FIGURE 61. (*contd.*).

30· (1033) R₁ = R₂ = R₃ = H

31· R₁ = R₂ = H, R₃ = Ac

32· R₁ = OAc, R₂ = H, R₃ = Ac

33· R₁ = OAc, R₂ = CH₃, R₃ = Ac

34· R = H

35· R = Ac

36·

37·

38· 13(14)d·b·

39·

40·

FIGURE 61. (*contd.*).

41·

42·

43·

44·

FIGURE 61. (*contd.*).

1· [276] $R_1 = R_2 = R_3 = R_4 = H$

2· $R_1 = Ac, R_2 = R_3 = R_4 = H$

3· [277] $R_1 = R_3 = R_4 = H, R_2 = OH$

4· [278] $R_1 = R_3 = R_4 = H, R_2 = OAc$

5· [279] $R_1 = R_2 = R_4 = H, R_3 = OH$

6· [284] $R_1 = R_2 = R_3 = H, R_4 = OAc$

7· $R_1 = R_2 = H, R_3 = R_4 = OH$

8· [292] $R_1 = R_2 = H, R_3 = OH, R_4 = OAc$

9· [289] $R_1 = R_2 = H, R_3 = OAc, R_4 = OH$

10· $R_1 = R_2 = R_3 = H, R_4 = OH$

11· $R_1 = OH, R_2 = H$

12· $R_1 = H, R_2 = OH$

FIGURE 62. Proposed conformation and twelve representative structures of the sandara-copimaranes isolated from *Garuleum* and *Osteospermum* (*156*). ^{1}H-NMR data for these structures are included in Table 16.

CHAPTER 7

References

1. Achilladelis, B., and J.R. Hanson: Studies in terpenoid biosynthesis. I. The biosynthesis of metabolites of *Tricothecium roseum*. Phytochem. **7**, 589 (1968).

2. Ahmed, A.A., D.A. Gage, J.S. Calderon, and T.J. Mabry: Labdane diterpenes from *Brickellia vernicosa*. Phytochem. **25**, 1385 (1986).

3. Alieva, S.A., Z.M. Putieva, E.S. Kondratenko, and N.K. Abubakirov: A diterpene glycoside—doronicoside D—from *Doronicum macrophyllum*. Khim. Prir. Soedin. 658 (1977).

4. Almqvist, S.-O., C.R. Enzell, and F.W. Wehrli: Carbon-13 NMR studies of labdane diterpenoids. Acta Chem. Scand **B 29**, 695 (1975).

5. Alvarez, L., R. Mata, G. Delgado, and A. Romo de Vivar: Sesquiterpene lactones from *Viguiera hypargyrea*. Phytochem. **24**, 2973 (1985).

6. Amaro, T.M., and M. Adrien R.: Diterpenoides del *Oxylobus glanduliferus* (Sch.-Bip.) Gray. Rev. Latinoamer. Quim. **13**: 110 (1982).

7. Anderson, A.B., R. McCrindle, and E. Nakamura: Diterpenoids of *Solidago arguta* Ait. The stereochemistry of *cis*-clerodanes. J. Chem. Soc. Chem. Commun. **1974**, 453.

8. Angeles, E., K. Folting, P.A. Grieco, J.C. Huffman, R. Miranda, and M. Salmon: Isolation and structure of stephalic acid, a new clerodane diterpene from *Stevia polycephala*. Phytochem. **21**, 1804 (1982).

9. Anthonsen, T.: A note on the constitution of the diterpene $C_{20}H_{28}O_3$ from *Solidago canadensis* L. Acta Chem. Scand. **20**, 904 (1966).

10. Anthonsen, T., and G. Bergland: The diterpenoids of *Solidago missouriensis* Nutt. Acta Chem. Scand. **24**, 1860 (1970).

11. Anthonsen, T., and G. Bergland: The diterpenoids of some *Solidago* species. Acta Chem. Scand. **25**, 1924 (1971).

12. Anthonsen, T., and G. Bergland: Constitution and stereochemistry of diterpenoids from *Solidago missouriensis* Nutt. Acta Chem. Scand. **27**, 1073 (1973).

13. Anthonsen, T., and S. Chantharasakul: Isolation of *ent*-16-kauren-19-oic acid and *ent*-16-kauren-19-ol from *Abrotanella nivigena* Muell. Acta Chem. Scand. **25**, 1925 (1971).

14. Anthonsen, T., M.S. Henderson, A. Martin, R. McCrindle, and R.D.H. Murray: Furan-containing diterpenoids from *Solidago serotina* Ait. Acta Chem. Scand. **22**, 351 (1968).

15. Anthonsen, T., M.S. Henderson, A. Martin, R.D.H. Murray, R. McCrindle, and D. McMaster: Constituents of *Solidago* species. Part IV. Solidagoic acids

A and B, diterpenoids from *Solidago gigantea* var. *serotina*. Can J. Chem. **51**, 1332 (1973).

16. Anthonsen, T., P.H. McCabe, R. McCrindle, and R.D.H. Murray: The constitution and stereochemistry of solidagenone. Acta Chem. Scand. **21**, 2289 (1967).

17. Anthonsen, T., P.H. McCabe, R. McCrindle, and R.D.H. Murray: Constituents of *Solidago* species I. The constitution and stereochemistry of diterpenoids from *Solidago canadensis* L. Tetrahedron **25**, 2233 (1969).

18. Anthonsen, T., and R. McCrindle: The constitution of diterpenoids from *Solidago elongata* Nutt. Acta Chem. Scand. **23**, 1068 (1969).

19. Apsimon, J.W., W.G. Craig, P.V. DeMarco, D.W. Mathieson, and W.B. Whalley: The n.m.r. spectra of some diterpenes. Chem. Comm. **1966**, 361.

20. Apsimon, J.W., W.G. Craig, P.V. DeMarco, D.W. Mathieson, and W.B. Whalley: The chemical shift—III. The NMR spectra of some diterpenes. Tetrahedron **23**, 2375 (1967).

21. Arnone, A., R. Mondelli, and J. St. Pyrek: 13C NMR Spectroscopy of natural substances. IV. 13C NMR Studies of trachylobane diterpenes: complete carbon assignment. Org. Magn. Reson. **12**, 429 (1979).

22. Arriaga-Giner, F.J., E. Wollenweber, I. Schober, P. Dostal, and S. Braun: 2β-Hautriwaic acid, a clerodane type diterpenoid and other terpenoids from three *Baccharis* species. Phytochem. **25**, 719 (1986).

23. Bahsas, A., and J. Triana: Inulaefol, nuevo furanoditerpeno del *Eupatorium inulaefolium* HBK. VI. Simposium Internacional Quimica de Productos Naturales, Monterrey, N. L., Mexico. April 23–28. **1979**, 15.

24. Bailey, J.A., G.G. Vincent, and R.S. Burden: Diterpenes from *Nicotiana glutinosa* and their effect on fungal growth. J. of General Microbiology. **85**, 57 (1974).

25. Banerjee, S., M. Grenz, J. Jakupovic, and F. Bohlmann: Some alicyclic terpenoids from the tribe Anthemideae. Planta Med. **1985**, 177 (1985).

26. Banerjee, S. and J. Jakupovic: Further diterpenes from *Palafoxia rosea*. Rev. Latinoamer. Quim. **17**, 202 (1986).

27. Banerjee, S., J. Jakupovic, and R.M. King: Weitere derivate des β-jonons aus *Haplopappus fremontii*. Pharmazie **41**, 157 (1986).

28. Barua, R.N., R.P. Sharma, G. Thyagarajan, W. Herz, and S.V. Govindan: New melampolides and darutigenol from *Sigesbeckia orientalis*. Phytochem. **19**, 323 (1980).

29. Baruah, R.N., C. Zdero, F. Bohlmann, R.M. King, and H. Robinson: Some sesqui- and diterpenes from the tribe Eupatorieae. Phytochem. **24**, 2641 (1985).

30. Bastard, J., D. Khac Duc, M. Fetizon, M.J. Francis, P.K. Grant, R.T. Weavers, C. Kaneko, G. Vernon Baddeley, J.-M. Bernaussau, I.R. Burfitt, P.M. Wovkulich, and E. Wenkert: CMR spectroscopy of labdanic diterpenes and related substances. J. Nat. Prod. **47**, 592 (1984).

31. Beale, M.H., J.R. Bearder, J. Macmillan, A. Matsuo, and B.O. Phinney: Diterpene acids from *Helianthus* species and their microbiological conversion by *Gibberella fugikuroi*, Mutant B1–41a. Phytochem. **22**, 875 (1983).

32. Bennett, R.D., E.R. Lieber, and E. Heftmann: Biosynthesis of steviol from (-)-kaurene. Phytochem. **6**, 1107 (1967).

33. Berti, G., O. Livi, and D. Segnini: Cistodiol and Cistodioic acid, diterpenoids with a *cis*-fused clerodane skeleton. Tetrahedron Letters **1970**, 1401.

34. Bhat, S.V., P.S. Kalyanaraman, H. Kohl, and N.J. De Souza: Inuroyleanol and 7-ketoroyleanone, two novel diterpenoids of *Inula royleana* DC. Tetrahedron **31**, 1001 (1975).
35. Bittner, M., and W.H. Watson: Phytochemical study of *Haplopappus ciliatus*. Rev. Latinoamer. Quim. **13**, 24 (1982).
36. Bittner, M.L., V. Zabel, W.B. Smith, and W.H. Watson: A new dihydro-*trans*-clerodane diacid from *Haplopappus ciliatus*. Phytochem. **17**, 1797 (1978).
37. Bjeldanes, L.F., and T.A. Geissman: Constituents of *Helianthus cilaris*. Phytochem. **11**, 327 (1972).
38. Bohlmann, F.: Neues über die chemie der Compositen. Naturwissenschaften **67**, 588 (1980).
39. Bohlmann, F., and W.-R. Abraham: Neue diterpene und weitere inhaltsstoffe aus *Helichrysum calliconum* und *Helichrysum heterolasium*. Phytochem. **18**, 889 (1979).
40. Bohlmann, F., and W.-R. Abraham: Neue diterpene aus *Helichrysum acutatum*. Phytochem. **18**, 1754 (1979).
41. Bohlmann, F. and W.-R. Abraham: Sesquiterpene und acetylenverbindungen aus *Cineraria*-arten. Phytochem. **17**, 1629 (1978).
42. Bohlmann, F., W.-R. Abraham, R.M. King, and H. Robinson: Diterpenes from *Koanophyllon* species. Phytochem. **20**, 1903 (1981).
43. Bohlmann, F., W.-R. Abraham, H. Robinson, and R.M. King: A new labdane derivative and geranylphloroglucinols from *Achyrocline alata*. *Phytochem.* **19**, 2475 (1980).
44. Bohlmann, F., W.-R. Abraham, H. Robinson, and R.M. King: Heliangolides and other constituents from *Bejaranoa semistriata*. Phytochem. **20**, 1639 (1981).
45. Bohlmann, F., W.-R. Abraham, and W.S. Sheldrick: Weitere diterpene mit helifulvan-gerüst und andere inhaltsstoffe aus *Helichrysum chionosphaerum*. Phytochem. **19**, 896 (1980).
46. Bohlmann, F., A. Adler, J. Jakupovic, R.M. King, and H. Robinson: A dimeric germacranolide and other sesquiterpene lactones from *Mikania* species. Phytochem. **21**, 1349 (1982).
47. Bohlmann, F., A. Adler, R.M. King, and H. Robinson. *ent*-Labdanes from *Mikania alvimii*. Phytochem. **21**, 173 (1982).
48. Bohlmann, F., A. Adler, A. Schuster, R.K. Gupta, R.M. King, and H. Robinson: Diterpenes from *Mikania* species. Phytochem. **20**, 1899 (1981).
49. Bohlmann, F., M. Ahmed, N. Borthakur, M. Wallmeyer, J. Jakupovic, R.M. King, and H. Robinson: Diterpenes related to grindelic acid and further constituents from *Grindelia* species. Phytochem. **21**, 167 (1982).
50. Bohlmann, F., M. Ahmed, J. Jakupovic, R.M. King, and H. Robinson: Labdane and dehydronerolidol derivatives from *Brickellia diffusa*. Phytochem. **21**, 691 (1982).
51. Bohlmann, F., M. Ahmed, J. Jakupovic, R.M. King, and H. Robinson: Dimeric sesquiterpene lactones and kolavane derivatives from *Gochnatia paniculata*. Phytochem. **22**, 191 (1983).
52. Bohlmann, F., M. Ahmed, J. Jakupovic, R.M. King, and H. Robinson: Three kolavane derivatives and 2β-angeloyloxy himachanolide from *Acritopappus longifolius*. Rev. Latinoamer. Quim. **15**, 16 (1984).
53. Bohlmann, F., M. Ahmed, R.M. King, and H. Robinson: Labdane and

eudesmane derivatives from *Ageratum fastigiatum*. Phytochem. **20**, 1434 (1981).

54. Bohlmann, F., M. Ahmed, H. Robinson, and R.M. King: A kolavane derivative from *Liatris scariosa*. Phytochem. **20**, 1439 (1981).

55. Bohlmann, F., N. Ates, and J. Jakupovic: Hirsutinolides from South African *Vernonia* species. Phytochem. **22**, 1159 (1983).

56. Bohlmann, F., N. Ates, R.M. King, and H. Robinson: Two sesquiterpenes from *Senecio* species. Phytochem. **22**, 1675 (1983).

57. Bohlmann, F., S. Banerjee, J. Jakupovic, M. Grenz, L.N. Misra, G. Schmeda-Hirschmann, R.M. King, and H. Robinson: Clerodane and labdane diterpenoids from *Baccharis* species. Phytochem. **24**, 511 (1985).

58. Bohlmann, F., S. Banerjee, C. Wolfrum, J. Jakupovic, R.M. King, and H. Robinson: Sesquiterpene lactones, geranylnerol and tremetone derivatives from *Ageratina* species. Phytochem. **24**, 1319 (1985).

59. Bohlmann, F., M. Bapuji, J. Jakupovic, R.M. King, and H. Robinson: Unusual diterpenes from *Brickellia eupatoriedes*. Phytochem. **21**, 181 (1982).

60. Bohlmann, F., M. Bapuji, R.M. King, and H. Robinson: Kolavenol derivatives from *Goyazianthus tetrastichus*. Phytochem. **21**, 939 (1982).

61. Bohlmann, F., R.N. Baruah, R.M. King, and H. Robinson: Alicyclic diterpenes from *Cronquistianthus bishopii*. Planta Med. **1985** 167 (1985).

62. Bohlmann, F., N. Borthakur, R.M. King, and H. Robinson: Dihydrodendroidinic acid from *Pleurocoronis pluriseta*. Phytochem. **20**, 2433 (1981).

63. Bohlmann, F., T. Burkhardt, and C. Zdero: Naturally Occurring Acetylenes. Academic Press. London and New York. 1973.

64. Bohlmann, F., V. Castro, and J. Jakupovic: Germacra-1(10),4-dien-*cis*-6,12-olides and elemanolides from *Montanoa atriplicifolia*. Phytochem. **22**, 1223 (1983).

65. Bohlmann, F., T. Chau-Thi, P. Singh, and J. Jakupovic: Alicyclic diterpenes from *Solidago* species. Planta Med. **1985**, 487 (1985).

66. Bohlmann, F., and H. Czerson: Neue labdan- und pimaren- derivate aus *Palafoxia rosea*. Phytochem. **18**, 115 (1979).

67. Bohlmann, F., A.K. Dhar, J. Jakupovic, R.M. King, and H. Robinson: Two sesquiterpene lactones with an additional propiolactone ring from *Disynaphia halimifolia*. Phytochem. **20**, 1077 (1981).

68. Bohlmann, F., K. Dhar, J. Jakupovic, R.M. King, and H. Robinson: A caryophylene derivative from *Fleischmannia pycnocephaloides*. Phytochem. **20**, 1425 (1981).

69. Bohlmann, F., L.N. Dutta, W. Dorner, R.M. King, and H. Robinson: Zwei neue guajanolide sowie weitere longipinenester aus *Stevia*-arten. Phytochem. **18**, 673 (1979).

70. Bohlmann, F., L. Dutta, H. Robinson, and R.M. King: Neue labdan-derivate aus *Chrysothamnus nauseusus*. Phytochem. **18**, 1889 (1979).

71. Bohlmann, F., D. Ehlers, and C. Zdero: Einige neue furanoeremophilane aus *Senecio* arten. Phytochem. **17**, 467 (1978).

72. Bohlmann, F., and U. Fritz: Neue diterpene und acetylenverbindungen aus *Nidorella*-arten. Phytochem. **17**, 1769 (1978).

73. Bohlmann, F., U. Fritz, H. Robinson, and R.M. King: Sesquiterpene and diterpene derivatives from *Solidago* species. Phytochem. **19**, 2655 (1980).

74. Bohlmann, F., U. Fritz, H. Robinson, and R.M. King: Isosesquicaren aus

Haplopappus tenuisectus. Phytochem. **18**, 1749 (1979).

75. Bohlmann, F., T. Gerke, J. Jakupovic, N. Borthakur, R.M. King, and H. Robinson: Diterpene lactones and other constituents from *Wedelia* and *Aspilia* species. Phytochem. **23**, 1673 (1984).

76. Bohlmann, F., and M. Grenz: Notiz über die isolierung von hautriwa-säure aus *Conyza ivaefolium* Less. Chem. Ber. **105**, 3123 (1972).

77. Bohlmann, F., and M. Grenz: Constituents from representatives of the *Eupatorium* group. Chem. Ber. **110**, 1321 (1977).

78. Bohlmann, F., and M. Grenz: Atisiren-derivate aus *Garuleum sonchifolium* (DC.)T. Norl. Chem. Ber. **111**, 1509 (1978).

79. Bohlmann, F., and M. Grenz: Diterpene vom sandarocopimaren-typ aus *Chrysanthemoides*-arten. Phytochem. **18**, 683 (1979).

80. Bohlmann, F., M. Grenz, A.K. Dhar, and M. Goodman: Labdane derivatives and flavones from *Gutierrezia dracunculoides.* Phytochem. **20**, 105 (1981).

81. Bohlmann, F., M. Grenz, J. Jakupovic, R.M. King, and H. Robinson: New labdane derivatives and other consituents from *Fleischmannia viscidipes.* Rev. Latinoamer. Quim. **15**, 1 (1984).

82. Bohlmann, F., M. Grenz, and H. Schwarz: Neue diterpene aus *Hinterhubera imbricata* Cuatr. Chem. Ber. **106**, 2579 (1973).

83. Bohlmann, F., M. Grenz, P. Wegner, and J. Jakupovic: Clerodan-derivate und neuartige diterpene aus *Conyza scabrida* DC. Liebigs Ann. Chem. **1983**, 2008.

84. Bohlmann, F., R.K. Gupta, R.M. King, and H. Robinson: A clerodane lactone and a tremetone derivative from *Bahianthus viscidus.* Phytochem. **20**, 331 (1981).

85. Bohlmann, F., R.K. Gupta, H. Robinson, and R.M. King: Labdane derivatives and a himachalanolide from *Acritopappus longifolius.* Phytochem. **20**, 275 (1981).

86. Bohlmann, F., L. Hartono, and J. Jakupovic: A diterpene related to erythroxydiol from *Helichrysum refluxum.* Phytochem. **24**, 611 (1985).

87. Bohlmann, F., L. Hartono, C. Zdero, and J. Jakupovic: Constituents of the genus *Oxylobus.* Phytochem. **24**, 1111 (1985).

88. Bohlmann, F., and H. Hoffmann: Further amides from *Echinecea purpurea.* Phytochem. **22**, 1173 (1983).

89. Bohlmann, F., and J. Jakupovic: Über neue chromene und andere inhaltsstoffe von *Lagascea rigida.* Phytochem. **17**, 1677 (1978).

90. Bohlmann, F., and J. Jakupovic: Neue labdan-derivate und andere inhaltsstoffe aus *Relhania acerosa.* Phytochem. **18**, 631 (1979).

91. Bohlmann, F., and J. Jakupovic: Neue labdan-derivate und sesquiterpene aus *Silphium*-arten. Phytochem. **18**, 1987 (1979).

92. Bohlmann, F., J. Jakupovic, M. Ahmed, M. Grenz, H. Suding, H. Robinson, and R.M. King: Germacranolides and diterpenes from *Viguiera* species. Phytochem. **20**, 113 (1981).

93. Bohlmann, F., J. Jakupovic, M. Ahmed, M. Wallmeyer, H. Robinson, and R.M. King: Labdane derivatives from *Hemizonia* species. Phytochem. **20**, 2383 (1981).

94. Bohlmann, F., J. Jakupovic, N. Ates, A. Schuster, J. Pickardt, R.M. King, and H. Robinson: Steiractinolide, eine neue gruppe von sesquiterpenlactonen. Liebigs Ann. Chem. **1983**, 962 (1983).

95. Bohlmann, F., J. Jakupovic, A.K. Dhar, R.M. King, and H. Robinson: Heliangolides and diterpenes from *Hartwrightia floridana*. Phytochem. **20**, 843 (1981).

96. Bohlmann, F., J. Jakupovic, A.K. Dhar, R.M. King, and H. Robinson: Two sesquiterpene and three diterpene lactones from *Acanthospermum australe*. Phytochem. **20**, 1081 (1981).

97. Bohlmann, F., J. Jakupovic, M. Hashemi-Nejad, and S. Huneck: Clerodane diterpenoids from *Aster alpinus*. Phytochem. **24**, 608 (1985).

98. Bohlmann, F., J. Jakupovic, R.M. King, and H. Robinson: Neue *ent*-atisiren- und *ent*-kaurensäure-derivate aus *Helianthus*-arten. Phytochem. **19**, 863 (1980).

99. Bohlmann, F., J. Jakupovic, R.M. King, and H. Robinson: New germacranolides, guaianolides and rearranged guaianolides from *Lasiolaena santosii*. Phytochem. **20**, 1613 (1981).

100. Bohlmann, F., J. Jakupovic, R.M. King, and H. Robinson: New labdane derivatives from *Madia sativa*. Phytochem. **21**, 1103 (1982).

101. Bohlmann, F., J. Jakupovic, H. Robinson, and R.M. King: Neue diterpene aus *Schkuhria*-arten. Phytochem. **19**, 881 (1980).

102. Bohlmann, F., J. Jakupovic, H. Robinson, and R.M. King: Diterpenes and other constituents of *Morithamnus crassus*. Phytochem. **19**, 2769 (1980).

103. Bohlmann, F., J. Jakupovic, and A. Schuster: Germacranolides from *Perymenium klattianum* and *Perymeniopsis ovalifolia*. Phytochem. **24**, 495 (1985).

104. Bohlmann, F., J. Jakupovic, and A. Schuster: 8-Hydroxypegolettiolide, a sesquiterpene lactone with a new carbon skeleton and further constituents from *Pegolettia senegalensis*. Phytochem. **22**, 1637 (1983).

105. Bohlmann, F., J. Jakupovic, A. Schuster, R.M. King and H. Robinson: Germacranolide, hydroxyverbenon und *ent*-kaur-15(16)-en-17,19-disäure aus *Helianthus occidentalis* var. *dowellianus*. Planta Med. **1984**, 202 (1984).

106. Bohlmann, F., J. Jakupovic, A. Schuster, R.M. King, and H. Robinson: Guaianolides and homoditerpenes from *Lasiolena morii*. Phytochem. **21**, 161 (1982).

107. Bohlmann, F., J. Jakupovic, A. Schuster, R.M. King, and H. Robinson: New melampolides, kaurene derivatives and other constituents from *Ichthyothere* species. Phytochem. **21**, 2317 (1982).

108. Bohlmann, F., J. Jakupovic, A. Schuster, R.M. King, and H. Robinson: Homogeranylnerol derivatives and a melampolide from *Smallanthus glabratus*. Phytochem. **24**, 1309 (1985).

109. Bohlmann, F., J. Jakupovic, C. Zdero, R.M. King, and H. Robinson: Neue melampolide und *cis,cis*-germacranolide aus vertretern der subtribus Melampodiinae. Phytochem. **18**, 625 (1979).

110. Bohlmann, F., W. Knauf, M. Grenz, and M.A. Lane: Ein neues diterpen aus *Xanthocephalum linearifolium*. Phytochem. **18**, 2040 (1979).

111. Bohlmann, F., W. Knauf, R.M. King, and H. Robinson: Ein neues diterpen und weitere inhaltsstoffe aus *Baccharis*-arten. Phytochem. **18**, 1011 (1979).

112. Bohlamnn, F., and K.-H. Knoll: Neue pimardien-derivate aus *Othonna*-arten. Phytochem. **15**, 1072 (1976).

113. Bohlmann, F., K.-H. Knoll, R.M. King, and H. Robinson: Neue α-santalen- und labdan-derivate aus *Ayapana amygdalina*. Phytochem. **18**, 1997 (1979).

114. Bohlmann, F., K.-H. Knoll, H. Robinson, and R.M. King: Neue kauren-

derivate und melampolide aus *Smallanthus uvedalia.* Phytochem. **19**, 107 (1980).

115. Bohlmann, F., K.-H. Knoll, H. Robinson, and R.M. King: Neue eudesmanolide aus *Steiractinia mollis.* Phytochem. **19**, 971 (1980).

116. Bohlmann, F., W. Kramp, M. Grenz, H. Robinson, and R.M. King: Diterpenes from *Baccharis* species. Phytochem. **20**, 1907 (1981).

117. Bohlmann, F., W. Kramp, J. Jakupovic, H. Robinson, and R.M. King: Diterpenes from *Baccharis* species. Phytochem. **21**, 399 (1982).

118. Bohlmann, F., and N. Le Van: Über neue diterpene aus *Dimorphotheca pluvialis* Moench. Chem. Ber. **109**, 1446 (1976).

119. Bohlmann, F., and N. Le Van: Sesquiterpenlactone und polyine aus der gattung *Arctotis.* Phytochem. **16**, 487 (1977).

120. Bohlmann, F., and N. Le Van: Neuen kaurensäure-derivate aus Wedelia-arten. Phytochem. **16**, 579 (1977).

121. Bohlmann, F., and N. Le Van: Neue guajanolide aus *Podachaenium eminens.* Phytochem. **16**, 1304 (1977).

122. Bohlmann, F., and N. Le Van: Neue sesqui- und diterpene aus *Bedfordia salicina.* Phytochem. **17**, 1173 (1978).

123. Bohlmann, F. and N. Le Van: Neue kaurensäure-derivate und germacranolide aus *Montanoa pteropoda.* Phytochem. **17**, 1957 (1978).

124. Bohlmann, F., and M. Lonitz: Neue sadaracopimardien-derivate, sesquiterpene und sesquiterpenlactone aus *Zexmenia*-arten. Chem. Ber. **111**, 843 (1978).

125. Bohlmann, F., and M. Lonitz: Über die inhaltsstoffe von *Zexmenia gnaphaloides* und die synthese von valerenan-derivaten. Chem. Ber. **113**, 2410 (1980).

126. Bohlmann, F., G.-W. Ludwig, J. Jakupovic, R.M. King, and H. Robinson: A daucanolide and further farnesene derivatives from *Ageratum fastigiatum.* Phytochem. **22**, 983 (1983).

127. Bohlmann, F., and P.K. Mahanta: Kaurenic acid derivatives from *Adenostemma caffrum.* Phytochem. **17**, 814, (1978).

128. Bohlmann, F., and P.K. Mahanta: Ein neue diterpensäure aus *Centipeda orbicularis.* Phytochem. **18**, 1067 (1979).

129. Bohlmann, F., L.N. Misra, and J. Jakupovic: Pseudoguaianolides and carabrone derivatives from *Loxothysanus sinuatus.* Phytochem. **24**, 1021 (1985).

130. Bohlmann, F., L.N. Misra, J. Jakupovic, R.M. King, and H. Robinson: Guaianolides, heliangolides, diterpenes and cycloartenol derivatives from *Balsamorhiza sagittata. Phytochem.* **24**, 2029 (1985).

131. Bohlmann, F., A.A. Natu, and P.K. Mahanta: Neue diterpene und germacranolide aus *Mikania*-arten. Phytochem. **17**, 483 (1978).

132. Bohlmann, F., and N. Rao: Neue hydroxyacetophenon-derivate aus *Espeletia schultzii.* Chem. Ber. **106**, 3035 (1973).

133. Bohlmann, F., H. Robinson, and R.M. King: Chemotaxonomy of the genus *Plagiocheilus.* Phytochem. **19**, 2235 (1980).

134. Bohlmann, F., E. Rosenberg, R.M. King, and H. Robinson: Neue labdanderivate aus *Aristeguietia buddleaefolia.* Phytochem. **19**, 977 (1980).

135. Bohlmann, F., C. Scheidges, R.M. King, and H. Robinson: Five labdane derivatives from *Koanophyllon conglobatum.* Phytochem. **23**, 1190 (1984).

136. Bohlmann, F., C. Scheidges, C. Zdero, R.M. King, and H. Robinson: *Ent*-labdanes from *Baccharis sternbergiana*. Phytochem. **23**, 1109 (1984).

137. Bohlmann, F., G. Schmeda-Hirschmann, and J. Jakupovic: Heliangolides and germacranolides from *Disynaphia multicrenulata*. Phytochem. **23**, 1435 (1984).

138. Bohlmann, F., G. Schmeda-Hirschmann, and J. Jakupovic: Neue melampolide aus *Acanthosperum australe*. Planta Med. **1984**, 37 (1984).

139. Bohlmann, F., G. Schmeda-Hirschmann, and J. Jakupovic: Nor-*ent*-labdan derivate aus *Austroeupatorium inulaefolium*. Planta Med. **1984**, 199 (1984).

140. Bohlmann, F., P. Singh, and J. Jakupovic: Further ineupatorolide-like germacranolides from *Inula cuspidata*. Phytochem. **21**, 157 (1982).

141. Bohlmann, F., P. Singh, J. Jakupovic, R.M. King, and H. Robinson: Eudesmanolides from *Dimerostemma brasilianum*. Phytochem. **21**, 1343 (1982).

142. Bohlmann, F., P. Singh, J. Jakupovic, H. Robinson, and R.M. King: An epoxygermacranolide and further consitituents from *Mikania* species. Phytochem. **21**, 705 (1982).

143. Bohlmann, F., P. Singh, R.K. Singh, K.C. Joshi, and J. Jakupovic: A diterpene with a new carbon skeleton from *Solidago altissima*. Phytochem. **24**, 1114 (1985).

144. Bohlmann, F., H. Suding, J. Cuatrecasas, R.M. King, and H. Robinson: Neue diterpene aus der subtribus Espeletiinae. Phytochem. **19**, 267 (1980).

145. Bohlmann, F., H. Suding, J. Cuatrecasas, H. Robinson, and R.M. King: Tricyclic sesquiterpenes and further diterpenes from *Espeletiopsis* species. Phytochem. **19**, 2399 (1980).

146. Bohlmann, F., A. Suwita, R.M. King, and H. Robinson: Neue *ent*-labdan-derivate aus *Austroeupatorium chaparense*. Phytochem. **19**, 111 (1980).

147. Bohlmann, F., A. Suwita, and T.J. Mabry: New labdane derivatives and further constituents of *Brickellia* species. Phytochem. **17**, 763 (1978).

148. Bohlmann, F., A. Suwita, H. Robinson, and R.M. King: Six guaianolides from *Stylotrichium rotundifolium*. Phytochem. **20**, 1887 (1981).

149. Bohlmann, F., K. Umemoto, and J. Jakupovic: Pseudoguaianolides related to confertin from *Stevia isomeca*. Phytochem. **24**, 1017 (1985).

150. Bohlmann, F., M. Wallmeyer, and J. Jakupovic: A new seco-labdane derivative from *Athrixia elata*. Phytochrem. **21**, 1806 (1982).

151. Bohlmann, F., M. Wallmeyer, J. Jakupovic, and J. Ziesche: Diterpenes and sesquiterpenes from *Osteospermum* species, Phytochem. **22**, 1645 (1983).

152. Bohlmann, F., M. Wallmeyer, R.M. King, and H. Robinson: 2-Oxo-labda-8(17),13-dien-15-ol from *Ophryosporus chilca*. Phytochem. **23**, 1513 (1984).

153. Bohlmann, F., and P. Wegner: *Ent*-beyer-15-ene derivatives from *Nidorella anomala*. Phytochem. **21**, 1175 (1982).

154. Bohlmann, F., and P. Wegner: Three diterpenes from *Conyza podocephala*. Phytochem. **21**, 1693 (1982).

155. Bohlmann, F., P. Wegner, and J. Jakupovic: Unusual diterpenes and sesquitepene xylosides from *Nidorella hottentotica*. Phytochem. **21**, 1109 (1982).

156. Bohlmann, F., G. Weickgenannt, and C. Zdero: Neue diterpene aus der tribus Calendulaceae (sic.). Chem. Ber. **106**, 826 (1973).

157. Bohlmann, F., and C. Zdero: Notiz über ein weiteres diterpen aus *Osteospermum subulatum* DC. Chem. Ber. **108**, 362 (1975).

158. Bohlmann, F., and C. Zdero: Über inhaltsstoffe der gattung *Brickellia*. Chem. Ber. **109**, 1436 (1976).

159. Bohlmann, F., and C. Zdero: Über ein neues diterpen aus Melampodium perfoliatum (Carv.)A. Gray. Chem. Ber. **109**, 1670 (1976).

160. Bohlmann, F., and C. Zdero: Neue terpen-inhaltsstoffe aus *Verbesina*-arten. Phytochem. **15**, 1310 (1976).

161. Bohlmann, F., and C. Zdero: Inhaltsstoffe der gattung *Polymnia*. Phytochem. **16**, 492 (1977).

162. Bohlmann, F., and C. Zdero: Ein neue diterpensäure aus *Perymenium ecuadoricum*. Phytochem. **16**, 786 (1977).

163. Bohlmann, F., and C. Zdero: Ein neues clerodan-derivat sowie weitere inhaltsstoffe aus der gattung *Macowania*. Phytochem. **16**, 1583 (1977).

164. Bohlmann, F., and C. Zdero: Neue norkauren- und thymol derivate aus *Athrixia*-arten. Phytochem. **16**, 1773 (1977).

165. Bohlmann, F., and C. Zdero: Diterpenes with a new carbon skeleton from *Printzia laxa*. Phytochem. **17**, 487 (1978).

166. Bohlmann, F., and C. Zdero: Ein neues kaurensäure und ein euparin-derivat aus *Oyedaea*-arten. Phytochem. **18**, 492 (1979).

167. Bohlmann, F., and C. Zdero: Neue phloroglucin—derivate aus *Helichrysum natalitium* und *Helichrysum bellum*. Phytochem. **18**. 641 (1979).

168. Bohlmann, F., and C. Zdero: Korrektur der konstitutionen von diterpenen aus *Palafoxia rosea*. Phytochem. **18**, 2038 (1979).

169. Bohlmann, F., and C. Zdero: Sandaracopimarene derivatives from *Senecio subrubriflorus*. Phytochem. **21**, 1697 (1982).

170. Bohlmann, F. and C. Zdero: Stevisalicinon, ein neuer Diterpentyp, sowie weitere inhaltsstoffe aus *Stevia*-arten. Liebigs Ann. Chem. 1764 (**1985**).

171. Bohlmann, F., C. Zdero, W.-R. Abraham, A. Suwita, and M. Grenz: Neue diterpene und neue dihydrochalkon-derivate sowie weitere inhaltsstoffe aus *Helichrysum*-arten. Phytochem. **19**, 873 (1980).

172. Bohlmann, F., C. Zdero, and M. Ahmed: New sesquiterpene lactones, geranyllinalol derivatives and other constituents from *Geigeria* species. Phytochem. **21**, 1679 (1982).

173. Bohlmann, F., C. Zdero, D. Berger, A. Suwita, P. Mahanta, and C. Jeffrey: Neue furanoeremophilane und weitere inhaltsstoffe aus südafrikanischen *Senecio-arten*. *Phytochem.* **18**, 79 (1979).

174. Bohlmann, F., C. Zdero, J. Cuatrecasas, R.M. King, and H. Robinson: Neue sesquiterpene und norditerpene aus vertretern der gattung *Libanothamnus*. Phytochem. **19**, 1145 (1980).

175. Bohlmann, F., C. Zdero, L. Fiedler, H. Robinson, and R.M. King: A labdane derivative from *Chromolaena collina* and a *p*-hydroxyacetophenone derivative from *Stomatanthes corumbensis*. Phytochem. **20**, 1141 (1981).

176. Bohlmann, F., C. Zdero, and M. Grenz: New sesquiterpenes of the genus *Othonna*. Chem. Ber. **107**, 3928 (1974).

177. Bohlmann, F., C. Zdero, and M. Grenz: Weitere inhaltsstoffe aus vertretern der *Eupatorium*-gruppe. Chem. Ber. **110**, 1034 (1977).

178. Bohlmann, F., C. Zdero, M. Grenz, A.K. Dhar, H. Robinson, and R.M. King:

Five diterpenes and other constituents from nine *Baccharis* species. Phytochem. **20**, 281 (1981).

179. Bohlmann, F., C. Zdero, R.K. Gupta, R.M. King, and H. Robinson: Diterpenes and teranorditerpenes from *Acritopappus* species. Phytochem. **19**, 2695 (1980).

180. Bohlmann, F., C. Zdero, E. Hoffmann, P.K. Mahanta, and W. Dorner: Neue diterpene und sesquiterpene aus südafrikanischen *Helichrysum*-arten. Phytochem. **17**, 1917 (1978).

181. Bohlmann, F., C. Zdero, and S. Huneck: Diterpenes from *Heteropappus altaicus*. Phytochem. **24**, 1027 (1985).

182. Bohlmann, F., C. Zdero, J. Jakupovic, N. Ates, R.M. King, and H. Robinson: Steiractinolide aus *Aspilia* und *Wedelia*-arten. Liebigs Ann. Chem. **1983**, 1257 (1983).

183. Bohlmann, F., C. Zdero, J. Jakupovic, T. Gerke, M. Wallmeyer, R.M. King, and H. Robinson: Neue sesquiterpenlactone und rosan-derivate aus *Trichogonia*-arten. Liebigs Ann. Chem. **1984**, 162.

184. Bohlmann, F., C. Zdero, J. Jakupovic, R.M. King, and H. Robinson: Diterpenes from *Acritopappus confertus*. Phytochem. **22**, 2243 (1983).

185. Bohlmann, F., C. Zdero, R.M. King, and H. Robinson: Neue labdan-derivate aus *Carterothamnus anomalochaeta*. Phytochem. **18**, 621 (1979).

186. Bohlmann, F., C. Zdero, R.M. King, and H. Robinson: Ein neues labdan-derivat und neue thymol-derivate aus *Bishovia boliviensis*. Phytochem. **18**, 1234 (1979).

187. Bohlmann, F., C. Zdero, R.M. King, and H. Robinson: Neue labdan-derivate aus *Gutierrezia*-arten. Phytochem. **18**, 1533 (1979).

188. Bohlmann, F., C. Zdero, R.M. King, and H. Robinson: Ein neues germacran-8,12-olid und neue diterpene aus *Polymnia canadensis*. Phytochem. **19**, 115 (1980).

189. Bohlmann, F., C. Zdero, R.M. King, and H. Robinson: Germacranolides, a guaianolide with a β-lactone ring and further constituents from *Grazielia* species. Phytochem. **20**, 1069 (1981).

190. Bohlmann, F., C. Zdero, R.M. King, and H. Robinson: Thirteen kolavane derivatives from *Symphyopappus* species. Phytochem. **20**, 1657 (1981).

191. Bohlmann, F., C. Zdero, R.M. King, and H. Robinson: Sesquiterpenes, guaianolides and diterpenes from *Stevia myriadenia*. Phytochem. **21**, 2021 (1982).

192. Bohlmann, F., C. Zdero, R.M. King, and H. Robinson: New germacranolides and other constituents from *Trichogoniopsis morii*. Phytochem. **21**, 2035 (1982).

193. Bohlmann, F., C. Zdero, R.M. King, and H. Robinson: Eudesmanolides and kaurane derivatives from *Wedelia hookeriana*. Phytochem. **21**, 2329 (1982).

194. Bohlmann, F., C. Zdero, R.M. King, and H. Robinson: New labdan-derivate aus *Aristeguietia pseudoarborea*. Liebigs Ann. Chem. **1983**, 2127.

195. Bohlmann, F., C. Zdero, R.M. King, and H. Robinson: Epoxycannabinolid und diterpene mit neuem kohlenstoffgerüst aus *Villanova titicaensis*. Liebigs Ann. Chem. **1984**, 250.

196. Bohlmann, F., C. Zdero, R.M. King, and H. Robinson: Kingidiol, a kolavane derivative from *Baccharis kingii*. Phytochem. **23**, 1511 (1984).

197. Bohlmann, F., C. Zdero, R.M. King, and H. Robinson: Gutierrezial and

further diterpenes from *Gutierrezia sarothrae*. Phytochem. **23**, 2007 (1984).

198. Bohlmann, F., C. Zdero, R.M. King, and H. Robinson: A further geranylnerol derivative from *Helianthopsis* species. Phytochem. **24**, 1108 (1985).

199. Bohlmann, F., C. Zdero, and P. Mahanta: Neue diterpene aus *Dimorphotheca*- und *Viguiera*-arten. Phytochem. **16**, 1073 (1977).

200. Bohlmann, F., C. Zdero, H. Robinson, and R.M. King: Ein neues germacrenderivat sowie ein diterpenmalonat aus *Baccharis* arten. Phytochem. **18**, 1993 (1979).

201. Bohlmann, F., C. Zdero, H. Robinson, and R.M. King: 15α-Methylacryloyloxy-*ent*-kaurenic acid from *Ichthyothere* species. Phytochem. **20**, 522 (1981).

202. Bohlmann, F., C. Zdero, H. Robinson, and R.M. King: A diterpene, a sesquiterpene quinone and flavanones from *Wyethia helenioides*. Phytochem. **20**, 2245 (1981).

203. Bohlmann, F., C. Zdero, H. Robinson, and R.M. King: Labdane derivatives from *Planaltoa lychnophoroides*. Phytochem. **21**, 465 (1982).

204. Bohlmann, F., C. Zdero, H. Robinson, and R.M. King: germacranolides from *Lychnophora* species. Phytochem. **21**, 1087 (1982).

205. Bohlmann, F., C. Zdero, G. Schmeda-Hirschmann, J. Jakupovic, V. Castro, R.M. King, and H. Robinson: Heliangolide, trachyloban und villanovan-derivate aus *Viguiera*-arten. Liebigs. Ann. Chem. **1984**, 495 (1984).

206. Bohlmann, F., C. Zdero, and S. Schoneweiss: Über die Inhaltsstoffe aus *Stevia*-arten. Chem. Ber **109**, 3366 (1976).

207. Bohlmann, F., C. Zdero, and B. L. Turner: Guaianolides and heliangolides from *Hymenopappus newberryi*. Phytochem. **23**, 1055 (1984).

208. Bohlmann, F., C. Zdero, R. Zeisberg, and W.S. Sheldrick: Helifulvanolsaure -ein neues diterpen mit anomalem kohlenstoffgerust aus *Helichrysum fulvum*. Phytochem. **18**, 1359 (1979).

209. Bohlmann, F., and J. Ziesche: Neue germacren-derivate aus *Senecio*-arten. Phytochem. **18**, 1489 (1979).

210. Bohlmann, F., and J. Ziesche: Neue diterpene aus *Gnaphalium*-arten. Phytochem. **19**, 71 (1980).

211. Bohlmann, F., J. Ziesche, R.M. King, and H. Robinson: Neuemelampolide aus *Smallanthus fruticosus*. Phytochem. **19**, 973 (1980).

212. Bohlmann, F., J. Ziesche, R.M. King, and H. Robinson: Seven furanoeremophilanes from three *Senecio* species. Phytochem. **19**, 2675 (1980).

213. Bohlmann, F., J. Ziesche, R.M. King, and H. Robinson: Eudesmanolides and diterpenes from *Wedelia trilobata* and an *ent*-kaurenic acid derivative from *Aspilia parvifolia*. Phytochem. **20**, 751 (1981).

214. Bohlmann, F., J. Ziesche, R.M. King, and H. Robinson: Eudesmanolides and other constituents from *Dimerostemma asperatum*. Phytochem. **20**, 1335 (1981).

215. Bohlmann, F., J. Ziesche, R.M. King, and H. Robinson: Eudesmanolides, guaianolides, germacranolides and elemanolides from *Zinnia* species. Phytochem. **20**, 1623 (1981).

216. Bohlmann, F., J. Ziesche, R.M. King, and H. Robinson: A pseudoguaianolide and a hydroxygeranylnerol from *Kingianthus paradoxus*. Phytochem. **20**, 1146 (1981).

217. Bohlmann, F., P. Zitzkowski, A. Suwita, and L. Fiedler: *cis*-Kolaveninsäure und weitere inhaltsstoffe aus vertretern der tribus Eupatorieae. Phytochem. **17**, 2101 (1978).

218. Brieskorn, C.H., and E. Pohlmann: Diterpene vom kaurantyp aus der composite *Espeletia schultzii* (Wedd.). Tetrahedron Lett., 566 (1968).

219. Brieskorn, C.H., and E. Pohlmann: Kaurandien-(9(11).16)-saure-(19) und 15α-acetoxy-kauren-(16)-saure-(19). Chem. Ber. **102**, 2621 (1969).

220. Brunn, T., L.M. Jackman, and E. Stenhagen: Grindelic and oxygrindelic acids. Acta. Chem. Scand. **16**, 1675 (1962).

221. Calderon, J.S., and J.R. Puig: Diterpenos tipo labdano de *Stevia origanoides*. X. Simposium Internacional Quimica de Productos Naturales, Monterrey, N. L., Mexico. April 28–30, 1983. p. 6.

222. Calderon, J.S., Quijano, L., Cristia, M., Gomez, F., and T. Rios: Labdane diterpenes from *Brickellia veronicaefolia*. Phytochem. **22**, 1783 (1983).

223. Calderon, J.S., L. Quijano, M. Garduno, F. Gomez, and T. Rios: 2α-iso-valeroyloxy-eperuic acid, a diterpene from *Eupatorium petiolare*. Phytochem. **22**, 2617 (1983).

224. Canonica, L., B. Rindone, and C. Scolastico: Structure and stereochemistry of psiadiol, a new diterpenoid. Tetrahedron Lett. 2639 (1967).

225. Canonica, L., B. Rindone, and C. Scolastico: A new diterpene with pimarane skeleton. Tetrahedron Lett. 4801 (1969).

226. Carlquist, S.: Structure and ontogeny of glandular trichomes of *Madinae* (Compositae). Amer. J. Bot. **45**, 675 (1958).

227. Castro, V. and J. Jakupovic: Two further 6,12-*cis*-germacranolides from *Montanoa tomentosa* subsp. *xanthiifolia*. Phytochem. **24**, 2449 (1985).

228. Castro, V. and J. Jakupovic: 4β,19-Epoxy-norkaurene and other diterpenes from *Mikania banisteriae*. Phytochem. **24**, 2450 (1985).

229. Ceccarelli, N., R. Lorenzi, and A. Alpi: Kaurene and kaurenol biosynthesis in cell-free systems of *Phaseolus coccineus* suspensor. Phytochem. **18**, 1657 (1979).

230. Cheng, P.C., C. Hufford, and N.J. Doorenbos: Isolation of 11-hydroxylated kauranic acids from *Adenostemma lavenia*. J. Nat. Prod. **42**, 183 (1979).

231. Connolly, J.D., F. Orsini, F. Pelizzoni, and G. Ricca: 13C NMR spectra of cassane diterpenoids. The stereochemistry of the caesalpins. Org. Magn. Reson. **17**, 163 (1981).

232. Craig, J.C., M.L. Mole, S. Billets, and F. El-Feraly: Isolation and identification of the hypoglycemic agent, carboxyatractylate from *Xanthium strumarium*. Phytochem. **15**, 1178 (1976).

233. Cronquist, A.: The Compositae revisited. Brittonia **29**, 137 (1977).

234. Croteau, R., and M.A. Johnson: Biosynthesis of terpenoids in glandular trichomes. In: E. Rodriguez (ed.). Biology and Chemistry of Plant Trichomes. Plenum Press (1984).

235. Cruse, W.B.T., M.N.G. James, and Ali A. Al-Shamma, J.K. Beal, and R.W. Doskotch: The molecular structure of gutierolide, a novel chloro-diterpenoid lactone. J. Chem. Soc. Chem. Commun. 1278 (1971).

236. Cuatrecasas, T.: Prima Flora Colombiana. 3. Compositae—Astereae. Webbia **24**, 1 (1969).

237. Cuevas, L.A., F. Garcia Jimenez, and A. Romo de Vivar: Estructura de la estenolobina. Rev. Latinoamer. Quim. **3**, 22 (1972).

238. Cutler, H.G.: Carboxyatractyloside: A compound from *Xanthium strumarium* and *Atractylis gummifera* with plant growth inhibiting properties. The probable "Inhibitor A." J. Nat. Prod. **46**, 609 (1983).

239. Danieli, B., E. Bombardelli, A. Bonati, and B. Gabetta: Structure of the diterpenoid carboxyatractyloside. Phytochem. **11**, 3501 (1972).

240. Delgado, G., L. Alvarez, and A. Romo de Vivar: Terpenoids and a flavan-3-ol from *Viguiera quinqueradiata*. Phytochem. **23**, 675 (1984).

241. Delgado, G., L. Alvarez, and A. Romo de Vivar: 15-Oxo-zoapatlin, a diterpene lactone from *Viguiera maculata*. Phytochem. **23**, 2674 (1984).

242. Delgado, G., H. Candenas, G. Pelaez, and A. Romo de Vivar: Terpenoids from *Viguiera excelsa* and *Viguiera oaxacana*. J. Nat. Prod. **47**, 1042 (1984).

243. Delgado, G., and A. Romo de Vivar: *ent*-Kaurenoid methyesters from *Viguiera stenoloba*, structural revision of stenolobin and its biomimetic conversion to zoapatlin. Chem. Lett. 1237 (1984).

244. Delgado, G., A. Romo de Vivar, J. Cardenas, R. Pereda-Miranda, and E. Huerta: *ent*-Beyerene and *ent*-atisene-diterpenes from *Viguiera insignis*. Phytochem. **23**, 2285 (1984).

245. Delgado, G., A. Romo de Vivar, A. Ortega, J. Cardenas, and E. O. Schlemper: Diterpenoids from *Viguiera insignis*. Phytochem. **22**, 1227 (1983).

246. Dell, B., and A.J. McComb: Resin production and glandular hairs in *Beyeria viscosa* (Labill.) Miq. (Euphorbiaceae). Aust. J. Bot. **25**, 195 (1974).

247. Dell, B., and A.J. McComb: Glandular hairs, resin production, and habitat of *Newcastelia viscida* E. Pritzel (Dicrastylidaceae). Aust. J. Bot. **23**, 373 (1975).

248. Dominguez, X.A.: Eupatorieae—chemical review, In V.H. Heywood, J.B. Harborne and B.L. Turner (eds.). The biology and chemistry of the Compositae. vol. 1, 487–502, Academic Press, London (1977).

249. Dominguez, X.A., D. Butruille, I. Sandler, and G. Vazquez: Identification de la solidagenona en *Solidago sempervirens*. Rev. Latinoamer. Quim. **6**, 159 (1975).

250. Dominquez, X.A., C. Cisneros, E. Guajardo, T., R. Villarreal, and A. Zamudio: Diterpenoids from *Palafoxia rosea*. Phytochem. **14**, 1665 (1975).

251. Dominquez, X.A., C. Cisneros, R. Guajardo, R. Villarreal, V. Zebel, and W.H. Watson: Jesromotetrol, a new diterpene from *Palafoxia rosea* (Compositae). Rev. Latinoamer. Quim. **9**, 99 (1978).

252. Dominquez, X.A., and S. Jimenez Jimennez: Aislamiento y estructuras del julslimtetrol, *nor*-julslimdiolona y la isocumambrina, metabolitos secundarius del *Croptilon divaricatum* (Compuesta). Rev. Latinoamer. Quim. **3**, 177 (1973).

253. Dominguez, X.A., J. Verde S., N.E. Guerra R., E. Ellenmaurer, and J. Jakupovic: A *nor*-diterpene and other constituents from *Isocoma coronopifolia*. Phytochem. **25**, 2893 (1986).

254. Edwards, O.E., G. Feniak, and M. Los: Diterpenoid quinones of *Inula royleana* D.C. Can. J. Chem. **40**, 1540 (1962).

255. Edwards, O.E., and M.N. Rodger: The alkaloids of *Inula royleana*. Can. J. Chem. **37**, 1187 (1959).

256. Eguren, L., A. Perales, J. Fayos, B. Rodriquez, G. Savona, and F. Piozzi: New neoclerodane diterpenoid containing an oxetane ring isolated from *Teucrium chamaedrys*. X-ray structure determination. J. Org. Chem. **47**, 4157 (1982).

257. El-Emary, N.A., G. Kusano, and T. Takemoto: Kaurenoids from *Cacalia bulbifera*. Phytochem. **14**, 1660 (1975).

258. Elliger, C.A., D.F. Zinkel, B.G. Chan, and A.C. Waiss, Jr.: Diterpene acids as larval growth inhibitors. Experientia **32**, 1364 (1976).

259. El-Naggar, S.F., R.W. Doskotch, T.M. O'Dell, and L. Girard: Antifeedant diterpenes for the gypsy moth larvae from *Kalmia latifolia*: Isolation and characterization of ten grayanoids. J. Nat. Prod. **43**, 617 (1980).

260. Evans, F.J., and C.J. Soper: The tigliane, daphnane and ingenane diterpenes, their chemistry, distribution and biological activities. A review. Lloydia **41**, 193 (1978).

261. Evans, F.J., and S.E. Taylor: Pro-inflammatory, tumour-promoting and anti-tumour diterpenes of the plant families Euphorbiaceae and Thymelaeaceae. Fortschritte Chem. Org. Naturst. **44**, 1 (1984).

262. Fahn, A.: Secretory Tissues in Plants. Chap. 9 (158–222), Tissues secreting lipophilic substances. Academic Press, London, New York, and San Francisco.

263. Ferguson, G., W.C. Marsh, R. McCrindle, and E. Nakamura: Stereochemistry of clerodanes. X-ray structure of a key diterpenoid from *Solidago arguta*. J. Chem. Soc. Chem. Commun. 299 (1975).

264. Ferguson, G., R. McCrindle, S.T. Murphy, and M. Parvez: Further diterpenoid constituents of *Helianthus annuus* L. Crystal and molecular structure of methyl *ent*-15β-hydroxy-trachyloban-19-oate. J. Chem. Res. **5**, 200 (1982).

265. Fernandez-Gadea, F., M.L. Jimeno, and B. Rodriguez: Complexation of hydroxylated *ent*-kaurane diterpenoids with boric acid as an aid to the assignment of the 13C NMR spectra. Org. Magn. Reson. **22**, 515 (1984).

266. Fischer, N.H., E.J. Olivier, and H.D. Fischer: The biogenesis and chemistry of sesquiterpene lactones, In: W. Hertz, H. Grisebach and G. W. Kirby (eds.). Progress in the chemistry of organic natural products, 48–390. Spring-Verlag, New York (1979).

267. Fujita, E., K. Fuji, Y. Nagao, M. Node, and M. Ochiai: Bull. Inst. Chem. Res., Kyoto Univ., **61**, 142 (1983).

268. Gage, D.A.: Chemical data and their bearing upon generic delineations in the Eupatorieae. Taxon **34**, 61 (1985).

269. Gage, D.A.: A Chemosystematic Study of the genus *Piptothrix* (Compositae, Eupatorieae). Doctoral Dissertation. The University of Texas at Austin (1986).

270. Ganguly, S.N., T. Ganguly, and S.M. Sircar: Gibberellins of *Enhydra fluctuans*. Phytochem. **11**, 3433 (1972).

271. Gao, F., M. Leidig, and T.J. Mabry: *ent*-Labdane derivatives from *Gutierrezia grandis*. Phytochem. **24**, 1541 (1985).

272. Gao, F., M. Leidig, and T.J. Mabry: Diterpene glycosides from *Gutierrezia sphaerocephala*. Phytochem. **25**, 1371 (1986).

273. Gao, F., and T.J. Mabry: An acyclic diterpene from *Viguiera deltoidea*. Phytochem. **24**, 3061 (1985).

274. Gao, F., and T.J. Mabry: Ten *cis*-clerodane-type diterpene lactones from *Gutierrezia texana*. Phytochem. **26**, 209 (1987).

275. Gao, F., M. Miski, D.A. Gage, and T.J. Mabry: Terpenoid constituents of *Viguiera dentata*. J. Nat. Prod. **48**, 316 (1985).

276. Gao, F., M. Miski, D.A. Gage, J.A. Norris, and T.J. Mabry: Terpenoids from *Viguiera potosina*. J. Nat. Prod. **48**, 489 (1985).

277. Gao, F., H. Wang, and T.J. Mabry: Sesquiterpene lactones and flavonoids from *Helianthus* species. J. Nat. Prod. **50**, 23 (1987).

278. Garcia, E.E., E. Guerreiro, and P. Joseph-Nathan: *ent*-Pimaradiene diterpenes from *Gochnatia glutinosa*. Phytochem. **24**, 3059 (1985).

279. Geissman, T.A., and D.H.G. Crout: Organic Chemistry of Secondary Plant Metabolism, Chap. 10 (Diterpenoid Compounds) pp. 291–311. Freeman, Cooper and Company, San Francisco.

280. Gershenzon, J., and T.J. Mabry: Sesquiterpene lactones from a Texas population of *Helianthus maximiliani.* Phytochem. **23**, 1959 (1984).

281. Gershenzon, J., and T.J. Mabry: Furanoheliangolides from *Helianthus schweinit-zii.* Phytochem. **23**, 2557 (1984).

282. Gershenzon, J., T.J. Mabry, J.D. Korp, and I. Bernal: Germacranolides from *Helianthus californicus.* Phytochem. **23**, 2561.

283. Gianello, J.C., and O.S. Giordano: Barticulidiol, un nuevo furano diterpeno aislado de *Baccharis articulata* Lam (Persoon). Rev. Lationamer. Quim. **13**, 76 (1982).

284. Givovich, A., A. San-Martin, and M. Castillo: Neo-clerodane diterpenoids from *Baccharis incarum.* Phytochem. **25**, 2829 (1986).

285. Glasby, J.S.: Encyclopaedia of the Terpenoids, Vols. 1 and 2. Wiley-Interscience Publication, Chichester and New York (1982).

286. Gomez, G.F., J.S. Calderon, L., Quijano, and T. Rios: Acido brickelico C, un neuvo diterpeno aislado de *Brickellia* sp. VIII. Simposium Internacional Quimica de Productos Naturales, Monterrey, N. L., Mexico. March 26–28, 1981.

287. Gomez, F., L. Quijano, J.S. Calderon, and T. Rios: 3α-Angeloyloxy-2α-hydroxy-cativic acid, a new diterpene from *Brickellia paniculata.* Phytochem. **22**, 1292 (1983).

288. Gonzalez, A.G., J.M. Artega, B.M. Fraga, M.G. Hernandez, and J. Fayos: The structure of jhanilactone. Experientia **34**, 554 (1978).

289. Gonzalez, A.G., J.M. Artega, J.L. Breton, and B.M. Fraga: Five new labdane diterpene oxides from *Eupatorium jhanii.* Phytochem. **16**, 107 (1977).

290. Gonzalez, A.G., J.J. Mendoza, J.G. Luis, A.G. Ravelo, X.A. Dominguez, and G. Cano: Diterpenes from *Palafoxia texana.* Phytochem. **24**, 3056 (1985).

291. Goswami, A., R.N. Barua, R.P. Sharma, J.N. Baruah, P. Kulanthaivel, and W. Herz: Clerodanes from *Solidago virgaurea.* Phytochem. **23**, 837 (1984).

292. Grau, J.: Astereae—systematic review. In: The Biology and Chemistry of the Compositae (Heywood, V.H., J.B. Harborne, and B.L. Turner, eds.). New York and London: Academic Press. 1977.

293. Guerreiro, E., J. Kavka, and O.S. Giordano: Acido 1-hidroxigrindelico en *Grindelia pulchella.* Rev. Latinoamer. Quim. **13**, 72 (1982).

294. Guerreiro, E., J. Kavka, J.R. Saad, M.A. Oriental, and O.S. Giordano: Acidos diterpenicos en *Grindelia pulchella* and *G. chiloensis* Cabr. Rev. Latinoamer. Quim. **12**, 77 (1981).

295. Guerrero, C., and A. Romo de Vivar: Estructura y estereoquimica de la bacchofertina; diterpeno aislado de *Baccharis conferrta* H.B.K. Rev. Latinoamer. Quim. **4**, 178 (1973).

296. Gutierrez, A.B., J.C. Oberti, P. Kulanthaivel, and W. Herz: Sesquiterpene lactones and diterpenes from *Mikania periplocifolia.* Phytochem. **24**, 2967 (1985).

297. Hahn, D.W., E.W. Ericson, M.T. Lai, and A. Probst: Antifertility activity of *Montanoa tomentosa* (zoapatle). Contraception **23**, 133 (1981).

298. Hänsel, R., E.-M. Cybulski, B. Çubukçu, A.H. Meriçli, F. Bohlmann, and C. Zdero: Neue pyron-derivate aus *Helichrysum*-arten. Phytochem. **19**, 639 (1980).

299. Hanson, J.R.: The biosynthesis of the diterpenes. In: Progress in the Chemistry of Organic Natural Products (Herz, W., H. Grisebach, and G. W. Kirby, eds.). Wien and New York: Springer-Verlag. 1971.

300. Hanson, J.R.: The bicyclic diterpenes. In: Progress in Phytochemistry (Reinhold, L., and Y. Liwschitz, eds.). London: Interscience Publishers of John Wiley and Sons. 1972.

301. Hanson, J.R.: The di- and sesquiterpenes. In: Chemistry of Terpenes and Terpenoids (Newman, A.A., ed.). London and New York: Academic Press. 1972.

302. Hanson, J.R. (ed.): Terpenoids and Steroids (Specialist Periodical Reports), Vol. 12. London: The Royal Society of Chemistry. 1983.

303. Hanson, J.R.: The biosynthesis of C_5-C_{20} terpenoid compounds. Natural Products Reports **1**, 443 (1984).

304. Hanson, J.R.: Diterpenoids. Natural Products Reports **1**, 171 (1984).

305. Hanson, J.R.: Diterpenoids. Natural Products Reports **1**, 339 (1984).

306. Hanson, J.R.: Diterpenoids. Natural Products Reports **1**, 533 (1984).

307. Hanson, J.R.: Diterpenoids. Natural Products Reports **3**, 307 (1986).

308. Hanson, J.R.: The chemistry of the tetracyclic diterpenoids—X. Some beyerene 2 and 3-alcohols. Tetrahedron **26**, 2711 (1970).

309. Hanson, J.R., G. Savona, and M. Siverns: 13C Nuclear magnetic resonance spectra and microbiological hydroxylation of 7α- and 7β-hydroxykaurenolide. J.C.S. Perkin I **1974**, 2002.

310. Hanson, J.R., M. Siverns, F. Piozzi, and G. Savona: The 13C Nuclear magnetic resonance spectra of kauranoid diterpenes. J.C.S. Perkin I **1976**, 114.

311. Hanson, J.R., and A.F. White: Studies in terpenoid biosynthesis II. The biosynthesis of steviol. Phytochem. **7**, 595 (1968).

312. Harborne, J.B.: Inuleae—chemical review. In: The Biology and Chemistry of the Compositae (Heywood, V.H., J.B. Harborne and B.L. Turner, eds.). New York and London: Academic Press. 1977.

313. Hedden, P., and J.E. Grabbe: Kaurenolide biosynthesis in a cell-free system from *Cucurbita maxima* seeds. Phytochem. **20**, 1011 (1981).

314. Hegnauer, R.: Compositae. In: Chemotaxonomie der Pflanzen, Vol. 3. Basel: Birkhäuser Verlag. 1964.

315. Hegnauer, R.: The chemistry of the Compositae. In: The Biology and Chemistry of the Compositae (Heywood, V.H., J.B. Harborne, and B.L. Turner, eds.). New York and London: Academic Press. 1977.

316. Heiser, C.B., D.M. Smith, S.B. Clevenger, and W.C. Martin: The North American Sunflowers (*Helianthus*). Bull. Torr. Bot. Club **22**, 1 (1969).

317. Henderson, M.S., R. McCrindle, and D. McMaster: Constituents of *Solidago* species. Part V. Non-acidic diterpenoids from *Solidago gigantea* var. *serotina*. Can. J. Chem. **51**, 1346 (1973).

318. Henderson, M.S., R.D.H. Murray, R. McCrindle, and D. McMaster: Constituents of *Solidago* species. Part III. The constitution of diterpenoids from *Solidago juncea* Ait. Can. J. Chem. **51**, 1322 (1973).

319. Herz, W.: Astereae—chemical review. In: The Biology and Chemistry of the Compositae (Heywood, V.H., J.B. Harborne, and B.L. Turner, eds.). New York and London: Academic Press. 1977.

320. Herz, W., S.V. Bhat, and R. Murari: The diterpene darutigenol from *Palafoxia arida*. Phytochem. **17**, 1060 (1978).

321. Herz, W., and M. Bruno: Heliangolides, kauranes and other constituents of *Helianthus heterophyllus*. Phytochem. **25**, 1913 (1986).

322. Herz, W., D. Gage, and N. Kumar: Damsinic acid and ambrosanolides from vegetative *Ambrosia hispida*. Phytochem. **20**, 1601 (1981).

323. Herz, W., S.V. Govindan, and J.F. Blount: Tetracyclic analogues of the rosane lactones from *Eupatorium album*. J. Org. Chem. **44**, 2999 (1979).

324. Herz, W., S.V. Govindan, and K. Watanabe: Diterpenes of *Helianthus rigidus* and *H. salicifolius*. Phytochem. **21**, 946 (1982).

325. Herz, W., and P. Kulanthaivel: Eudesmanolides and *ent*-pimaranes from *Liatris laevigata*. Phytochem. **22**, 715 (1983).

326. Herz, W., and P. Kulanthaivel: *Ent*-kaurenes and trachylobanes from *Helianthus radula*. Phytochem. **22**, 2543 (1983).

327. Herz, W., and P. Kulanthaivel: *Ent*-pimaranes, *ent*-kauranes, heliangolides and other constituents of three *Helianthus* species. Phytochem. **23**, 1453 (1984).

328. Herz, W., and P. Kulanthaivel: *Ent*-kauranes and 10α-methyl-eudesman-8αH,12-olides from *Wedelia calycina* and *Wedelia hispida*. Phytochem. **23**, 2271 (1984).

329. Herz, W., and P. Kulanthaivel: Trihydroxy-C_{18}-acids and a labdane from *Rudbeckia fulgida*. Phytochem. **24**, 89 (1985).

330. Herz, W., and P. Kulanthaivel: Diterpenes and sesquiterpene lactones from *Mikania congesta*. Phytochem. **24**, 1761 (1985).

331. Herz, W., and P. Kulanthaivel: Diterpenes from *Viguiera porteri*. J. Nat. Prod. **48**, 676 (1985).

332. Herz, W., P. Kulanthaivel, and K. Watanabe: *Ent*-kauranes and other constituents of three *Helianthus* species. Phytochem. **22**, 2021 (1983).

333. Herz, W., and N. Kumar: Sesquiterpene lactones from *Helianthus grosseserratus*. Phytochem. **20**, 99 (1981).

334. Herz, W., A.-M. Pilotti, A.-C. Söderholm, I.K. Shuhama, and W. Vichnewski: New *ent*-clerodane-type diterpenoids from *Baccharis trimera*. J. Org. Chem. **42**, 3913 (1977).

335. Herz, W., and R.P. Sharma: Ligantrol and ligantrol monoacetate, two new linear polyoxgenated diterpenes from *Liatris elegans*. J. Org. Chem. **40**, 192 (1975).

336. Herz, W., and R.P. Sharma: New hydroxylated *ent*-kauranoic acids from *Eupatorium album*. J. Org. Chem. **41**, 1021 (1976).

337. Herz, W., K. Watanabe, P. Kulanthaivel, and J.F. Blount: Cycloartanes from *Lindheimera texana*. Phytochem. **24**, 2645 (1985).

338. Hochmannova, J., L. Novotny, and V. Herout: Hydrocarbons from *Petasites albus* (L.)Gaertn. rhizomes. Coll. Czech. Chem. Commun. **27**, 2711 (1962).

339. Hoeneisen, M. and M. Silva: Quimica y actividad biologica de algunas Compositae chilenas. Rev. Latinoamer. Quim. **17**, 24 (1986).

340. Hoffmann, J.J., B.E. Kingsolver, S.P. McLaughlin, and B.N. Timmerman:

Production of resins by arid-adapted Astereae. In: Recent Advances in Phytochemistry (Timmermann, B.N., C. Steelink, and F.A. Loewus, eds.), Vol. 18, p. 251. New York and London: Plenum Press. 1984.

341. Hoffmann, J.J., S.P. McLaughlin, S.D. Jolad, K.H. Schram, M.S. Tempesta, and R.B. Bates: Constituents of *Chrysothamnus paniculatus* (Compositae): Chrysothame, a new diterpene, and 6-oxogrindelic acid. J. Org. Chem. **47**, 1725 (1982).

342. Hosozawa, S., N. Kato, and K. Munakata: Antifeeding active substances for insects in *Caryopteris divaricata* Maxim. Agr. Biol. Chem. **38**, 823 (1974).

343. Hubert, T.D., D.F. Wiemer: Ant-repellent terpenoids from *Melampodium divaricatum*. Phytochem. **24**, 1197 (1985).

344. Huneck, S., Y. Asakawa, Z. Taira, A.F. Cameron, J.D. Connolly, and D.S. Rycroft: Gymnocolin, a new *cis*-clerodane diterpenoid from the liverwort, *Gymnocolea inflata*. Crystal structure analysis. Tetrahedron Lett. **24**, 115 (1983).

345. Hutchison, M., P. Lewer, and J. Macmillan: Carbon-13 nuclear magnetic resonance spectra of eighteen derivatives of *ent*-kaur-16-en-19-oic acid. J. Chem. Soc. Perkin Trans. I **1984**, 2363.

346. Jakupovic, J., S. Banerjee, F. Bohlmann, R.M. King, and H. Robinson: New diterpenes from *Chiliotrichium rosmarinifolium* and Nardophyllum lanatum. Tetrahedron **42**, 1305 (1986).

347. Jakupovic, J., R.N. Baruah, F. Bohlmann, R.M. King, and H. Robinson: New alicyclic diterpenes and *ent*-labdanes from *Guterrezia solbrigii*. Tetrahedron **41**, 4537 (1985).

348. Jakupovic, J., R.N. Baruah, C. Zdero, F. Eid, V.P. Pathak, T.V. Chau-Thi, F. Bohlmann, R.M. King, and H. Robinson: Further diterpenes from plants of the Compositae, subtribe Solidagininae. Phytochem. **25**, 1873 (1986).

349. Jakupovic, J., E. Ellmauerer, F. Bohlmann, R.M. King, and H. Robinson: *ent*-Labdanes from *Austrobrickellia patens*. Phytochem. **25**, 1927 (1986).

350. Jakupovic, J., E. Ellmauerer, F. Bohlmann, A. Whittemore, and D.A. Gage: Diterpenes from *Eupatorium turbinatum*. Phytochem. **25**, 2677 (1986).

351. Jakupovic, J., J. Kuhnke, A. Schuster, M.A. Metwally and F. Bohlmann: Phloroglucinol derivatives and other constituents from South American *Helichrysum* species. Phytochem. **25**, 1133 (1986).

352. Jeffries, P.R., J.R. Knox, K.R. Price, B. Scaf: Constituents of the tumor-inhibitory extract of *Olearia muelleri*. Aust. J. Chem. **27**, 221 (1974).

353. Jiang, J.B., M.J. Urbanski, and Z.G. Hajos: Total synthesis of dioxane analogues related to zoapatanol. J. Org. Chem. **48**, 2001 (1983).

354. Jiménez, E.M., and M. Gonzalez de la Parra: nuevo alcohol diterpenico aislado de *Piqueria trinervia* Cav. Rev. Latinoamer. Quim. **14**, 20 (1983).

355. Kaneda, N., H. Kohda, K. Yamasaki, O. Tanaka, and K. Nishi: Paniculosides I-V. Diterpene glucosides from *Stevia ovata* Lag. Chem. Pharm. Bull. **26**, 2266 (1978).

356. Kanojia, R.M., M.P. Wachter, S.D. Levine, R.E. Adams, R. Chen, E. Chin, M.L. Cotter, A.F. Hirsch, R. Huettemann, V.V. Kane, P. Ostrowski, and C.J. Shaw: Isolation and structural elucidation of zoapatanol and montanol, novel oxepane diterpenoids from the Mexican plant zoapatle (*Montanoa tomentosa*). J. Org. Chem. **47**, 1310 (1982).

357. Khac Manh Duc, D., M. Fetizon, S. Lazare, P.K. Grant, M.J. Nicholls,

H.T.L. Liau, M.J. Francis, J. Poisson, J.-M. Bernassau, N.F. Roque, P.M. Wovkulich, and E. Wenkert: 13C NMR spectroscopy of tetracarbocyclic diterpenes and related substances. Tetrahedron **37**, 2371 (1981).

358. Kim, J.H., K.D. Han, K. Yamasaki, and O. Tanaka: Darutoside, a diterpenoid from *Siegesbeckia pubescens* and its structure revision. Phytochem. **18**, 894 (1979).

359. Kinghorn, A.D., D.D. Soejarto, N.P.D. Nanayakhara, C.M. Compadre, H.C. Makapugay, J.M. Hovanec-Brown, P.J. Medon, and S.K. Kamath: A phytochemical screening procedure for sweet *ent*-kaurene glycosides in the genus *Stevia*. J. Nat. Prod. **47**, 439 (1984).

360. Knudsen, F., W. Vilegas, F. Oliveira, and N.F. Roque: Pimaradiene diterpenes from *Mikania triangularis*. Phytochem. **25**, 1240 (1986).

361. Kobayashi, M., S. Horikawa, I.H. Degrandi, J. Ueno, and H. Mitsuhashi: Dulcosides A and B, new diterpene glycosides from *Stevia rebaudiana*. Phytochem. **16**, 1405 (1977).

362. Kohda, H., R. Kasai, K. Yamasaki, K. Murakami, and O. Tanaka: New sweet diterpene glycosides from *Stevia rebaudiana*. Phytochem. **15**, 981 (1976).

363. Kohda, H., O. Tanaka, and K. Nishi: Diterpene—glycosides of *Stevia paniculata* Lag.: Structures of aglycones. Chem. Pharm. Bull. **24**, 1040 (1976).

364. Kusumoto, S., T. Okazaki, A. Ohsuka, and M. Kotake: The structure of solidagonic acid, a bitter principle of goldenrod. Tetrahedron Lett. **40**, 4325 (1968).

365. Kusumoto, S., T. Okazaki, A. Ohsuka, and M. Kotake: The structure and stereochemistry of solidagonic acid. Bulletin of the Chemical Society of Japan. **42**, 812 (1969).

366. Lassak, E.V., and J.T. Pinhey: The constituents of some *Helichrysum* species (family Compositae). Aust. J. Chem. **21**, 1927 (1968).

367. Lee, E.F., J. Gershenzon, and T.J. Mabry: Terpenoids of *Helianthus nuttallii*. J. Nat. Prod. **47**, 1021 (1984).

368. Le Quesne, P.W., G.A. Cooper-Driver, M. Villani, M.N. Do, P.A. Morrow, and D.A. Tonkin: Biologically active diterpenes from *Solidago* species. Page 271 in: A. Rahman and P. W. Le Quesne (eds.). New Trends in Natural Products Chemistry. Studies in Organic Chemistry, Vol. 26. Elsevier Scientific Publishers. Amsterdam (1986).

369. Lequesne, P.W., V. Honkan, K.D. Onan, P.A. Morrow, and D. Tonkyn: Oxidized kaurene derivatives from leaves of *Solidago missouriensis* and *S. rigida*. Phytochem. **24**, 1785 (1985).

370. Le Van, N., and T.V. Cuong Pham: Viscidic acid A and B, two *ent*-labdane derivatives from *Chrysothamnus viscidiflorus*. Phytochem. **19**, 1971 (1980).

371. Levan, N., and N.J. Fischer: Three new melampolide sesquiterpenes, polymatin A, B, and C, from *Polymnia maculata*. Phytochem. **18**, 851 (1979).

372. Levine, S.D., D.W. Hahn, M.L. Cotter, F.C. Greenslade, R.M. Kanojia, S.A. Pasquale, M. Wachter, and J.L. McGuire: The Mexican plant zoapatle (*Montanoa tomentosa*) in reproductive medicine. Past, present and future. Journal of Reproductive Medicine **26**, 524 (1981).

373. Lloyd, H.A., S.L. Evans, and H.M. Fales: Terpene alcohols of *Helichrysum*

dendroideum. (1978).

374. Lloyd, H.A., and H.M. Fales: Terpene alcohols of *Helichrysum dendroideum*. Tetrahedron Lett. **48**, 4891 (1967).

375. Luteijn, J.M., A. Van Veldhuizen, and A. De Groot: The assignment of ^{13}C NMR shift data in clerodanes and related structures. Organic Magnetic Resonance **19**, 95 (1982).

376. Mahanto, S.B., A.K. Sen, P.C. Mazumbar, and K. Yamasaki: Diterpenes of *Conyza stricta*, identification of conyzic acid, seconidoresedasäure and strictic acid. Phytochem. **20**, 850 (1981).

377. Manabe, S., N. Enoki, and C. Nishino: Application of the CD homoallylic benzoate method as a chiroptical tool for determination of absolute configuration. Tetrahedron Lett. **26**, 2213 (1985).

378. Manabe, S. and C. Nishino: Stereochemistry of *cis*-clerodane diterpenes. Tetrahedron **42**, 3461 (1986).

379. Manchand, P.S., J.F. Blount, T. McCabe, and J. Clardy: Structure of evillosin, a novel labdane diterpenoid lactone from *Eupatorium villosum* Sw. J. Org. Chem. **44**, 1322 (1979).

380. Mangoni, L., and M. Belardini: Sui componenti della *Grindelia robusta*. Nota II. Acido 6-oxo-grindelico. Gass. Chim. Ital. **92**, 983 (1962).

381. Mangoni, L., and M. Belardini: Sui componenti della *Grindelia robusta*. Nota III. Acido $7\alpha,8\alpha$-oxido-diidrogrindelico. Gass. Chim. Ital. **92**, 995 (1962).

382. Marcelle, G.B., N. Bunyapraphatsara, G.A. Cordell, H.H.S. Fong, K.C. Nicolaou, and R.E. Zipkin: Studies on Zoapatle I. The extraction of zoapatle (*Montanoa tomentosa*) and the identification of 21-normontanol as the initial decomposition product of zoapatanol. J. Nat. Prod. **48**, 739 (1985).

383. Mathur, S.B., and C.M. Fermin: Terpenes of *Mikania mongenansis*. Phytochem. **12**, 226 (1973).

384. Mathur, S.B., P. Garcia Tello, C.M. Fermin, and V. Mora-Arellano: Terpenoids of *Mikania monagasensis* and their biological activities. Rev. Latinoamer. Quim. **6**, 201 (1975).

385. McCrindle, R., and E. Nakamura: Constituents of *Solidago* species. Part VI. The constitution of diterpenoids from a chemically distinct variety of *Solidago serotina*. Can. J. Chem **52**, 2029 (1974).

386. McCrindle, R., E. Nakamura, and A.B. Anderson: Constituents of *Solidago* species. Part VII. Constitution and stereochemistry of the *cis*-clerodanes from *Solidago arguta* Ait. and of related diterpenoids. J. Chem. Soc. Perkin Trans. I. **15**, 1590 (1976).

387. Melek, F.R., D.A. Gage, J. Gershenzon, and T.J. Mabry: Sesquiterpene lactone and diterpene constituents of *Helianthus annuus*. Phytochem. **24**, 1537 (1985).

388. Merxmüller, H., P. Leins, and H. Roessler: Inuleae—systematic review. In: The Biology and Chemistry of the Compositae (Heywood, V.H., J.B. Harborne, and B.L. Turner, eds.), Vol. 1. London: Academic Press.

389. Michie, M.J., and W.W. Reid: Biosynthesis of complex terpenes in the leaf cuticle and trichomes of *Nicotiana tabacum*. Nature **218**, 578 (1968).

390. Misra, L.N., J. Jakupovic, F. Bohlmann, and G. Schmeda-Hirschmann: Isodaucane derivatives, norsesquiterpenes and clerodanes from *Chromolaena laevigata*. Tetrahedron **41**, 5353 (1985).

391. Misra, R., R.C. Pandey, and Sukh Dev: The absolute stereochemistry of hardwickiic acid and its congeners. Tetrahedron Letters **1968**, 2681.

392. Mitscher, L.A., G.S.R. Rao, T. Veysoglu, S. Drake, and T. Haas: Isolation and identification of trachyloban-19-oic and (-)-kaur-16-en-19-oic acids as antimicrobial agents from the prairie sunflower, *Helianthus annuus*. J. Nat. Prod. **46**, 745 (1983).

393. Miyakado, M., N. Ohno, H. Yoshioka, T.J. Mabry, and T. Whiffin: Gymnospermin: A new labdantriol from *Gymnosperma glutinosa*. Phytochem. **13**, 189 (1974).

394. Morales Méndez, A., A. Usubillaga, A.K. Banerjee, and T. Nakano: Studies on the constituents of *Espeletia weddellii*. Planta Med. **24**, 243 (1973).

395. Mossa, J.R., J.M. Cassady, M.D. Antoun, S.R. Byrn, A.T. McKenzie, J.F. Kozlowski, and P. Main: Saudin, a hypoglycemic diterpenoid with a novel 6,7-secolabdane carbon skeleton from *Cluytia richardiana*. J. Org. Chem. **50**, 916 (1985).

396. Muradian, J., M. Motidome, and P.C. Ferreira: Flavonols and (-)-kaur-16-en-19-oic acid from *Mikania hirsutissima* DC. var. *hirsutissima*. Rev. Latinoamer. Quim. **8**, 88 (1977).

397. Murakami, T., T. Isa, and T. Satake: Eine neuuntersuchung der inhaltsstoffe von *Siegesbeckia pubescens* Makino. Tetrahedron Lett. **50**, 4991 (1973).

398. Nair, A.G.R., S. Subramanan, F. Bohlmann, S. Schöneweiss, and T.J. Mabry: A new diterpene galactoside from *Acanthospermum hispidum*. Phytochem. **15**, 1776 (1976).

399. Nakano, T., and A. Martin: Synthesis of labda-8(17),13(14)-diene-15,16-olide and 15,16-epoxy-labda-8(17),13(16)-14-triene and their rearrangement to clerodane derivatives. Tetrahedron **38**, 1217 (1982).

400. Nicholas, H.J.: Biosynthesis and metabolism of [^{14}C] sclareol. Biochem. Biophys. Acta **84**, 80 (1964).

401. Nishino, C., S. Manabe, M. Kazui, and T. Matsuzaki: Piscicidal *cis*-clerodane diterpenes from *Solidago altissima* L.: Absolute configurations of 5α,10α-*cis*-clerodanes. Tetrahedron Lett. **25**, 2809 (1984).

402. Niwa, M., and S. Yamamura: Stereostructures of several solidagolactones (elongatolides). Tetrahedron Lett. **22**, 2789 (1981).

403. Nurmukhamedova, M.R., N.D. Abdullaev, and G.P. Sidyakin: Diterpenoids of *Pulicaria salviifolia*. II. Structure of salvicin. Khim. Prir. Soedin. 299 (1986).

404. Nurmukhamedova, M.R., S.Z. Kasymov, N.D. Abdullaev, G.P. Sidyakin, and M.R. Yagudaev: Diterpenoids of *Pulicaria salviifolia* I. Structures of salvin and salvinin. Khim. Prir. Soedin. 201 (1985).

405. Oberti, J.C., A.B. Pomilio, and E.G. Gros: Diterpenes and sterols from *Wedelia glauca*. Phytochem. **19**, 2051 (1980).

406. Oberti, J.C., V.E. Sosa, P. Kulanthaivel, and W. Herz: *Ent*-norlabdane triols from *Austroeupatorium inulaefolium*. Phytochem. **23**, 2003 (1984).

407. Oehlschlager, A.C., and G. Ourisson: A comparison of *in vivo* and *in vitro* skeletal transformations of diterpenes. In: Terpenoids in Plants (Pridham, J.B., ed.). London and New York: Academic Press. 1967.

408. Oelrichs, P.P., P.J. Vallely, J.K. MacLeod, and I.A.S. Lewis: Isolation of a new potential antitumor compound from *Wedelia asperrima*. J. Nat. Prod. **43**,

414 (1980).

409. Ohno, N., J. Geshenzon, P. Newman, and T.J. Mabry: Diterpene carboxylic acids and a heliangolide from *Helianthus angustifolius*. Phytochem. **20**, 2393 (1981).

410. Ohno, N., and T.J. Mabry: Sesquiterpene lactones and diterpene carboxylic acids in *Helianthus niveus* subspecies *canescens*. Phytochem. **19**, 609 (1980).

411. Ohno, N., T.J. Mabry, V. Zabel, and W.H. Watson: Tetrachyrin, a new rearranged kaurenoid lactone, and diterpene acid from *Tetrachryon orizabaensis* and *Helianthus debilis*. Phytochem. **18**, 1687 (1979).

412. Ohsuka, A., S. Kusumoto, and M. Kotake: The steric structures of diterpene carboxylic acids from *Solidago altissima* L. Nippon Kagaku Kaishi **1973**, 590.

413. Okazaki, T., A. Ohsuka, and M. Kotake: Structure of diterpene lactones from *Solidago altissima* L. Nippon Kagaku Kaishi **1973**, 584.

414. Oriental, M.A., E. Guerreiro, and O.S. Giordano: Diterpene acids from *Grindelia aegialitis*. Rev. Latinoamer. Quim. **15**, 73 (1984).

415. Ortega, A., R. Martinez, and C.L. Garcia: Los diterpenos de *Stevia salicifolia* Cav. Estructura del stevinsol and salicifoliol. Rev. Latinoamer. Quim. **11**, 45 (1980).

416. Ortega, A., F.J. Morales, and M. Salmon: Kaurenic acid derivatives from *Stevia eupatoria*. Phytochem. **24**, 1850 (1985).

417. Oshima, Y., G.A. Cordell, and H.H.S. Fong: Oxepane diterpenes from *Montanoa tomentosa*. Phytochem. **25**, 2567 (1986).

418. Oshima, Y., J. Saito, and H. Hikino: Sterebins A, B, C, and D, bisnorditerpenoids of *Stevia rebaudiana* leaves. Tetrahedron **42**, 6443 (1986).

419. Oshima, Y., S.-M. Wong, C. Konno, G.A. Cordell, D.P. Waller, D.D. Soejarto, and H.H.S. Fong: Studies on Zoapatle, II. Leucantholide, a novel sesquiterpene lactone from *Montanoa leucantha* ssp. *leucantha*. J. Nat. Prod. **49**, 313 (1986).

420. Ourisson, G.: Some aspects of the distribution of diterpenes in plants. In: Chemistry in Botanical Classification (Nobel Foundation, Stockholm), Nobel Symposium **25**. New York and London: Academic Press. 1973).

421. Pakrashi, S.C., P.P. Ghosh Dastidar, and E. Ali: Newer (-)-kaurene derivatives from *Enhydra fluctuans* Lour. Indian J. Chem. **9**, 84 (1971).

422. Pakrashi, S.C., P.P. Ghosh Dastidar, and S.K. Gupta: Diterpenoids from *Enhydra fluctuans*. Phytochem. **9**, 459 (1970).

423. Pandey, U.C., A.K. Singhal, N.C. Barua, R.P. Sharma, J.N. Baruah, K. Watanabe, P. Kulanthaivel, and W. Herz: Stereochemistry of strictic acid and related furanoditerpenes from *Conyza japonica* and *Grangea maderaspatana*. Phytochem. **23**, 391 (1984).

424. Patra, A., A.K. Mitra, S. Biswas, C.D. Gupta, T.K. Chatterjee, K. Basu, and A.K. Barua: Carbon-13 NMR spectra of some labdane diterpenoids. Org. Magn. Reson. **17**, 301 (1981).

425. Pezzuto, J.M.: Chemistry, metabolism and biological activity of steviol (*ent*-13-hydroxykaur-16-en-19-oic acid), the aglycone of stevioside. Page 371 in A. Rahman and P.W. Le Quesne (eds.). New Trends in Natural Products Chemistry. Studies in Organic Chemistry, Vol. 26. Elsevier Scientific Publishers. Amsterdam (1986).

426. Pinhey, J.T., R.F. Simpson, and J.L. Batey: The constituents of *Olearia*

heterocarpa. The structure of olearin, a diterpene dilactone of the cascarillin group. Aust. J. Chem. **25**, 2621 (1972).

427. Pinto, A.C., M.L. Patitucci, R.S. Dasilva, P.P.S. Queiroz, and A. Kelecom: Pimarane and cleistanthane diterpenes from Velloziaceae: absolute configuration and biomimetic conversion. Tetrahedron **39**, 3351 (1983).

428. Piozzi, F., S. Passannanti, and M.P. Paternostro: Kaurenoid diterpenes in *Espeletia grandiflora*. Phytochem. **10**, 1164 (1971).

429. Proksch, P., and E. Rodriguez: Chromenes and benzofurans of the Asteraceae, their chemistry and biological significance. Phytochem. **22**, 2335 (1983).

430. Purushothaman, K.K., A. Sarada, A. Saraswathy, and J.D. Connolly: Sempervirenic acid, a diterpene acid from *Solidago sempervirens*. Phytochem. **22**, 1042 (1983).

431. Quijano, L., J.S. Calderon, F. Gomez G., V. Rosario M., and T. Rios: Oxepane diterpenoids and sesquiterpene lactones from "zoapatle" (*Montanoa tomentosa*), a Mexican plant with oxytocic activity. Phytochem. **24**, 2337 (1985).

432. Quijano, L., J.S. Calderon, F. Gomez G., V. Rosario, and T. Rios: Acyclic precursors of the uterotonic oxepane diterpenoids of "zoapatle" (*Montanoa tomentosa*). Phytochem. **24**, 2741 (1985).

433. Quijano, L., J.S. Calderon, F. Gomez, J.L. Vega, and T. Rios: Diterpenes from *Stevia monardaefolia*. Phytochem. **21**, 1369 (1982).

434. Quijano, L., and N.H. Fischer: Sesquiterpene and diterpene lactones from *Melampodium longipilum*. Phytochem. **23**, 829 (1984).

435. Quijano, L., and N.H. Fischer: Two oxepane diterpene lactones from *Melampodium diffusum*. Phytochem. **23**, 833 (1984).

436. Quijano, L., F.R. Fronczek, and N.H. Fischer: Sesquiterpene and diterpene lactones from *Melampodium leucanthum* and the molecular structure of 4(5)-dihydromelampodin-B. Phytochem. **24**, 1747 (1985).

437. Railton, I.D., B. Fellows, and C.A. West: *Ent*-kaurene synthesis in chloroplasts from higher plants. Phytochem. **23**, 1261 (1984).

438. Robinson, H., and R.M. King: Eupatorieae—systematic review. In: The Biology and Chemistry of the Compositae (Heywood, V.H., J.B. Harborne, and B.L. Turner, eds.), Vol. 1. London: Academic Press. 1977.

439. Rogers, L.J., S.P.J. Shah, and T.W. Goodwin: Compartmentation of biosynthesis of terpenoids in green plants. Photosynthetica **2**, 184 (1968).

440. Rose, A.F.: Grindelane diterpenoids from *Chrysothamnus nauseosus*. Phytochem. **19**, 2689 (1980).

441. Rose, A.F., K.C. Jones, W.F. Haddon, and D.L. Dreyer: Grindelane diterpenoid acids from *Grindelia humilis*: Feeding deterrency of diterpene acids towards aphids. Phytochem. **20**, 2249 (1981).

442. Rossomando, P.C., O.S. Giordano, J. Espiñeira, and P. Joseph-Nathan: A diterpene acid from *Baccharis tucumanensis*. Phytochem. **24**, 787 (1985).

443. Rustaiyan, A., E. Simozar, A. Ahmadi, M. Grenz, and F. Bohlmann: A hardwickiic acid derivative from *Pulicaria gnaphalodes*. Phytochem. **20**, 2772 (1981).

444. Ruzicka, L.: The isoprene rule and the biogenesis of terpenic compounds. Experientia **9**, 357 (1953).

445. Ruzicka, L., A. Eschenmoser, and H. Heusser: Biogenesis of steroids and terpenic compounds. Experientia **9**, 362 (1953).

446. St. Pyrek: Neutral diterpenoids of *Helianthus annuus*. J. Nat. Prod. **47**, 822 (1984).

447. Sakamoto, I., K. Yamasaki, and O. Tanaka: Application of ^{13}C NMR spectroscopy to chemistry of plant glycosides: rebaudiosides-D and -E, new sweet diterpene-glucosides of *Stevia rebaudiana* Bertoni. Chem. Pharm. Bull. **25**, 3437 (1977).

448. Salmon, M., A. Ortega, G. Garcia de la Mora, and E. Angeles. A diterpenic acid from *Stevia lucida*. Phytochem. **22**, 1512 (1983).

449. San-Martin, A., A. Givovich, and M. Castillo: Neo-clerodane diterpenoids from *Baccharis incarum*. Phytochem. **25**, 264 (1986).

450. San-Martin, A., J. Rovirosa, R. Becker, and M. Castillo: Diterpenoids from *Baccharis tola*. Phytochem. **19**, 1985 (1980).

451. San-Martin, A., J. Rovirosa, and M. Castillo: Diterpenoids from *Baccharis tola*. Phytochem. **22**, 1461 (1983).

452. San-Martin, A., J. Rovirosa, C. Labbe, A. Givovich, M. Mahu and M. Castillo: Neo-clerodane diterpenoids from *Baccharis rhomboidalis*. Phytochem. **25**, 1393 (1986).

453. Schmeda-Hirschmann, G., J. Jakupovic, V.P. Pathak, and F. Bohlmann: Balansolide and other sesquiterpene lactones from *Bejaranoa balansae*. Phytochem. **25**, 2167 (1986).

454. Schmeda-Hirschmann, G., C. Zdero, R.N. Baruah, and F. Bohlmann: Further sesquiterpene lactones from *Calea* and *Viguiera* species. Phytochem. **24**, 2019 (1985).

455. Schteingart, C.D., and A.B. Pomilio: Terpenoids from *Wedelia bupthalmiflora*. Phytochem. **20**, 2589 (1981).

456. Schteingart, C.D. and A.B. Pomilio: Atractyloside, toxic compound from *Wedelia glauca*. J. Nat. Prod. **47**, 1046 (1984).

457. Scott, A.I., F. McCapra, F. Comer, S.A. Sutherland, D.W. Young, G.A. Sim, and G. Ferguson: Stereochemistry of the diterpenoids—IV. Structure and stereochemistry of some polycyclic diterpenoids. Tetrahedron **20**, 1339 (1964).

458. Seaman, F.: Sesquiterpene lactones as taxonomic characters in the Asteraceae. Bot. Rev. **48**, 121 (1982).

459. Seaman, F., A.J. Malcolm, and N.H. Fischer: Tomexanthin, an oxepane diterpene from *Montanoa tomentosa*. Phytochem. **23**, 464 (1984).

460. Sen, A.K., S.B. Mahato, and N.L. Dutta: Conyzic acid, a new diterpene resin from *Conyza stricta*. Indian J. Chem. **13**, 504 (1975).

461. Sholichin, M., K. Yamasaki, R. Miyama, R. Yahara, and O. Tanaka: Labdane-type diterpenes from *Stevia rebaudiana*. Phytochem. **19**, 326 (1980).

462. Silva, M., and P.G. Sammes: A new diterpenic acid and other constituents of *Haplopappus foliosus* and *H. angustifolius*. Phytochem. **12**, 1755 (1973).

463. Singh, P., M.C. Sharma, K.C. Joshi, and F. Bohlmann: Diterpenes derived from clerodanes from *Pulicaria angustifolia*. Phytochem. **24**, 190 (1985).

464. Soejarto, D.D., A.D. Kinghorn, and N.R. Farnsworth: Potential sweetening agents of plant origin. III. Organoleptic evaluation of *Stevia* leaf herbarium samples for sweetness. J. Nat. Prod. **45**, 590 (1982).

465. Sorensen, N.A.: Polyacetylenes and conservation of chemical characters in the Compositae. In: The Biology and Chemistry of the Compositae (Heywood,

V.H., J.B. Harborne, and B.L. Turner, eds.), Vol. 1. London: Academic Press. 1977.

466. Stapel, G., H.G. Menssen, and G. Snatzke: Isolierung und strukturaufklärung von zwei diterpenen aus *Baccharis articulata*. Planta Med. **38**, 366. (1980).

467. Stuessy, T.F.: Heliantheae—systematic review. In: The Biology and Chemistry of the Compositae (Heywood, V.H., J.B. Harborne, and B.L. Turner, eds.), Vol. 1. London: Academic Press. 1977.

468. Swain, T., and C.A. Williams: Heliantheae—chemical review. In: The Biology and Chemistry of the Compositae (Heywood, V.H., J.B. Harborne, and B.L. Turner, eds.), Vol. 1. London: Academic Press. 1977.

469. Tabacik, C., and M. Bard: Etude chimio-taxonomique dans le genre *Cistus*. Phytochem. **10**, 3093 (1971).

470. Tandon, S., and R.P. Rastogi: Strictic acid, a novel diterpene from *Conyza stricta*. Phytochem. **18**, 494 (1979).

471. Timmermann, B.N., J.J. Hoffmann, S.D. Jolad, R.B. Bates, and T.J. Siahaan: Diterpenoids and flavonoids from *Grindelia discoidea*. Phytochem. **25**, 723 (1986).

472. Timmermann, B.N., J.J. Hoffmann, S.D. Jolad, R.B. Bates, and T.J. Siahaan: Labdane diterpenoids from *Grindelia discoidea*. Phytochem. **25**, 1389 (1986).

473. Timmermann, B.N., J.J. Hoffmann, S.D. Jolad, and K.H. Schram: Grindelane diterperoids from *Grindelia squarrosa* and *G. camporum*. Phytochem. **24**, 1031 (1985).

474. Timmermann, B.N., J.J. Hoffmann, S.D. Jolad, K.H. Schram, R.E. Klenck, and R.B. Bates: Constituents of *Chrysothamnus paniculatus* (Compositae). 2. Chrysolic acid, a new labdane-derived diterpene with an aromatic B ring. J. Org. Chem. **47**, 4114 (1982).

475. Timmermann, B.N., D.J. Luzbetak, J.J. Hoffmann, S.D. Jolad, K.H. Schram, R.B. Bates, and R.E. Klenck: Grindelane diterpenoids from *Grindelia camporum* and *Chrysothamnus paniculatus*. Phytochem. **22**, 523 (1983).

476. Tomassini, T.C.B., and M.E.O. Matos: On the natural occurrence of 15α-tiglinoyloxy-kaur-16-en-19-oic acid. Phytochem. **18**, 663 (1979).

477. Tonn, C.E., P.C. Rossomando, and O.S. Giordano: Batudioic acid and flavonoids from *Baccharis tucumanensis*. Phytochem. **21**, 2599 (1982).

478. Tsankova, E., and F. Bohlmann: A monoterpene from *Aster bakeranus*. Phytochem. **22**, 1285 (1983).

479. Usubillaga, A., J. de Hernandez, N. Perez, and M. Kiriakidis: Kauranoid diterpenes in *Espeletia* species. Phytochem. **12**, 2999 (1973).

480. Usubillaga, A., and A. Morales: Kaurenic acids in *Espeletia* species. Phytochem. **11**, 1856 (1972).

481. Usubillaga, A., and A. Nakano: Kauranoid diterpenes in *Ruilopezia margarita*. Planta Med. **35**, 331 (1978).

482. Vichnewski, W., H. de Freitas Leitao Filho, R. Murari, and W. Herz: Cinnamoylgrandifloric acid from *Mikania oblongifolia*. Phytochem. **16**, 2028 (1977).

483. Vichnewski, W., R. Murari, and W. Herz: New clerodanes from *Symphiopappus itatiayensis*. Phytochem. **18**, 129 (1979).

484. Vis, E., and H.G. Fletcher: Stevioside. IV. Evidence that stevioside is a sophoroside. J. Am. Chem. Soc. **78**, 4709 (1956).

485. Vlad, P.F., N.D. Ungur, and M.N. Koltsa: Cyclization and rearrangements of diterpenoids. I. Synthesis of tetracyclic diterpenoids with a new carbon skeleton from labdanes. Tetrahedron **39**, 3947 (1983).

486. Von Carstenn-Lichterfelde, C., C. Pascual, J. Pons, R.M. Rabanal, B. Rodriguez, and S. Valverde: Carbon-13 NMR spectra of ent-beyerane and ent-beyerene derivatives. Tetrahedron Lett. **1975**, 3569.

487. Waddell, T.G., C.B. Osborne, R. Collins, M.J. Levine, M.C. Cross, J.V. Silverton, H.M. Fales, and E.A. Sokoloski: Erigerol, a new labdane diterpene from *Erigeron philadelphicus*. J. Org. Chem. **48**, 4450 (1983).

488. Wagenitz, G.: Systematics and phylogeny of the Compositae (Asteraceae). Plant Syst. Evol. **125**, 29 (1976).

489. Wagner, H., R. Seitz, V.M. Chari, and H. Lotter: Bacchotrcuneatin A and B, zwei neue diterpenlactone aus *Baccharis tricuneata* var. *tric.* Tetrahedron Lett. **1977**, 3039.

490. Wagner, H., R. Seitz, and H. Lotter: New furanoid ent-clerodanes from *Baccharis tricuneata*. J. Org. Chem. **43**, 3339 (1978).

491. Waiss, A.C., B.G. Chan, C.A. Elliger, V.H. Garrett, E.C. Carlson, and B. Beard: Larvicidal factors contributing to host-plant resistance against sunflower moth. Naturwissenschaften **64**, 341 (1977).

492. Warning, U., F. Bohlmann, H. Sanchez V., E. Del Rio S., and X.A. Dominguez: New constituents of *Baccharis salicifolia*. Rev. Latinoamer. Quim. **17**, 199 (1986).

493. Watanabe, K., N. Ohno, H. Yoshioka, J. Gershenzon, and T.J. Mabry: Sesquiterpene lactones and diterpenoids from *Helianthus argophyllus*. Phytochem. **21**, 709 (1982).

494. Wehrli, F.W., and T. Nishida: The use of carbon-13 nuclear magnetic resonance spectroscopy in natural products chemistry. Fortschritte d. Chem. org. Naturst. **36**, 1 (1979).

495. Wenkert, E.: Structural and biogenetic relationships in the diterpene series. Chem. and Ind. **1955**, 282.

496. Wenkert, E., and Z. Kumazawa: Manool-derived diterpenic hydrocarbons and related products. Chemical Communications **1968**, 140.

497. West, C.A.: Biosynthesis of diterpenes. In: Biosynthesis of Isoprenoid Compounds (Porter, J.W., and S.L. Spurgeon, eds.), Vol. 1, p. 376. New York: John Wiley and Sons. 1981.

498. West, C.A., M.W. Dudley, and M.T. Dueber: Regulation of terpenoid biosynthesis in higher plants. In: Topics in the Biochemistry of Natural Products (Swain, T., and G.R. Waller, eds.). New York and London: Plenum Press.

499. Wollenweber, E., and V.H. Dietz: Occurrence and distribution of free flavonoid aglycones in plants. Phytochem. **20**, 869 (1981).

500. Yamamura, Y., I. Masatoki, M. Niwa, Hasegawa, S. Ohba, and Y. Saito: The isolation and structure of tricyclosolidagolactone, a new diterpene from *Solidago altissima*. Tetrahedron Lett. **22**, 739 (1981).

501. Yamasaki, K., H. Kohda, T. Kobayashi, N. Kaneda, R. Kasai, O. Tanaka, and K. Nishi: Application of 13C nuclear magnetic resonance spectroscopy to chemistry of glycosides: structures of paniculosides-I, -II, -III, -IV, and -V, diterpene glucosides of *Stevia paniculata* Lag. Chem. Pharm. Bull. **25**, 2895 (1977).

502. Yamasaki, K., H. Kohda, T. Kobayashi, R. Kasai, and O. Tanaka: Structures of *Stevia* diterpene-glucosides: Application of 13C NMR. Tetrahedron Lett. **1976**, 1005.

503. Zdero, C., F. Bohlmann, R.M. King, and H. Robinson: Diterpene glycosides and other constituents from Argentinian *Baccharis* species. Phytochem. **25**, 2841 (1986).

504. Zurcher, R.F.: Protonenresonanzspektroskopie und steroidstruktur. I. Das C-19-methylsignal in funktion der substituenten. Helv. Chim. Acta **44**, 1380 (1961).

505. Zurcher, R.F.: Protonenresonanzspektroskopie und steroidstruktur. II. Die lage der C-18- und C-19-methylsignale in abhängigkeit von den substituenten am steroidgerüst. Helv. Chim. Acta **46**, 2054 (1963).

Appendix

APPENDIX. Alphabetical Listing of Molecular Substituents including Ester Sidechains.

Name	Code	Structure
(1,2-dihydroxyethyl)acrylate	(1,2-OH-Et)Acr	
(2α-acetoxyethyl)acrylate	(2-AcEt)Acr	
(2α-hydroxyethyl)acrylate	(2-OH-Et)Acr	
2,3-diacetyloxy-2-methylbutyrate	2-Mebut-2,3-Ac	
2,3-dihydroxy-2-methylbutyrate	2-Mebut-2,3-OH	
2,3-dihydroxyisobutyrate	i-But-2,3-OH	

APPENDIX. (*contd.*).

Name	Code	Structure
2,3-epoxy-2-methylbutyrate	Epoxyang	
2,3-epoxy-n-butyrate	Epoxy-n-But	
2-hydroxy-2-methylbutyrate	2-Mebut-2-OH	
2-hydroxy-3-acetyloxy-2-methylbutyrate	2-Mebut-2-OH-3-Ac	
2-hydroxy-3-chloro-isobutyrate	i-But-2-OH-3-Cl	
2-hydroxy-3-oxo-2-methylbutyrate	2-Mebut-2-OH-3=O	
2-hydroxy-isobutyrate	i-But-2-OH	
2-methyl-butanoate	2-Mebut	

APPENDIX. (*contd.*).

Name	Code	Structure
3-acetoxy-isovalerate	i-Val-3-Ac	
3-furoate	Fur	
3-hydroxy-2-ethoxy-isobutyrate	i-But-3-OH-2-OEt	
3-hydroxy-2-methylbutyrate	2-Mebut-3-OH	
3-hydroxy-isovalerate	i-Val-3-OH	
3-methyl-2,3-dehydro-valerate	deh-Val-3-Me	
3-methyl-coumarate	3-Me-Coum	
4,5-dihydroxytiglate	Tig-4,5-OH	

APPENDIX. *(contd.)*.

Name	Code	Structure
4-[3-hydroxy-stearoyloxy]-tiglate	Tig-4-Stear	
4-acetyloxy-angelate	Ang-4-Ac	
4-hydroxy-5-[5-hydroxy-tiglinoyloxy]-tiglate	Tig-4-OH-5-Tig-5-OH	
4-hydroxy-5-acetyloxytiglate	Tig-4-OH-5-Ac	
4-hydroxy-5-tiglinoyloxy-tiglate	Tig-4-OH-5-Tig	
4-hydroxy-angelate	Ang-4-OH	
4-hydroxy-isobutyrate	i-But-4-OH	

APPENDIX. (*contd.*).

Name	Code	Structure
4-hydroxy-methacrylate	Mac-4-OH	
4-hydroxy-sarracinate	Sar-4-OH	
4-hydroxy-tiglate	Tig-4-OH	
5-[5-hydroxy-tiglinoyloxy]-tiglate	Tig-5-OTig-5-OH	
5-acetyloxy-tiglate	Tig-5-Ac	
5-hydroxy-2,3-dihydro-3-sulfydryl-sarracinate	dih-Sar-5-OH-3-SH	
5-hydroxy-tiglate	Tig-5-OH	
acetate	Ac	

APPENDIX. (*contd.*).

Name	Code	Structure
acetyloxy	OAc	
acetylsarracinate	Sarac	
angelate	Ang	
anthranoate	Anth	
arabinoside	Arab	
arachinate	Arac	$[CH_3(CH_2)_{18}CO-]$
behenate (CO(CH2)20Me	Behen	$[CH_3(CH_2)_{20}CO-]$
benzoate	Benz	

APPENDIX. *(contd.)*.

Name	Code	Structure
benzoyloxy	O-Ben	
bernsteinoate (= succinate)	O-bern	
cis-3-S-methylacrylate	cis-Acr-S-Me	
cis-cinnamate	cis-Cinn	
cis-coumarate	cis-Coum	
coumarate	Coum	
epoxymethacrylate	Epoxymac	
epoxysarracinate	Epoxysar	

APPENDIX. (*contd.*).

Name	Code	Structure
formyloxy	OCHO	
β-Galactoside	Gal	
β-D-[6-acetoxyglucopyranoside]	Glu-6-Ac	
β-D-glucopyranoside	Glu	
glycyl	Gly	
hydrogen	H	
hydroxy	OH	
hydrocinnamate	H-Cinn	

Appendix. (*contd.*).

Name	Code	Structure
isobutyrate	i-But	
isovalerate	i-Val	
laurate	Dodec	$[CH_3(CH_2)_{10}CO-]$
linoleate	Linol	
linolenate	Linolen	
malonate	Mal	
methacrylate	Mac	
methoxy	OMe	
methyl	Me	

APPENDIX. (*contd.*).

Name	Code	Structure
methylene	CH2	
methylvalerate	MeVal	
myristate	Tetradec	$[CH_3(CH_2)_{12}CO-]$
p-hydroxy-hydrocinnamate	p-OH-H-Cinn	
p-hydroxy-phenylacetate	p-OH-Phe-Ac	
palmitate	Hexadec Palm	$[CH_3(CH_2)_{14}CO-]$
peroxy	O-OH	
peroxy-bridge	-O-O-	
phenylacetate	Phe-Ac	

APPENDIX. (*contd.*).

Name	Code	Structure
piperidine	Pip	
propionate	Pro	
sarracinate	Sar	
senecioate	Sen	
stearate	Octadec Stear	$[CH_3(CH_2)_{16}CO-]$
stearoyloxy-isovalerate	i-Val-5-O-Stear	$(CH_2)_{15}CH_3$
succinate	Succ	
tiglate	Tig	

APPENDIX. (*contd.*).

Name	Code	Structure
<u>trans</u>-3-S-methylacrylate	Trans-Acr-S-**Me**	
<u>trans</u>-cinnamate	trans-Cinn	

Compound Index

---

* Acetylation products of compound 94. These compound numbers were applied to other structures, but these earlier uses were not removed.

Species Index

General Index